D1236839

CAMBRIDGE STUDIES IN ECOLOGY

Megaherbivores

The influence of very large body size on ecology

Also in the series

Hugh G. Gaugh, Jr. *Multivariate analysis in community ecology*
Robert Henry Peters *The ecological implications of body size*
C. S. Reynolds *The ecology of freshwater phytoplankton*
K. A. Kershaw *Physiological ecology of lichens*
Robert P. McIntosh *The background of ecology: concept and theory*
Andrew J. Beattie *The evolutionary ecology of ant–plant mutualisms*
F. I. Woodward *Climate and plant distribution*
Jeremy J. Burdon *Diseases and plant population biology*
N. G. Hairston *Community ecology and salamander guilds*
Janet I. Sprent *The ecology of the nitrogen cycle*
H. Stolp *Microbial ecology: organisms, habitats and activities*

Megaherbivores
The influence of very large body size on ecology

R. NORMAN OWEN-SMITH

University of the Witwatersrand

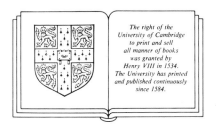

The right of the
University of Cambridge
to print and sell
all manner of books
was granted by
Henry VIII in 1534.
The University has printed
and published continuously
since 1584.

CAMBRIDGE UNIVERSITY PRESS

Cambridge

London   New York   New Rochelle

Melbourne   Sydney

Published by the Press Syndicate of the University of Cambridge
The Pitt Building, Trumpington Street, Cambridge CB2 1RP
32 East 57th Street, New York, NY 10022, USA
10 Stamford Road, Oakleigh, Melbourne 3166, Australia

© Cambridge University Press 1988

First published 1988

Printed in Great Britain at the University Press, Cambridge

British Library cataloguing in publication data
Owen-Smith, R. Norman
  Megaherbivores
  1. Large herbivores. Ecology
  I. Title
  599.7

Library of Congress cataloguing in publication data
Owen-Smith, R. Norman.
  Megaherbivores: the influence of very large body size on ecology
  R. Norman Owen-Smith.
    p.   cm. – (Cambridge studies in ecology)
  Bibliography: p.
  Includes index.
  ISBN 0 521 36020 X
  1. Herbivora–Size.   2. Herbivora–Ecology.   3. Ungulata–Size.
4. Ungulata–Ecology.   5. Mammals–Size.   6. Mammals–Ecology.
!. Title.   II. Series.
  QL737.U4095 1988
  599.7–dc19
  88-6941

ISBN 0 521 36020 X

SE

'Hints for an agenda: How big is it and how fast does it happen?'
(G. Evelyn Hutchinson 1975)

Contents

Preface   *xi*

1 Prologue   1

2 Morphology, evolutionary history and recent distribution   6
Introduction   6
Morphology   6
Evolutionary origins and relationships   16
Paleontological diversity   20
Distribution of extant species   21

3 Food and other habitat resources   30
Introduction   30
Food   30
Water and other habitat needs   45
Comparisons with smaller ungulates   50

4 Space–time patterns of habitat use   53
Introduction   53
Temporal patterning of activities   53
Utilization of space   61
Comparisons with smaller ungulates   67

5 Body size and nutritional physiology   69
Introduction   69
Metabolic requirements   70
Gut anatomy   71
Food intake and digestion   72

6 Body size and feeding ecology   82
Introduction   82
Diet quality   82
Foraging time   87
Home range extent   95
Trophic ecology of megaherbivores: summary   98

 7 **Social organization and behavior**   101
    Introduction   101
    Group structure   101
    Male dominance relations   109
    Courtship and mating   116
    Responses to predators   124
    Comparisons with smaller ungulates   131

 8 **Life history**   133
    Introduction   133
    Infancy and juvenilehood   133
    Adolescence and puberty   138
    Reproduction by females   144
    Reproduction by males   151
    Mortality and lifespan   152
    Comparisons with smaller ungulates   159

 9 **Body size and sociobiology**   160
    Introduction   160
    Grouping patterns   160
    Male dominance systems   167
    Female mate choice   177
    Summary   179

10 **Body size and reproductive patterns**   181
    Introduction   181
    Seasonality of reproduction   183
    Age at first conception   185
    Birth intervals   186
    Maternal investment in reproduction   190
    Offspring sex ratio   195
    Summary   198

11 **Demography**   200
    Introduction   200
    Population structure   200
    Population growth   212
    Population density and biomass   221
    Comparisons with smaller ungulates   225

12 **Community interactions**   226
    Introduction   226
    Impact on vegetation   226
    Effects on other large herbivores   239
    Comparisons with smaller ungulates   245

13 Body size and population regulation   246
Introduction   246
Demographic models   248
Interactions with vegetation   257
Dispersal   260
Summary   264

14 Body size and ecosystem processes   265
Introduction   265
Biomass levels   265
Energy flux   274
Nutrient cycling   277
Ecosystem stability and disturbance   278

15 Late Pleistocene extinctions   280
Introduction   280
Pattern of extinctions   281
Climatic change   284
Human predation   289
The role of megaherbivores   292
Summary   296

16 Conservation   297
Introduction   297
Conservation objectives   298
Problems of overabundance   299
Problems of overexploitation   307
Summary   308

17 Epilogue: the megaherbivore syndrome   309
Faunal patterns   309
Social and life history patterns   312
Demographic patterns   313
Community and ecosystem patterns   314

Appendixes   317
Appendix I   317
Appendix II   325

References   331

Index   364

Preface

Like the animals it describes, this book has had a long gestation. It started as a Ph.D. thesis on the white rhinoceros, grew into a monograph on rhinoceroses, expanded to include other similarly large herbivores, and then settled on the focus adopted in the pages that follow: the consequences of large size for the ecology of animals such as elephants, rhinoceroses and hippopotami, and by implication extinct species of similar size.

I hope that this work will be of interest to a variety of readers. Firstly, it is written for biologists interested in allometric scaling effects on ecological processes. The correlates of a body mass at the upper limit of the size range among mammalian herbivores are analyzed at various levels, including ecophysiology, behavioral ecology, demography, community interactions and ecosystem processes. Secondly, the book should be an aid to professional conservationists and wildlife managers concerned about the future survival of such large mammals. Scientific facts about these species must be given due cognizance if management actions are to achieve their desired objectives. Thirdly, it is directed towards paleobiologists interested in the ecological roles that similarly large mammals played in the faunas and ecosystems of the past. In particular potential causes of the extinctions of the so-called megafauna during the late Pleistocene are assessed. Finally, I hope that this book will be illuminating to all those who have marvelled at the ways of living of these largest among land animals, whether in the wild or on film. The features of their ecology and behavior are compared and contrasted with those of lesser ungulates.

I owe many debts of gratitude to people who have contributed help or ideas at various stages of the life history of this manuscript. It was a suggestion made by George A. Petrides, then Visiting Professor of Wildlife Management at the University of Pretoria, that first drew me into a study of the white rhinoceros. Rudolf C. Bigalke, then Principal Research Officer of the Natal Parks Board, made the study possible by organizing a temporary position for me with his organization, and by giving me free reign to delve into the social life of white rhinos. John T. Emlen, my doctoral supervisor at the University of Wisconsin, provided much inspiration and support both during the fieldwork period and over the subsequent period of thesis writing far from these animals and their environment. The impetus to delve into allometry was sparked by a conversation with Fred L. Bunnell at the University of British Columbia. R. Dale Guthrie of the University of Alaska opened my mind to the full richness of the large mammal fauna of the far north during the Pleistocene, and the intertwined puzzle of its disappearance.

Many personnel of the Natal Parks Board provided practical assistance. John Vincent saw to my needs during the initial six months, and arranged the aerial counts. Peter Hitchins developed the techniques used for installing radio transmitters on the rhinos, and generously provided much data on black rhinos. Greig Stewart initiated me into the vegetation of the area; Brian Downing and Orty Borquin helped identify many plants, and later Roger Porter provided further botanical assistance. Michael Keep provided veterinary assistance with post-mortems and rhino capture. Aid in catching the rhinos required for marking was provided by John Daniel, Ken Rochat, Mark Astrupp and Brian Thring. Park Wardens Ian Player, Nick Steele and Gordon Bailey helped in many ways, and imparted their wilderness ideals. Several of these people, and also Graham Root, assisted with aerial counts. Bob Crass and later Don Stewart provided valuable support from head office. David Rowe-Rowe organized the printing of photographs. Dawn Denyer saw to it that my grocery book went out each week. Mqhoyi Nkosi carried my equipment and imparted his bush knowledge, while later his son Bheki served as a keen and unobtrusive field assistant.

The book developed its final form under the helpful guidance of various editors and reviewers, including Norman Myers, Tim Clutton-Brock, Robin Dunbar, Robin Pellew and Martin Walters. Special thanks are due to Thomas Foose, for permission to quote extensively from his unpublished thesis, and to Richard Bell, for his vast generosity with comments and ideas on many sections of the book. A number of people provided information or helpful comments on various sections of the manuscript, including D.G. Ashton, Esmond Bradley Martin, Martin Brooks, Graeme Caughley, Johan du Toit, Peter Goodman, Russell Graham, Hans Grobler, Dale Guthrie, Andrew Hansen, Peter Hitchins, Andrew Laurie, Chris Lightfoot, Hanne Lindemann, Paul Martin, Thys Mostert and my wife Margaret Loffell.

The numerous illustrations were patiently drawn by Cheryl Hughes,

John Dallman, Cherry Allan, Carol Cardoso and Jenni Saley. Phillip Prim assisted with photographic work. Carol Sam helped with part of the typing, while my own fingers did the rest.

The white rhino study would not have been possible without the financial support provided by the Natal Parks Board, the Wisconsin Alumni Research Foundation, and US National Science Foundation grant no. GB-15304 to John T. Emlen. The Centre for Resource Ecology in the Department of Botany, University of the Witwatersrand, provided the ideal environment within which to develop into their final form the facts and ideas presented in this book.

Norman Owen-Smith
University of the Witwatersrand

1

Prologue

The subjects of this book are the animals that I will designate as megaherbivores. I define this term to include those plant-feeding mammals that typically attain an adult body mass in excess of one megagram, i.e. ten to the power six grams, 1000 kg, or one metric tonne. This demarcating criterion conveniently encompasses elephant, rhinoceros and hippopotamus among living forms, while giraffe slip marginally into the category. Such animals have been colloquially designated pachyderms; but a thick skin is a minor feature, and it is their very large body size that sets these few species apart from the numerous smaller species of unguligrade herbivore that occupy a wide variety of terrestrial ecosystems today. Paleontologists such as Martin (1967) have used the term 'megafauna' to encompass those species attaining a body mass exceeding about 45 kg (100 pounds). However this division is arbitrary and has no functional basis. In this book I show that there are distinctions between animals reaching a mass in excess of 1000 kg, and those of smaller size, in almost all aspects of ecology.

Of course, many whales attain a larger size than any terrestrial mammal, but all whales are carnivorous, feeding on other animals ranging from tiny shrimps to seals. However, among marine mammals there are also the strictly herbivorous sirenians (manatees and dugongs), which feed on submerged plants growing in shallow lagoons and coastal waters. Manatees may weigh up to 1600 kg, while the recently extinct Steller's sea cow weighed up to 6000 kg. I will say little further about these marine megaherbivores in this book, simply because it is difficult to compare their ecology meaningfully with that of terrestrial species, and because I lack familiarity with the literature on marine animals.

The terrestrial megaherbivores extant at the present time are a depauperate remnant of the much greater variety of such forms of animal represented in the faunas of the past until as recently as 11 000 years ago (the end of the Pleistocene). While today elephant, rhino and hippo are found only in Africa and tropical Asia, in previous eras through to the late Pleistocene there were comparably large mammals of a diversity of forms on all continents, occupying a range of ecosystems from arctic steppe and taiga to tropical rain forest and semidesert. During the mesozoic era certain reptilian herbivores (the so-called dinosaurs) attained immense body sizes comparable to or even exceeding those of the largest whales, although it has been suggested that the largest dinosaurs may have been semi-aquatic in their habits. These reptilian megaherbivores all became extinct over a relatively short period at the end of the Cretaceous period, some 100 million years BP. The disappearance of giant mammals from the Americas, Europe and the palearctic region of Asia at the end of the Pleistocene was even more dramatic in its suddenness, and coincided with the spread of humans through these regions.

The continued survival of most of the megaherbivore species currently in existence is somewhat precarious. The large appetites and destructive power of these beasts are incompatible with human agronomic objectives, so that their ranges have become restricted mostly to the island sanctuaries provided by national parks and wildlife reserves. Populations of elephants and rhinos have recently been depleted even within conservation areas, due to illegal hunting stimulated by the high prices fetched by ivory and rhino horn. All three Asian species of rhino are listed by the International Union for the Conservation of Nature as seriously endangered, as also is the Asian elephant. The northern race of the African white rhino is currently on the critical list, while numbers of black rhino and elephant are dwindling rapidly through most of Africa.

However, in African conservation areas where they have been effectively protected, populations of elephant, white rhino and hippo have increased to levels where they have induced vegetation changes that have appeared detrimental not only to their own food resources, but also to habitat conditions for other species of plant and animal. This has led to management intervention in the form of culling operations. These undertakings are controversial because of their interference with ecological processes, and also on humanitarian or animal welfare grounds. We still understand little about how populations of such potentially destructive animals were regulated under pristine conditions.

Thus surviving African species of megaherbivore are embarassingly successful, when protected from human depredation. This raises questions concerning the reasons for the demise of elephants, rhinos and other similarly large beasts from Europe and the Americas at the end of the Pleistocene. The relative importance of climatic change and human predation in these extinctions remains an unresolved problem. What light can the ecology of surviving species of megaherbivore shed on these questions?

What difference has the disappearance of these animals made to the ecosystems that had persisted with megaherbivores as integral constituents until 11 000 years ago? These questions raise some deeper issues with a bearing on the ecological problems now confronting us, *Homo sapiens*, as our populations expand and exert an increasingly strong influence on the functioning of natural ecosystems.

A number of scientific papers, and several books, have appeared recently on the subject of size and scaling in biology. A pioneering recognition of the pervasive importance of body size was by G. E. Hutchinson (1975), as noted in the quotation prefacing this book. During the gestation of the present manuscript, three books have been published focussing specifically on this topic. Peters (1983) documented the pervasive influence of body size on a wide range of physiological and ecological features, from metabolism and locomotion to abundance and productivity. Calder (1984) emphasized underlying dimensional and rate factors, together with aspects of life history. Schmidt-Nielsen's (1984) focus was specifically on aspects of physiology, from heartbeat and metabolic rate to movement and temperature regulation. While Calder's book was restricted largely to birds and mammals, Peters and Schmidt-Nielsen considered the complete range of organisms from protozoa upwards. These books, and other published papers, provide ample evidence that a variety of biological functions are scaled in magnitude or rate in relation to some exponential power of body mass.

In this book the focus is specifically on ecology, although underlying features of nutritional physiology are included. Taxonomic coverage is restricted to ungulates of the orders Perissodactyla and Artiodactyla, together with the subungulates of the order Proboscidea, among extant forms. As large mammalian herbivores these species share a basic ecological unity. The range in body size encompassed spans three orders of magnitude, from the smallest antelope, weighing 4–5 kg, to the largest elephant, weighing over 6000 kg. Geographically I will emphasize the species that share African savanna environments with modern day elephants, rhinos and hippos, since less information is available for Asian ungulates. By restricting coverage to such an ecologically, geographically and taxonomically coherent group of animals, it is possible to delve more deeply into the factors underlying body mass trends than was possible in the wide-ranging reviews cited above. Furthermore this is the group of animals with which I am personally familiar from my own field research. Such first hand contact is an aid in assessing and interpreting the results reported for particular species by other investigators using a variety of techniques.

My doctoral research on the behavioral ecology of the white or square-lipped rhinoceros (*Ceratotherium simum*) provided the initial ideas for this book. The white rhino is in many ways a seminal species. Unlike elephants and other rhinos, it is exclusively a grazer – a habit shared only with hippopotamus among surviving megaherbivores, but with a wider proportion of extinct species. In its grazing habits it has ecological affinities with a variety of bovids and equids that share similar savanna environments in Africa. It thus provides a unique example for unravelling the common influences of very large size on ecology from those deriving from phyletic inheritance or dietary restrictions.

The chapters of this book fall mainly into two kinds. Some review succinctly what is known about the ecology and behavior of living species of megaherbivore. Most of the ecological data from my white rhino study have not been published outside my thesis (Owen-Smith 1973). Thus I will use the white rhino as a special example, reporting aspects of its ecology in somewhat greater detail than is the case for other species, for which information has been extracted from the literature.

Other chapters analyze the relationships between particular features of the ecology of megaherbivores and body size. For these I draw on the now extensive body of facts published about the variety of ungulate species that still occur in Africa, and in some cases from data on ungulate species from other regions. In analyzing trends in aspects of ecology in relation to body mass for large herbivores, I will adopt a hypothetico-deductive approach. From a consideration of allometric relationships, a proposition will be made as to how a particular ecological attribute ought to vary with body mass. In general, hypotheses take the form that a certain proportional variation in body mass entails a corresponding proportional variation in an ecological feature. Such relationships imply power functions of the form $E = aM^b$, where E represents the ecological attribute, M body mass, a a constant, and b the power coefficient of the relation. These relations will be tested statistically by standard least-squares regression (Sokal & Rohlf 1969; Peters 1983). The confidence limits for the regression coefficient between log functions of E and M indicate whether or not the data refute the starting hypothesis. While the general trend with body mass indicates the constraining effect of increased body size on adaptations, points for individual species deviating markedly from the overall regression line suggest special adaptations releasing these species to some degree from the body size constraint. Special consideration will be given to the extent to which megaherbivores have compensated adaptively for the ecological restrictions imposed by their very large size.

The general questions underlying the body mass chapters are these:

1. To what extent do aspects of the ecology of megaherbivores differ in magnitude, and perhaps in kind, from those of smaller ungulates?
2. Do the ecological features shown by megaherbivores merely represent extrapolations of the trends with body size evident among large herbivores?
3. What features of ecology set upper limits to the body sizes attained by mammalian herbivores?
4. What are the special influences of megaherbivores on ecosystem processes?

Two chapters confront the two special problems posed by megaherbivores, from the perspective of the understanding gained from earlier chapters. Chapter 15 addresses the biological puzzle raised by the complete demise of megaherbivores outside Africa and tropical Asia during the late Pleistocene. I evaluate the relative role of climatic change and human overkill as causal factors in these extinctions, and advance a synthetic hypothesis taking into account features of the ecosystem impact of megaherbivores that have not been considered by other authors.

The effects of megaherbivores on ecosystem processes on the one hand, and their vulnerability to human disturbance on the other, create problems for their conservation. In chapter 16 I offer practical suggestions on how best to conserve and manage their populations within the limited confines of national parks and equivalent reserves, without negating the special features of the ecology of these large beasts. However their continued survival into the twenty-first century also raises political issues concerning how much space can be allocated to them alongside burgeoning human populations. To be successful these socio-political decisions must rest on a sound ecological knowledge of the sensitivities of these species, and also on an appreciation of the instrinsic value of conserving them.

In the final chapter I summarize the typifying ecological features of the megaherbivore phenotype. I hope that this book will help disseminate a more enlightened understanding of these great beasts, as an aid to effective action in ensuring their continued persistence on this planet; so that the final summary does not serve merely as an epitaph.

2

Morphology, evolutionary history and recent distribution

Introduction

Eight living species of terrestrial mammal fall into the megaherbivore category in terms of maximum body mass attained. These include two species of elephant, four rhinoceros species, and single species of hippopotamus and giraffe. In this chapter I describe the ecologically important features of their morphology, document their historic and present day geographic distributions, and outline their paleontological origins. This information serves as an essential background to the ecological topics that will form the subject of subsequent chapters.

Morphology

The most basic feature of significance to this book is size. How big do males and females of extant species of megaherbivore grow, in terms of height and weight? Size factors are frequently exaggerated in the literature, especially for large animals that are inconvenient to weigh. A distinction needs furthermore to be made between the asymptotic weights most typically reached by adult animals, and the maximum weights that might be reached by exceptional individuals. Weights may furthermore differ between different subspecies, and within populations in relation to prevailing resource abundance. Animals held in captivity may grow larger or smaller than their wild counterparts, depending on the adequacy of the diets that are provided to them.

Also of fundamental importance are the anatomic features functioning in the procurement and digestion of food. These include in particular the dentition and the structure of the digestive tract.

All large mammalian herbivores are dependent to some degree upon the agency of microbial symbionts for degradation of the cellulose in plant cell walls. To facilitate cell wall digestion, a gut compartment is needed where food passage can be delayed, and within which the conditions of neutral or slightly alkaline pH required by the bacteria and protozoa can be maintained. Mammalian herbivores fall into two categories in terms of the location of the fermentation chamber. Foregut fermenters have a compartment developed from the esophagus or anterior portion of the stomach. Distinct subdivisions may further compartmentalize this chamber, with narrow connecting openings delaying the passage of food material between them, as in ruminants such as bovines and deer. In such species absorption of fermentation products such as volatile fatty acids occurs before the food residues enter the acidic gastric chamber, where protein digestion is initiated. In hindgut fermenters (sometimes termed cecalids), fermentation occurs in the cecum, a blind sac branching from the junction of the small and large intestines, and/or in the large intestine or colon. Pockets and folds in the walls may retard passage of digesta through these hindgut chambers. Digestion and absorption of protein and soluble carbohydrates occurs before food residues undergo fermentation (Langer 1984).

Dental structure distinguishes browsers, feeding primarily on the leaves and stems of woody plants and dicotyledonous herbs, from grazers, eating primarily the leaves of grasses and sedges. Grazers exhibit finely textured surfaces on high-crowned molars for grinding the fine, fibrous and dusty leaves typical of graminoids. Browsers show lower-crowned molars with prominent cusps for dissecting the softer and thicker leaves and stems of dicotyledonous plants. Among ruminants there are further distinctions in the design of the digestive tract between grazers ('bulk and roughage feeders') and browsers (so-called 'concentrate selectors'). Grazers have a capacious fermentation chamber, narrow ostia and moderate surface papillation to cope with the slowly fermenting leaves of graminoids. Browsers have relatively smaller rumens with profuse papillae and larger connecting openings to process the faster digesting leaves of dicotyledonous plants (Hofmann 1973). Comparable differences probably exist among hindgut fermenters, but remain undocumented.

The African elephant (*Loxodonta africana*) is the largest living mammalian herbivore (Fig. 2.1). There are two subspecies: the bush elephant *L. a. africana*, and the forest elephant *L. a. cyclotis*. The forest subspecies is distinguished by downward pointing tusks, and appears to be slightly smaller in size than the bush elephant, though confirmatory weights are lacking (Kingdon 1979).

The largest recorded African elephant is an adult male shot in Angola. This animal measured 4.0 m high at the shoulder, and weighed an estimated 10 000 kg; but it was clearly exceptional. Elephants culled in Uganda and Zambia yielded an asymptotic shoulder height of 3.2 m, and masses of up to

Fig. 2.1 African elephant bull.

5500–6000 kg, among males. Females were somewhat smaller, weighing up to 2500–2800 kg on average, with a maximum of 3232 kg recorded. The asymptotic shoulder height was 2.7 m (maximum 2.9 m). Males show a growth surge between 20 and 30 years, so that full weight is attained relatively late in life. However, elephants in other parts of Africa may attain somewhat larger sizes. In the Gonarezhou Park in south-eastern Zimbabwe, female weights of up to 3800 kg were estimated from the hind leg mass. Zoo-kept females originating from Tanzania and Mozambique reached weights between 2800 kg and 3200 kg at 16 years, associated with shoulder heights of 2.5–2.9 m. A male of this group weighed 6600 kg and stood almost 3 m at the shoulder when destroyed at 25 years of age; while another captive male was still growing having reached a height of 3.25 m at 28 years. In the Kruger Park in South Africa, some exceptionally large bulls attained a shoulder height of 3.4 m (Bullock 1962; Hall-Martin 1987; Hanks 1972a; Lang 1980; Laws 1966; Sherry 1978).

The Asian or Indian elephant (*Elephas maximus*) (Fig. 2.2) is a little smaller than the African species. Captive females yielded shoulder heights of up to 2.35 m, while males measured up to 2.7 m. The mean weight

Fig. 2.2 Asian elephant (photo courtesy M. J. B. Green).

attained by females is reported to be 2720 kg, although a maximum weight of 4160 kg and shoulder height of 2.54 m is claimed. For males a maximum weight of 5400 kg, and shoulder height of 3.2 m, is reported; although the measured weight of one captive male was only 3600 kg (Hanks 1972a; Shoshani & Eisenberg 1982).

Elephants have the nasal region modified into a prehensile trunk serving as a food-gathering appendage. The lower incisor teeth have been transformed into enormous tusks. The molars replace one another in sequence, each tooth filling an arm of the short jaw so that no more than one and a half teeth are in operation on each half of the jaw. The teeth of Asian elephants are somewhat higher-crowned than those of the African elephant. In both species the enamel is folded into numerous plates, able to grind grasses as well as dicotyledonous browse. Female Asian elephants commonly lack tusks, but this pattern is relatively rare among African elephants. The stomach is simple, with the cecum not especially large relative to other parts of the gut, and the colon uncompartmentalized. Fermentation occurs both in the cecum and colon (Benedict 1936; Clemens & Maloiy 1982; Laws 1966; Maglio 1973).

The white or square-lipped rhinoceros (*Ceratotherium simum*) (Fig. 2.3) is generally regarded as third largest among living land mammals. However, there are no measured weights to confirm this. A young adult male (with the

Fig. 2.3 White rhino bull.

last molar not fully erupted), destroyed and sectioned at Umfolozi for the purpose of estimating drug dosages required for immobilization, weighed 2130 kg (John Clark personal communication 1965). Field weights are estimated by Natal Parks Board personnel to be 2000–2300 kg for adult males, and about 1600 kg for adult females. A weight of about 1800 kg is reported for a zoo-kept female (Foose 1982). Males attain shoulder heights of up to 1.8 m, females up to 1.77 m (Heller 1913; Kirby 1920).

Two subspecies of white rhino are distinguished. Animals of the northern race (*C. s. cottoni*) are differentiated from those of the southern race (*C. s. simum*) by the flatter dorsal profile of their skulls, and by their somewhat smaller teeth. Northern animals appear somewhat higher-legged and less long in the body, and lack the body hair which is present, although very sparsely, on southern animals (Cave 1962; Heller 1913).

The African black or hook-lipped rhinoceros (*Diceros bicornis*) (Fig. 2.4) is somewhat smaller than the white rhino. East African specimens weigh up to 1313 kg, while in South Africa weights of 708–1022 kg for males and 718–1132 kg for females were measured. Shoulder heights vary between 1.4 m and 1.65 m (Meinertzhagen 1938; Hitchins 1968, and personal communication). Seven subspecies of black rhino have been described, but some of

Fig. 2.4 Black rhino cow.

these are of uncertain status. The subspecies *D. b. bicornis* which inhabited the Cape was the largest. Although now extinct there, it has been suggested that animals surviving in northern Namibia may represent this form (Groves 1967; Hall-Martin 1985; Joubert 1970).

Both African rhino species lack incisors and canine teeth. The white rhino uses its broad lips to pluck grass, while in the black rhino the upper lip is modified into a finger-like projection which aids browsing. The white rhino has high-crowned molars and premolars, with fine surfaces adapted for grazing. In black rhinos the molars are lower-crowned with high cusps, in support of its browsing habits. White rhinos are further distinguished from black rhinos by their much longer heads, also an aid to grazing, and by the hump on the back of the neck formed by the hypertrophied nuchal ligament. Both species bear two horns on the snout. The horns of females tend to be longer and more slender than those of males. Some taxonomists regard the differences between the two species as trophic adaptations insufficient to warrant the generic distinction, and hence refer to the white rhino as *Diceros simus* (Cave 1962; Ellerman, Morrison-Scott & Hayman 1953).

There is no notable difference in skin color between the two species of

Fig. 2.5 Indian rhino (photo courtesy W. A. Laurie).

African rhino, this being influenced by the color of the soil last wallowed in. While it has been claimed that the appelation 'white' is a corruption of the Dutch word 'wijd', meaning wide, there is no basis for this conjecture. An early mention of the species in a document dated 1796–98 refers to the 'witte' (white) rhinoceros (du Plessis 1969); while Barrow (1801) mentions the supposed occurrence of a 'white rhinoceros' on the outskirts of the Cape settlement, distinguishing it by its 'pale carnation colour'. The most plausible explanation is that the first specimens to be encountered did in fact appear paler than the black rhinos inhabiting the fringe of the Cape settlement, probably as a result of wallowing in the calcareous soils typical of the northern Cape.

The Indian or great one-horned rhinoceros (*Rhinoceros unicornis*) (Fig. 2.5) is closely similar to the white rhino in size, although comparative measurements are sparse. Weights attained in the wild are estimated to be 2100 kg for males and 1600 kg for females; while a weight of about 1800 kg is reported for a captive female. Shoulder height may reach 1.8 m. The Javan or lesser one-horned rhinoceros (*Rhinoceros sondaicus*) is somewhat smaller, weighing up to about 1300 kg. Both of these species are characterized by a single horn on the snout, and by the retention of lower incisors, which have become modified into short tusks used in fighting. The prehensile upper lip of the Indian rhino aids feeding. The molars of the Indian rhino are

Fig. 2.6 Hippopotamus.

moderately high-crowned (although much less so than those of the white rhino), while the Javan rhino has molars that are relatively low-crowned and high-cusped, indicating a diet of browse (Foose 1982; Heller 1913; Laurie 1982; Thenius 1968).

The Sumatran or Asian two-horned rhinoceros (*Dicerorhinus sumatrensis*) attains a weight of about 800 kg and shoulder height of 1.2 m. It thus does not fall into the megaherbivore category, although I will report what is known about it for comparison with other rhinos. These animals are hairier than other living rhinos, and bear two fairly small horns on the snout. Incisors and canines are present in both jaws, while molars are low-crowned and high-cusped (Groves & Kurt 1972).

The digestive anatomy of rhinos resembles that of equids. In the black rhino the stomach is simple, the cecum voluminous and sacculated, and the colon also sacculated and compartmentalized. The chief site of fermentation is the cecum, with further fermentation occurring in the colon (Clemens & Maloiy 1982). Comparative descriptions are not available for other rhinos, although the basic pattern of the gut is similar.

For the hippopotamus (*Hippopotamus amphibius*) (Fig. 2.6) a maximum weight of 2660 kg has been reported for a male, which exceeds the weight attained by white rhinos; but this animal seems to have been exceptional.

The heaviest hippo shot during culling operations in Uganda was a female weighing 2025 kg, while in the Kruger Park in South Africa the heaviest animal culled was a male weighing 2005 kg. In Zambia the maximum recorded weight of animals culled in the Luangwa River was 1600 kg for males and 1565 kg for females. In the large sample of animals culled in Uganda the mean male weight was 1480 kg and the mean female weight 1365 kg. Hippos are semi-aquatic in habits, with a squat build and maximum shoulder height of about 1.4 m (Bere 1959; Marshall & Sayer 1976; Meinertzhagen 1938; Pienaar, van Wyk & Fairall 1966a).

Hippos retain incisor teeth, and have tusk-like canines used by males in fighting. The wide lips are used to pluck grass, and the molars are high-crowned. The stomach of a hippo is large and partly subdivided, creating some separation of contents between the anterior chambers and the posterior glandular section. Microbial fermentation occurs in the anterior sections. A cecum is lacking, and the large intestine is relatively undifferentiated. Despite forestomach fermentation, hippos do not remasticate food like ruminants (Clemens & Maloiy 1982; Langer 1976; van Hoven 1978).

Giraffe (*Giraffa camelopardalis*) (Fig. 2.7) attain maximum recorded weights of 1930 kg for a male and 1180 kg for a female. More typical weights are 1200 kg for adult males and 800 kg for adult females. Thus females generally do not reach the megaherbivore threshold as defined in this book. Giraffe are the tallest of living land mammals, with males reaching a maximum head height of 5.5 m, and females 4.5 m (Dagg & Foster 1976; Meinertzhagen 1938; Pellew 1984a).

Giraffe have a long, muscular tongue, which aids in gathering leaves into the mouth; and the dentition is typical of browsers. Giraffe are ruminants and like other ruminants chew the cud. The ruminoreticulum is relatively small, and the connecting ostia between compartments relatively large (Hofmann 1973).

The eight extant species of megaherbivore fall into four clusters in terms of body size: (i) the two elephants, with adult female weights of 2500–3800 kg; (ii) white rhino, Indian rhino and hippo, with adult female weights of 1400–2000 kg; (iii) black rhino and Javan rhino, with adult females weighing 1000–1300 kg; (iv) giraffe, with adult female weights of 800–1200 kg. The next largest mammalian herbivore is the Asian gaur, with adult males weighing as much as 940 kg. American bison bulls weigh up to about 900 kg, while African buffalo weigh up to 860 kg. Record specimens of these species could possibly reach 1000 kg, but such weights are exceptional. (See Appendix I for scientific names and maximum and mean weights for all large herbivores referred to in the book).

Fig. 2.7 Giraffe.

Elephants and rhinos are hindgut fermenters, while giraffe are true ruminants with forestomach fermentation. Hippos exhibit forestomach fermentation, but without the clearly divided compartments and remastication typifying ruminants. All bovids (cattle, antelope, sheep and goats) and cervids (deer) are ruminants. Other hindgut fermenters include zebras and other equids, and warthog plus other pigs (Langer 1984). White rhino and hippo have the dentition of a grazer; while the two elephants tend towards grazer dentition, without reaching the extreme dental specialization shown by the extinct mammoths. The Indian rhino is intermediate in its dental features, while black rhino, Javan rhino and giraffe have typical browsing dentition.

Evolutionary origins and relationships
Elephants

The elephants belong to the order Proboscidea, which can be traced back to the genus *Moeritherium*, found in Eocene deposits in Egypt. *Moeritherium* had the form of a stout-legged pig, and was probably semi-aquatic in its habits. Its descendants showed early tendencies towards elongation of the upper lip and nose, development of incisor teeth into tusks, and greatly increased body size. The deinotheres, which appeared in Africa during the Miocene and persisted in Eurasia through the Pliocene, had lower incisors which curved downwards and backwards in the form of huge hooks. In gomphotheres of this time the lower jaw was greatly lengthened, with tusks present in both upper and lower jaws (Maglio 1973; Coppens *et al.* 1978).

The modern day elephants are descended from gomphothere ancestors. The earliest recognized elephant, *Primelephas*, made its appearance in Africa during the late Miocene. It was ancestral to both of the extant genera, *Loxodonta* and *Elephas*, as well as to the mammoths (*Mammuthus*).

Loxodonta is the most conservative of these genera in its dental features, although early forms showed changes in the center of gravity of the jaw and height of the skull which developed somewhat later in other elephant lineages. It has remained exclusively African in its distribution. *Loxodonta atlantica*, which persisted until the mid Pleistocene, was larger than the modern African elephant, and also somewhat more progressive in its dentition. The living species, *L. africana*, first appeared during the late Pliocene, but from the paucity of fossil remains appears at that time to have been restricted to forest habitats. It does not become prominent in deposits at Olduvai Gorge in Tanzania and other savanna regions until late in the Pleistocene.

The genus *Elephas* appeared in both Africa and Eurasia during the mid Pleistocene. The African species *E. recki* exhibited molars which were higher crowned and more complexly folded than those of *L. atlantica*. *E. recki* and its descendant *E. iolensis*, similarly adapted for grazing, remained the most common elephant species in East African fossil deposits until about 35 000 years BP, when the lineage went extinct. The genus died out in Europe at about the same time, but in Asia it has persisted to the present day in the form of the Asian elephant *E. maximus*.

The genus *Mammuthus* originated in Africa during the mid Pliocene, but had disappeared from this continent by the early Pleistocene. Its main center of radiation was Eurasia, and mammoths entered North America fairly late in the Pleistocene. *Mammuthus* was the most advanced genus among the elephants in terms of its dental and cranial features – most particularly in its very high-crowned molars with their extremely complex pattern of fine enamel ridges, clearly specialized for grazing fibrous and abrasive forage. Other notable adaptations are the paired finger-like extensions on the tip of the trunk, combined with lateral wing-like extensions, which could have aided the gathering of grass (Guthrie 1982). With a shoulder height of over 4.0 m, the mid Pleistocene species *M. armenicus* was the largest of all proboscideans. The woolly mammoth *M. primigenius*, which persisted in both Europe and North America until the end of the last glacial, was somewhat smaller, with a shoulder height of about 2.8 m (Fig. 2.8).

Allied with the true elephants are the mastodonts of the family Mammutidae. Their prime distinguishing feature is the prominent mammary-gland-like cusps on their molar teeth, an indication of a primarily browsing diet. Mastodonts became extinct in Europe during the early Pleistocene, but persisted in North America until the end of the Pleistocene. Three genera of gomphotheres were also represented in South America during the Pleistocene. Among these *Stegomastodon* showed dental adaptations for grazing (E. Anderson 1984).

Rhinoceroses

A possible precursor of the Rhinocerotidae (order Perissodactyla) is the genus *Hyrachyus*, found in late Eocene deposits in North America and Asia. These were small, light-bodied animals, hardly different from the first horses and tapirs. Descendants soon exhibited a tendency towards large body size and towards the development of peculiar boneless horns, the presence of which is revealed by roughened areas on the fossilized skulls.

Baluchitherium, a rhinocerotid which occurred in Asia during the

Fig. 2.8 Mammoth (photo courtesy British Museum (Natural History)).

Oligocene and early Miocene, was the largest land mammal ever to walk the earth. It had a shoulder height of over 5 m, and was massively built. *Teleoceros*, a squat, heavy-bodied rhinoceros that lived in North America during the Miocene and early Pliocene, had a single small horn on the end of its snout; while the Miocene diceratheres exhibited two small nasal horns arranged side by side. *Elasmotherium*, from the Pleistocene steppes of Eurasia, had a gigantic single horn on its forehead. Rhinoceroses exhibited their greatest abundance during the latter half of the Tertiary era, and their generic richness had declined by the beginning of the Pleistocene (Colbert 1969).

The five living species of rhinoceros fall into three distinct subfamilies, which have had independent evolutionary histories for some time. The Asian one-horned rhinos belong to the subfamily Rhinocerotinae, which can be traced back to the Miocene genus *Gaindatherium*, found in the Sivalik hills of northern India. This subfamily is characterized by the single nasal horn, and by the retention of lower incisors, which have become modified into short tusks used in fighting. Of the two surviving species the Javan rhino is the more primitive, having altered little since the late Miocene some ten million years ago (Thenius 1969).

The Asian two-horned rhinos or Dicerorhinae can be traced back some forty million years to the tapir-sized *Dicerorhinus tagicus* from the Oligocene. The extant Sumatran rhino differs relatively little from its Oligocene predecessors. Four other species of this subfamily occurred in Europe during the course of the Pleistocene, including Merck's rhinoceros (*Dicerorhinus kirchbergensis*) as well as the woolly rhino (*Coelodonta antiquitatis*). The woolly rhino paralleled the white rhino in its high-crowned molars, absence of incisor and canine teeth, lengthened skull and other adaptations indicating a mainly graminoid diet. However it was somewhat smaller, attaining an estimated body mass of about 1100 kg. The giant *Elasmotherium* (of uncertain affinities) had enormous evergrowing molars with a complex enamel pattern, and was clearly also a grass-feeder (Fortelius 1982, and personal communication).

The African rhinos, or Dicerinae, are distinguished from the Dicerorhinae primarily by their lack of any ossification of the nasal septum. Both modern species lack incisor as well as canine teeth. The finding of the aberrant genus *Paradiceros* in Miocene deposits in Kenya suggests that the group had an independent African origin. The earliest known *Diceros* species, *D. pachygnathus* and *D. douariensis*, occurred in North Africa and adjacent parts of the Mediterranean region during the late Miocene. The modern black rhinoceros *D. bicornis* made its earliest appearance in Pliocene deposits dated at four million years BP, and by 2.5 million years BP its teeth had become indistinguishable from those of the modern representatives. The white rhinoceros lineage first appears in the form of the species *Ceratotherium praecox* in Pliocene deposits in Kenya dated at about seven million years BP. This ancestral form retained four incisors in the upper jaw, while its molars were not as high-crowned as those of the modern species. It was also somewhat larger than living white rhinos. *C. praecox* is abundant in fossil deposits at Langebaanweg in the Cape dated at 4–5 million years ago, so that its range was continent-wide. The modern species *C. simum* appears 3–4 million years ago in fossil deposits in East Africa, and is especially abundant in the Pleistocene deposits at Olduvai Gorge. The fossil evidence thus suggests a splitting of *Ceratotherium* from *Diceros* during the course of the early Pliocene (Hooijer 1969; Hooijer & Patterson 1972).

Hippopotamuses

The ancestors to the Hippopotamidae (order Artiodactyla) were the Anthracotheriidae, a widespread family of large, pig-like animals which occurred from the Oligocene to the late Pliocene. Genera such as *Merycopotamus* resembled modern hippos to a remarkable degree. True hippos appeared in Africa during the Pliocene, spreading later to southern Europe and Asia. The species *Hippopotamus antiquus* was especially common in Europe, including the British Isles, during Pleistocene interglacials. In Africa two species coexisted during the Pleistocene. The extinct *H. gorgops* had a longer and shallower skull, more elevated orbits and larger mandible than the modern hippo *H. amphibius*. The living pigmy hippo *Hexaprotodon liberiensis* is a tapir-sized animal, but extinct forms of *Hexaprotodon* attained sizes approaching those of living hippopotamus (Coryndon 1978).

Giraffes

Included among Giraffidae (order Artiodactyla) was the extinct genus *Sivatherium*, which occurred in both Africa and Eurasia during the Pliocene and early Pleistocene. Representatives attained shoulder heights of over 2.0 m, and with their massive build may have come marginally into the megaherbivore range. The earliest representatives of *Giraffa* appeared in Africa during the late Miocene. Giraffes occurred in southern Eurasia as well as in Africa during the Pliocene and Pleistocene (Churcher 1978).

Summary

Extant megaherbivores thus represent four mammalian families falling into three distinct orders. African and Asian rhinos are rather distantly related to each other, while the two elephant species are somewhat more closely allied. These species cannot be regarded as especially primitive in their evolutionary origins, except perhaps for the Javan rhino. All surviving species, except for the Javan rhino, date from the Pliocene radiations when most of the extant species of bovid and cervid also originated.

Paleontological diversity

Very large mammals were prominent in world faunas throughout the Tertiary and into the Quaternary. Apart from the proboscideans and rhinocerotids mentioned above, similar body sizes were attained by other forms of mammal. In the Americas giant ground sloths of the genera *Megatherium* and *Eremotherium* reached estimated masses of 3500 kg, while certain of the grazing mylodonts probably weighed over 1500 kg. The notoungulate genus *Toxodon* attained hippo-like proportions in South America; while the Australian marsupial *Diprotodon* was rhino-like in its build. The extinct South American camelid *Titanocamelops* reached a head height of 3.5 m, and was the local counterpart of the giraffe. Extinct forms of *Bison* such as *B. antiquus* and *B. latifrons* may have exceeded 1000 kg in maximum weight, but such sizes were probably attained only by the largest males. No bovid falls truly into the megaherbivore category.

During the course of the Pliocene and Pleistocene, megaherbivores were represented among fourteen mammalian families: the Diprotodontidae among the Marsupiala; the Megatheriidae and Mylodontidae in the Edentata; the Toxodontidae in the Notoungulata; the Gomphotheriidae, Deinotheriidae, Stegodontidae, Mammutidae and Elephantidae of the order Proboscidea; the Rhinocerotidae and Chalicotheriidae of the Perissodactyla; and the Anthracotheriidae, Hippopotamidae and Giraffidae among the Artiodactyla (Table 2.1). Throughout the course of the Pleistocene these families were represented by over 20 genera, and a somewhat greater number of species, worldwide (Table 2.2). Within regional faunas there were commonly between two and six species of megaherbivore, including perhaps two proboscideans, one or two rhinoceroses, a hippo-like animal, and perhaps a giant ground sloth or tall giraffe-like browser. Megaherbivores occurred in a complete range of ecosystems, from tropical forest and savanna, through deciduous and coniferous woodland, to the open grassy steppe of the subarctic during the ice ages.

The generic diversity of megaherbivores declined only slightly over the five million years of the Pliocene and Pleistocene. This was due mainly to a reduction in the number of proboscidean genera (Table 2.2). Between the late Pleistocene and the Holocene there was a dramatic reduction in diversity affecting all orders. Only three out of six orders persisted into the Holocene. While megaherbivores were formerly represented worldwide, surviving forms are restricted to the Old World tropics and subtropics of Africa and Asia.

Distribution of extant species

All surviving species of megaherbivore have become somewhat restricted in numbers and range in modern times compared with their early historic distribution.

Historically the African elephant was distributed continent-wide, from the environs of Cape Town to the fringe of the Sahara. In Roman times it evidently occurred also in the Mediterranean region. It is still represented today through much of this region, although its distribution has become somewhat fragmented. The species occupies habitats ranging from equatorial rain forest through various forms of savanna to semidesert in Namibia and Mali. Its center of greatest abundance seems to be the broad savanna woodland region extending through central and eastern Africa;

Table 2.1. *List of the genera of megaherbivores extant during Pliocene,*
Pleistocene and Recent times

Taxon	Time period	Geographic range
Marsupiala		
Diprotodontidae		
Diprotodon	Pleistocene	Australia
Edentata		
Megatheriidae		
Megatherium	Pleistocene	South America
Eremotherium	Pleistocene	South and North America
Mylodontidae		
Mylodon	Pleistocene	South America
Notoungulata		
Toxodontidae		
Toxodon	Pleistocene	South America
Proboscidea		
Gomphotheriidae		
Anancus	Miocene–M. Pleistocene	Africa, Eurasia
Cuvieronius	Pliocene–L. Pleistocene	South and North America
Haplomastodon	L. Pleistocene	South America
Stegomastodon	Pliocene–L. Pleistocene	South and North America
Deinotheriidae		
Amebelodon	Miocene–E. Pliocene	Europe, North America
Deinotherium	Miocene–E. Pleistocene	Africa, Asia, Europe
Gnathobelodon	Pliocene	Europe
Platybelodon	Miocene–E. Pliocene	Asia
Stegodontidae		
Stegodon	L. Pliocene–L. Pleistocene	Africa, Asia
Stegolophodon	L. Pliocene	Eurasia
Mammutidae		
Mammut	E. Miocene–L. Pleistocene	Eurasia, North America
Zygolophodon	Pliocene	Africa, Europe
Elephantidae		
Elephas	E. Pliocene–Recent	Africa, Asia, Europe
Loxodonta	M. Pliocene–Recent	Africa
Mammuthus	E. Pliocene–L. Pleistocene	Africa, Eurasia, North America
Primelephas	Miocene–E. Pliocene	Africa
Stegodibelodon	Miocene–E. Pliocene	Africa
Perissodactyla		
Rhinocerotidae		
Ceratotherium	M. Pliocene–Recent	Africa
Coelodonta	E.–L. Pleistocene	Eurasia
Dicerorhinus	Oligocene–Recent	Europe and Asia

Table 2.1. (*cont.*)

Taxon	Time period	Geographic range
Diceros	E. Pliocene–Recent	Africa, S. Europe
Elasmotherium	M.–L. Pleistocene	Asia
Rhinoceros	E. Pliocene–Recent	Asia
Sinotherium	Pliocene	Asia
Teleoceras	Miocene–E. Pliocene	Eurasia, North America
Chalicotheriidae		
Ancyclotherium	Pliocene–E. Pleistocene	Africa, S. Eurasia
Artiodactyla		
Anthracotheriidae		
Merycopotamus	Miocene–L. Pliocene	Africa, Asia
Hippopotamidae		
Hexaprotodon	Pliocene–Recent	Africa, Asia
Hippopotamus	Miocene–Recent	Africa, Eurasia
Giraffidae		
Giraffa	Miocene–Recent	Africa, Eurasia
Sivatherium	E. Pliocene–M. Pleistocene	Africa, Asia

Sources: Compiled from Martin and Guilday (1967), Kurtén (1968), Maglio & Cooke (1978), Kurtén and Anderson (1980), Simpson (1980), E. Anderson (1984) and Martin (1984a).
Note: E. Pleistocene = early Pleistocene
          M. Pleistocene = middle Pleistocene
          L. Pleistocene = late Pleistocene

although large numbers also occur in the forests of Zaire and Gabon, judging from the amount of ivory of the forest race coming onto the market. The current world population of African elephants totals under one million, and numbers are diminishing rapidly (Cumming & Jackson 1984; Douglas-Hamilton 1987; Kingdon 1982).

The Asian elephant was formerly distributed through most of tropical Asia, from India and Sri Lanka through to Malaysia, Indonesia and southern China. Today remnant populations persist in the wild in parts of Sri Lanka, south-eastern India, Assam, Burma, Thailand, Malaya and most of the larger islands of the Malaysian and Indonesian archipelago. The species occupies both forest and open woodland habitats (Eltringham 1982).

The hippopotamus was distributed in historic times from the Cape to the upper Nile River, wherever suitable water bodies occurred. The species still occurs today over much of this range, with its center of greatest abundance in the lakes of the western rift valley along the Uganda–Zaire border. In the

24   *Megaherbivores*

Table 2.2. *Numbers of megaherbivore genera represented in various continental faunas during different time periods*

Continent	Time period	Marsupiala	Edentata	Notoungulata	Proboscidea	Perissodactyla	Artiodactyla	Total
Africa	Pliocene	0	0	0	8	3	4	15
	Early Pleistocene	0	0	0	5	3	4	12
	Mid Pleistocene	0	0	0	3	2	4	9
	Late Pleistocene	0	0	0	2	2	2	6
	Recent	0	0	0	1	2	2	5
Asia	Pliocene	0	0	0	6	3	4	13
	Early Pleistocene	0	0	0	5	3	3	11
	Mid Pleistocene	0	0	0	3	4	2	9
	Late Pleistocene	0	0	0	2	4	1	7
	Recent	0	0	0	1	2	0	3
Europe	Pliocene	0	0	0	7	3	2	12
	Early Pleistocene	0	0	0	4	2	1	7
	Mid Pleistocene	0	0	0	2	2	1	5
	Late Pleistocene	0	0	0	2	2	1	5
	Recent	0	0	0	0	0	0	0
North America	Pliocene	0	0	0	3	1	0	4
	Early Pleistocene	0	1	0	4	0	0	5
	Mid Pleistocene	0	1	0	3	0	0	4
	Late Pleistocene	0	1	0	3	0	0	4
	Recent	0	0	0	0	0	0	0
South America	Pliocene	0	0	1	0	0	0	1
	Early Pleistocene	0	2	1	2	0	0	5
	Mid Pleistocene	0	2	1	2	0	0	5
	Late Pleistocene	0	3	1	3	0	0	7
	Recent	0	0	0	0	0	0	0
Australia	Pliocene	1	0	0	0	0	0	1
	Early Pleistocene	1	0	0	0	0	0	1
	Mid Pleistocene	1	0	0	0	0	0	1
	Late Pleistocene	1	0	0	0	0	0	1
	Recent	0	0	0	0	0	0	0
World	Pliocene	1	0	1	15	7	4	28
	Early Pleistocene	1	3	1	9	6	4	24
	Mid Pleistocene	1	3	1	8	6	4	23
	Late Pleistocene	1	3	1	8	6	2	21
	Recent	0	0	0	2	4	2	8

Sources: Compiled from Kurtén (1968), Maglio (1978), Kurtén and Anderson (1980), Simpson (1980), E. Anderson (1984), and Martin (1984a).

Pleistocene its range extended quite far northwards in Europe during interglacial periods (Kingdon 1979).

Giraffe remain widely distributed through savanna regions, from the Kalahari and Transvaal lowveld in the south to Mali and Somalia in the north. The historic range did not extend south of the Pongola River into Natal, but animals have been introduced into wildlife reserves in Zululand and are thriving there. In central Africa the species shows a distribution gap associated with the miombo woodlands of Zambia and southern Tanzania, apart from an isolated population in the Luangwa valley (Kingdon 1982).

The distribution of the Indian rhino was limited historically to north India and adjacent regions of Nepal. The habitats that it occupies consist of tall floodplain grassland and adjacent woodland. Today it occurs only in a few reserves in Assam and Nepal, the total population numbering about 1700 animals (Laurie 1982).

The Javan rhino was formerly widely distributed through most of south-east Asia from India to China and southwards through Indonesia. It generally occupied lowland forest. The current population is 55 animals restricted to the western tip of Java (Schenkel & Schenkel-Hulliger 1969b; Hoogerwerf 1970). The Sumatran rhino formerly ranged from Assam through to Vietnam and the islands of Borneo and Sumatra. It seems to favor more broken mountainous forest than the Javan rhino. About 700–800 animals remain, thinly scattered through Sumatra, Malaya, Borneo, Thailand and Burma (Borner 1979).

The historic distribution of the black rhinoceros was almost as wide as that of the African elephant, extending from the south-western Cape to Somaliland and the northern Cameroons–Ivory Coast border. However, the species was absent from the equatorial forest region of central Africa, its favored habitats being drier savanna and arid shrub steppe. Today, black rhinos survive in scattered population fragments through this range. The large populations that formerly occurred in the Luangwa Valley in Zambia and in southern Tanzania have been reduced to small remnants by poaching, and the largest surviving population is currently that in the Zambezi valley. Latest (1986) estimates indicate that only about 4500 black rhinos remain in Africa (Kingdon 1979; unpublished reports of the African Rhino and Elephant Specialist Group of the IUCN).

The white rhinoceros was distributed historically in the form of two discrete populations, separated by a gap of over 2000 km (Fig. 2.9). The species did not occur south of the Orange River in historic times, while in the east its southern limit was the region of the present-day Umfolozi Game Reserve. The northern boundary of the range of the southern race was the

Fig. 2.9 Historic distribution of the white rhino. Distributional limits of southern population adapted from Huntley (1967). Distributional limits of northern population from Lang (1920).

Zambezi River and the region of the Namibia–Angola border (Huntley 1967). White rhinos were particularly abundant in eastern Botswana and adjacent parts of the western Transvaal. For example Harris (1838) reported seeing 80 in a day's march north of the Magaliesberg Range towards the upper Limpopo River. Favored habitats seem to have been semi-arid savanna, although in Zimbabwe animals were commonly associated with drainage line grasslands (Kirby 1920; Selous 1899).

   The northern subspecies was found only to the west of the Nile River, from northern Uganda northwards to the vicinity of Shambe in the Sudan, and westwards into the Central African Republic. This northern range is associated mostly with mesic, *Combretum*-dominated savanna with tall grass prevalent except after fires. Although there are no historic records of white rhinos in East Africa, teeth found in Kenya and Tanzania suggest that the species occurred there during the Holocene. Rock engravings in Algeria indicate that the species was present in North Africa perhaps 5000–10 000 years ago (Heller 1913; Hooijer & Patterson 1972; Lang 1920, 1923).

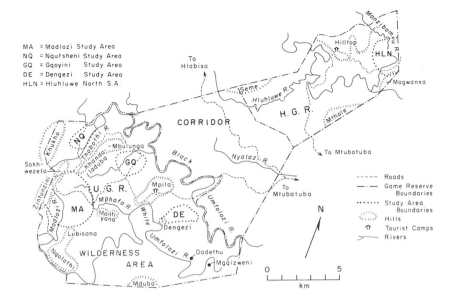

MA = Madlozi Study Area
NQ = Nqutsheni Study Area
GQ = Gqoyini Study Area
DE = Dengezi Study Area
HLN = Hluhluwe North S.A.

----- Roads
—·— Game Reserve Boundaries
······· Study Area Boundaries
Hills
↑ Tourist Camps
>— Rivers

Fig. 2.10 The Umfolozi–Corridor–Hluhluwe Game Reserve complex, showing locations of main features and of the intensive study areas.

The northern race of white rhino numbered about 2000 animals during the early 1960s. In 1980 an estimated 800 white rhinos still survived in Sudan and north-eastern Zaire; but the population has since been decimated by poaching for the horns, so that at the time of writing only 17 animals remain, all in the Garamba Park in Zaire (Schomber 1966; Hillman-Smith, Oyisenzoo & Smith 1986).

White rhinos were exterminated in southern Africa during the late nineteenth century, except for a few score surviving at the southern limit of the range between the Black and the White Umfolozi rivers in Zululand (although another remnant survived in the Nuanetsi region of Mozambique until the 1930s). The Umfolozi Game Reserve was proclaimed in 1898 to protect the last survivors, although protection did not become effective until 1920 (Sidney 1966; Vincent 1970).

Since then numbers have increased steadily to reach a peak of 2000 animals in the Umfolozi–Hluhluwe region in 1970. Starting in the 1960s, animals have been translocated from this population to restock other areas where the species formerly occurred. The current total population numbers about 4000 white rhinos, over half of which now occur outside the Umfolozi–Hluhluwe Reserve.

My study of the white rhino was carried out in the Umfolozi Game Reserve and neighboring Hluhluwe Reserve in Zululand (Fig. 2.10).

Fig. 2.11 View over the white rhino study area in the western region of Umfolozi Game Reserve.

Following an initial survey in 1966, the main study period extended from November 1968 to September 1971. Most intensive observations were carried out near the Madlozi outpost in the western region of the Umfolozi Reserve, where the highest densities of white rhinos existed (Fig. 2.11).

The Umfolozi Game Reserve covers an area of about 450 km², and is joined to the Hluhluwe Reserve by an intervening area of State Land known as the Corridor. The total extent of this unit, termed the Umfolozi–Hluhluwe complex, is about 950 km². The Umfolozi section consists of mostly gently rolling thorn savanna, underlaid by relatively fertile soils derived from Ecca shales and sandstones of the Karroo formation. In the Hluhluwe Reserve the country is more steeply rolling with open grassy hills and forest patches. The climate in Umfolozi is hot, with mean daily maximum temperatures of 32.6 °C in January (mid-summer) and 25.3 °C in July (mid-winter). The corresponding mean daily minima are 21.8 °C and 13.2 °C. The mean annual rainfall at Mpila Camp in Umfolozi is 700 mm (1959–1980), 70% of which falls during the six summer months of October to March. Rainfall increases northwards to reach a mean of 985 mm (1932–1980) at Hilltop Camp in Hluhluwe. The vegetation in Umfolozi is dominated by small trees and shrubs of the genus *Acacia* in the woody layer, and the grass *Themeda triandra* in the herb layer, although extensive areas of mixed short grasses occur.

In summary, most of the extant species of megaherbivore had a nearly continent-wide distribution in early historic times, wherever suitable savanna or forest environments occurred. The exceptions are the Indian rhino and, to a lesser degree, the white rhino. However, the white rhino was widely distributed through Africa during the Pleistocene, and there is evidence that its disappearance from East Africa was relatively recent. Other ungulate species with comparably wide historic distributions in Africa include warthog, bushpig, bushbuck and African buffalo; while those widely distributed through tropical Asia include water buffalo, sambar deer and muntjac.

All three Asian species of megaherbivore have become greatly restricted in geographic range, with the Javan rhino poised on the verge of extinction. The white rhino was hunted to near extinction in southern Africa during the last century, but has since recovered under protection. However the northern race of white rhino has been reduced to critically low numbers in the past few years. Black rhinos numbers are dwindling rapidly. While the current status of the African elephant is healthy overall, populations are undergoing steady attrition.

3

Food and other habitat resources

Introduction

The habitat resources of interest to this chapter are those that individual animals of a species depend upon for their survival. These include food sources, surface water, and refuges from weather extremes.

Food

For large herbivores dietary intake may be characterized either in terms of (i) the plant species eaten, (ii) the plant parts ingested, or (iii) the nutrient contents of the ingested material.

In terms of plant species, the basic classification is in terms of the graminoid:dicotyledon proportions (including non-graminaceous monocots with dicots). The leaves of grasses have higher contents of fibrous cell wall components, and digest more slowly, than the leaves of woody and herbaceous dicots. Silica bodies present in grass leaves further reduce digestibility and also abrade teeth. However, the leaves of woody dicots are ultimately less digestible than those of grasses, due to a higher proportion of indigestible lignin incorporated in the cell wall. Furthermore, the leaves of woody and herbaceous dicots frequently contain toxic or digestibility-reducing compounds, which are much less common in grasses.

In terms of plant parts, the proportions of foliage, stemmy material and fruits in the diet are of interest. Supporting tissues such as stems and bark tend to be high in indigestible fiber, while fruit pulp and seeds contain stores of soluble carbohydrates. Leaves contain the photosynthetic enzymes and are highest in protein and minerals (apart from calcium), although protein content declines as leaves age and fiber contents increase.

Nutrient content is most widely expressed in terms of the 'crude protein' (nitrogen $\times$ 6.25) concentration in the dry matter. Energy availability is dependent upon the digestibility of the structural carbohydrates (cellulose and hemicellulose) forming, together with lignin, the cell walls. However, the overall dry matter digestibility tends to be closely related to the crude protein content (Owen-Smith 1982).

In most environments food abundance and quality change seasonally due to the phenology of plant growth. Food quality is highest early in the growing season due to the prevalence of new leaves and shoots, while food abundance peaks later in the growing season. During the dormant season many woody plants shed their leaves, while grasses withdraw nutrients and leave standing dead leaves. However fruits and seeds may provide a high quality supplement during the early part of the dormant period.

Nutritional balance depends most directly not on the potential food abundance in the vegetation, but on the rate of food ingestion and on the nutritional value of the ingested material. Defecation rates provide an indication of food passage rates, and, indirectly, of daily food intake (allowing for digestibility).

Diet composition
Elephants

African elephants exhibit much variation in grass:browse proportions in the diet (Table 3.1). Under open grassland conditions, such as prevail in the Murchison Falls and Queen Elizabeth National Parks in Uganda, grass may form 60–95% of the diet year-round. In wooded savannas in Kenya, Uganda, Zambia, Zimbabwe and Tanzania, grass occupies between 40% and 70% of the feeding time during the wet season, but only 2% to 40% during dry season months. Bulls tend to select slightly higher proportions of grass than cows. Grass is insignificant in the diet of elephants inhabiting forests in Ghana and Ivory Coast, where woody browse and fruits are the main food components. Fruits and seed pods are also actively sought out by savanna elephants when available (Barnes 1982a; Buss 1961; Field 1971; Field & Ross 1976; Guy 1976a; Laws & Parker 1968; Lewis 1986; Merz 1981; Napier Bax & Sheldrick 1963; Short 1981; Williamson 1975a).

When feeding on grasses, African elephants favor leaves and inflorescences during the wet season. Commonly eaten species include short grasses such as *Cynodon* as well as taller grasses like *Panicum*, *Setaria*, *Themeda* and *Hyparrhenia* species. During the dry season elephants select the leaf bases and roots of genera including *Andropogon*, *Cymbopogon* *Hyparrhenia* and *Setaria*, kicking tussocks free of the ground with their feet with leaves and stems being discarded uneaten (Fig. 3.1) (Field 1971; Field & Ross 1976; Lewis 1986).

When browsing during the wet season, African elephants strip leaves,

Table 3.1. *Grass: browse proportions in the diet of African elephants in different regions*

Area	Method	Season	Graminoids (%)	Forbs (%)	Woody plants (%)	Reference
Uganda						
Murchison Falls	Stomach contents	Wet	97	{ 3 }		1
		Dry	95	{ 5 }		1
Queen Elizabeth	Fecal contents	Wet	90.5	{ 9 }		1
	Feeding time	Wet	59	30	11	1
		Dry	66	18	16	1
Kidepo	Feeding time	Wet	57	22	21	2
		Dry	28.5	13	59	2
	Fecal contents	Wet	73	{ 27 }		2
		Dry	66	{ 34 }		2
Tanzania						
Ruaha	Feeding time	Wet	66	9	25	3
		Dry	2.5	2	92	3
Zimbabwe						
Sengwa	Feeding time	Wet	39	—	16	4
		Dry	7	—	32	4
Zambia						
Luangwa	Feeding time	Dry	31	0	62	5

Source references: 1 – Field 1971; 2 – Field and Ross 1976; 3 – Barnes 1982; 4 – Guy 1976a; 5 – Lewis 1986.

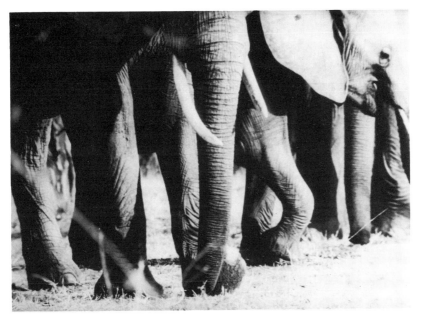

Fig. 3.1 African elephants kicking out grass plants with their feet (Luangwa Valley, Zambia).

and break off branchlets to consume the terminal twigs (Fig. 3.2). Bark may be stripped by drawing small branches through the mouth. From *Acacia* trees the woody material ingested outweighs the foliage. In the Serengeti elephants obtain most of their browse from drainage line thickets, and make relatively greater use of forbs in more open habitats. In some areas small woody plants less than 1 m in height are ignored, while in other areas they are uprooted and eaten whole. When feeding on thorny *Acacia xanthophloea* scrub, elephants flatten the thorns between a tusk and the base of the trunk, and may bite off and discard the more prickly distal end (Croze 1974a and b).

In the dry season elephants feed more on bark, woody stems and roots. Bark stripping commonly occurs just before trees flower or leaf out, i.e. when the bark is likely to be rich in sap. Burning of grasslands may cause elephants to exert greater pressure on trees, but animals tend to move out of burnt areas to concentrate their feeding in unburnt marshes or evergreen forest patches. Under severe drought conditions wood, bark and roots may occupy 70–80% of the feeding time. When pressed for food animals also feed on the soft, pithy stems of trees such as baobab (*Adansonia digitata*) and chestnut (*Sterculia* spp) trees, eventually destroying the plant.

Fig. 3.2 African elephant browsing on mopane foliage (Luangwa Valley, Zambia).

Favored genera of trees and shrubs include *Acacia, Azima, Baphia, Brachystegia* (certain species only), *Combretum, Colophospermum, Terminalia* and *Uapaca*. Genera eaten rarely or not at all include *Boscia, Burkea, Capparis, Diospyros, Melia* and *Protea*. When feeding on *Colophospermum mopane*, elephants prefer to browse regrowth from previously damaged trees. In the dry season they discard the resinous leaves of mopane and *Commiphora* and consume woody branchlets, bark and roots (Anderson & Walker 1974; Barnes 1982a; Buss 1961; Croze 1974a; Douglas-Hamilton 1972; Field & Ross 1976; Guy 1976a; Jachmann & Bell 1985; Laws, Parker & Johnstone 1975; Lewis 1986; Napier, Bax & Sheldrick 1963; Pienaar, van Wyk & Fairall 1966b; Western & Lindsay 1984; Williamson 1975a).

In Zimbabwe 70% of all browsing occurred below a level of 1.2 m, while in Malawi the preferred feeding level was about 1.5–2 m above ground. The maximum feeding reach with the trunk is about 6 m, and trees taller than 6 m may be pushed over bringing higher branches within reach. However, elephants do not always feed on trees they have felled. In Zimbabwe elephant bulls pushed over, as a year-round average, 6 trees per day, compared with 2.6 per day by cows. In Serengeti in Tanzania the tree pushing rate by bulls averaged only 0.7 per day, and only 30% of tree pushing attempts were successful, trees greater than about 0.25 m in diameter commonly withstanding attempts to push them over. The rate of food intake obtained from the herb layer is considerably higher than that secured from pushed over trees. Trees pushed over in the Kasungu National Park in Malawi showed a height mode of 4–5 m for favored species, but 2–3 m for species generally rejected as food. Since pushed over trees commonly coppice from the base, this selective damage could lead to increased availability of food at an accessible feeding level (Croze 1974a; Guy 1976a; Jachmann & Bell 1985).

Elephants feeding in forest patches in Uganda are attracted to regeneration in the patches opened by timber management. They favor stems under 250 mm in diameter and commonly break off leader shoots, thereby maintaining the secondary growth. Important timber species such as *Khaya* (mahogany), *Chrysophyllum*, *Cordea* and *Maesopsis* are favored as food, so that the course of forest succession is deflected to less desirable species (Laws 1970; Wing & Buss 1970). In Ivory Coast, elephants do most of their feeding in secondary rain forest, but depend on primary forest for certain fruiting trees (Merz 1981).

The daily food intake of African elephants has been estimated either from the mass of the stomach contents, assuming a mean turnover time of 12 hours; or by extrapolating from the feeding rate and daily feeding time. Both methods give similar results, indicating a mean daily food intake of about 1.0–1.2% of body mass per day for males and non-lactating females, and 1.2–1.5% of body mass per day for lactating females (dry mass/ livemass). Food intake appears somewhat higher in the wet season than in the dry season on a wet mass basis, but the difference would probably be reduced if measured as dry mass. Crude protein concentrations in the stomach contents of elephants culled in Uganda varied between 6% and 14% during the wet season, and 5% to 8% during the dry season. Animals from the Queen Elizabeth Park, where there was a higher proportion of browse in the diet, showed higher protein contents than animals from the Murchison Falls Park, where the diet consisted mostly of grass (Guy 1975; Laws, Parker & Johnstone 1975; Malpas 1977; McCullagh 1969).

African elephants defecate about 14–20 times per day in the wet season, and about 10 times per day during the dry season. An adult produces between 6 kg and 11 kg of feces per defecation on average, depending on its size. On average, the total quantity of feces produced per day would amount to about 150 kg wet mass, or 35 kg dry mass. If the daily food intake of an adult male is about 60 kg dry matter per day, this suggests a dry matter digestibility of about 40% (Barnes 1982a; Coe 1972; Dougall & Sheldrick 1964; Guy 1975; Wing & Buss 1970; Wyatt & Eltringham 1974).

Asian elephants devoted 89% of their feeding time to grass in the Ruhunu Park in Sri Lanka; while in the Gal Oya Park, also in Sri Lanka, grass formed just over 50% of the diet. Even in the forests of Malaya, grass occupied up to one third of feeding time. Most of the grass eaten consisted of short grasses, entire plants being consumed. Grass plants shorter than about 60 mm were kicked out of the ground with the feet, then gathered with the trunk. Tall swamp grasses were eaten during the winter dry season. Availability of freshly growing grass seemed to be the main factor controlling seasonal movements.

At Ruhunu, elephants foraged preferentially in open scrub or scrub-forest, and made relatively little use of evergreen forest vegetation. Gap-colonizing species of woody plant were favored over shade-tolerant species. In both areas in Sri Lanka, utilization of woody plants was largely by branch breaking, foliage plus bark being stripped from the broken off stem. Plants with a stem girth between 20 and 160 mm were favored. The rarity of bark stripping from main stems may be due to the lack of tusks in most of the elephants observed. Tree pushing was also rare. Small woody plants were commonly eaten whole. With spiny shrubs the trunk was drawn up the stem to flatten the spines before plucking. In Malayan forests elephants eat mostly quick growing pioneer species, and the dominant trees of the Dipterocarpaceae are rejected. Favored feeding areas in forest regions are the open glades bordering rivers (McKay 1973; Mueller-Dombois 1972; Olivier 1982; Vancuylenberg 1977).

Asian elephants defecate between 12 and 18 times per day in the wild. A captive adult male fed hay produced 114 kg (wet mass) of feces per day (Benedict 1936; Vancuylenberg 1977).

Hippopotamus

Hippos are area selective grazers with a preference for short green grass. Grasses and sedges form 95–99% of the food eaten by hippos in different regions, the remainder being made up by forbs, which increase in representation during the dry season. Short or leafy species of grass are favored, and coarser grasses tend to slip between the lips and so are neglected. Favored genera include *Panicum, Cynodon, Brachiaria, Sporobolus, Themeda, Digitaria, Heteropogon, Urochloa, Hemarthria, Echinochloa* and *Cyperus*, based on observations made in Uganda, Zaire and Natal. Aquatic herbs such as *Pistia* are eaten in small quantities. The large pods of *Kigelia* are reportedly eaten in Zambia.

Hippos pluck grass close to ground level, and their feeding promotes the development of short grass lawns bordering the pools serving as their daytime refuges. Most feeding occurs within a kilometer or less of water during the wet season, but in the dry season animals may move 2–3 km or more from rivers or lakes (Field 1970; Lock 1972; Mackie 1976; O'Connor & Campbell 1986; Olivier & Laurie 1974; Scotcher, Stewart & Breen 1978).

A hippo's stomach contents amount to 13–15% of body mass on a wet mass basis and represent two nights feeding. Estimated daily food intake on a dry mass basis is 1.1% of body mass for males and 1.3% for females, with lactating females having higher stomach fills than pregnant females (Laws 1968b).

Giraffe

Giraffe are almost exclusively browsers, feeding on the leaves and shoots of trees and shrubs. Herbaceous material, including climbers, vines and taller forbs (but no grass), forms between 0.2% and 7% of the year-round diet in different areas. Flowers, fruits and pods are favored when available. Females consume a higher proportion of herbaceous plants than males.

Giraffe favor deciduous trees and shrubs during the wet season, but rely increasingly on evergreen or semi-evergreen species as the dry season advances. During the late dry season their feeding tends to be concentrated in the forest or thicket fringes flanking rivers and dry watercourses. Important food genera include *Acacia, Capparis, Combretum, Kigelia, Securinega* and *Ziziphus* during the wet season, and *Albizia, Balanites, Boscia, Colophospermum, Diospyros, Euclea, Grewia* and *Melia* during the dry season. Favored fruits include the pods of various leguminous trees, the huge woody pods of *Kigelia* (Fig. 3.3), and hard baseball sized fruits of *Gardenia* and *Strychnos* (Field & Ross 1976; Hall-Martin 1974; Kok & Opperman 1980; Leuthold & Leuthold 1972; Lightfoot 1978; Pellew 1984b; Sauer, Theron & Skinner 1977).

Giraffe seek out the new unhardened shoots of *Acacia* species, and their feeding stimulates increased shoot production by these plants. From hardened shoots they strip leaves from branch ends with their tongues, or bite off the shoot end. The proportion of woody material ingested increases in the dry season, forming 15% of the rumen contents at this time, compared with 5% during the wet season. At Tsavo in Kenya giraffe performed 67% of their feeding below 2 m during the wet season, decreasing to 37% during the dry season; but at Kyle in Zimbabwe only 20% of feeding was below 2 m. At Serengeti in Tanzania bulls did 75% of their feeding above

Fig. 3.3 Giraffe plucking the large woody pod of *Kigelia pinnata* (Luangwa Valley, Zambia).

4 m, while for cows 78% of feeding was done below 3 m (Leuthold & Leuthold 1972; Lightfoot 1978; Pellew 1983a, 1984b).

Based on feeding rate and daily feeding time, the food intake of giraffe is estimated to be 2.1% of body mass for females, and 1.6% for males, on a dry mass basis. The crude protein content of the material in the rumen varies between about 19% in the wet season and 14.5% in the dry season for adult females, with values for adult males being about 3–4% lower (Field & Blankenship 1973; Hall-Martin & Basson 1975; Pellew 1984c).

Rhinoceroses

For Indian rhinos in Nepal, tall canelike grass species form the main food source year-round, in particular species of *Saccharum*. These are most favored when young in spring. Short grasses, such as *Cynodon*, and herbs make up the bulk of the diet during the monsoon period, with aquatic herbs such as *Ceratophyllum* and *Hydrilla* especially favored. Woody browse forms about 2.5% of the diet during the monsoon, increasing to 22% during the winter period. Fallen fruits are also eaten. When grazing on short grasses, Indian rhinos fold back the projecting upper lip (Laurie 1982).

Sumatran rhinos in the Gunung Leuser Park in Sumatra feed mainly on

Fig. 3.4 Black rhino browsing on low *Acacia* scrub (Umfolozi, South Africa).

small trees or saplings, consuming twigs, small branches and leaves, and also certain fruits. Herbs and lianas formed only 1% of the plants recorded as eaten. To reach the higher shoots of woody saplings, animals bend or break the stem by walking over the plant and pressing down on the trunk with the body. Sumatran rhinos snapped with ease plants with stem diameters of up to 50 mm (Borner 1979). Javan rhinos are also browsers, feeding mainly on the twigs and branches of saplings (Schenkel & Schenkel-Hulliger 1969b).

Black rhinos are predominantly ground feeders, concentrating on forbs and low-growing woody scrub (Fig. 3.4). Grass occupies no more than 1–5% of the feeding time even in open grassland habitats such as Ngorongoro Crater, Tanzania, and Masai Mara Park, Kenya. In the semi-arid steppe of Tsavo East National Park, Kenya, grasses were unrepresented in feeding records. Forbs and dwarf shrubs, especially legumes such as *Indigofera*, *Tephrosia*, *Trifolium*, *Lathyrus*, *Aeschynomene*, and *Caesalpinia*, are the favored food source, and occupy between one third and three-quarters of feeding time during the wet season.

Woody browse becomes more important in the diet following fires which remove the herb layer. Woody plants are also relatively more important in

Fig. 3.5 White rhino grazing short grass (Umfolozi, South Africa).

Namibia and at Addo in the southern Cape where herbs are less plentiful. When browsing the prehensile upper lip is used to pull twigs into the mouth, and these are then bitten off with the molars. The preferred browsing level is between 0.5 and 1.2 m, with the maximum reach being 1.5 m. Black rhinos bite off shoot ends 100–250 mm in length and up to 10 mm in thickness, and may consume 30–60% of the above-ground biomass of plants under 0.5 m in height. Species lacking thorns may be defoliated by running the lips over the twigs. Important food sources include genera such as *Acacia*, *Combretum*, *Croton*, *Dichrostachys*, *Grewia* and *Terminalia*. Common woody species rejected as food at Tsavo include *Boscia*, *Commiphora* and *Dobera*. *Colophospermum mopane* is eaten in limited amounts during the wet season in Namibia. Stem succulents such as species of *Euphorbia* become an important food source during the dry season if available. At Tsavo such plants formed up to 70% of the food intake despite their high latex content. Woody plants generally increase in representation in the dry season. During the hot, dry mid-summer period at Addo, leaf succulents such as *Portulacaria* made up over 40% of the food. The horns may be used to break down higher branches, with main stems up to 170 mm in diameter being snapped. Bark may also be stripped from certain species, for example

Table 3.2. *Trends in the food selection of white rhinos in relation to grassland condition*

Figures represent percentage of total plant bites.

Grass form	Grass condition					
	Early green[a]	Late green	Mainly green	Mainly brown	Brown	Year[b]
Short grass species	56.9	54.7	45.4	20.0	18.3	46.3
Climax grass species	26.8	23.9	27.2	56.5	66.6	33.4
Shade grass species	10.4	16.9	17.3	18.8	7.8	13.8
Miscellaneous grass species	4.8	4.2	7.4	2.0	5.2	5.3
Sedges	0.2	0.6	0.4	0	0.8	0.2
Forbs	0.7	0.2	1.9	3.7	1.3	1.0
Mean grass height (leaf table, mm):						
Before grazing	75	130	100	170	240	
After grazing	30	55	35	60	110	
N (total plant bites)	1634	2230	988	249	1002	6103

Notes: [a] Pre December 31.
[b] Weighted mean.

Euphorbia tirucalli (Goddard 1968, 1970a; Hall-Martin, Erasmus & Botha 1982; Hitchins 1979; Joubert & Eloff 1971; Mukinya 1977).

Black rhinos defecate 4–5 times in 24 hours. Coprophagy has been recorded during the dry season when legumes were sparsely available (Goddard 1968; Joubert & Eloff 1971).

White rhinoceros. To determine the diet composition of white rhinos, I observed feeding animals at close quarters (20–40 m range), and then inspected the feeding site after the animal had moved on. Within an area defined by what I could touch with my fingers while standing with legs straddled, I counted the number of plants of each species that had been freshly grazed. A unit plant was defined by the spread between my extended middle finger and thumb, which distance closely approximates the measured bite width of a white rhino (200 mm). Feeding records were divided according to the general grass condition, in terms of its greenness at the time of grazing, as influenced by prior rainfall.

Short grasses were the most important food source during the wet season while the grass remained green or mainly green (Fig. 3.5, Table 3.2). Favored short grass species included *Panicum coloratum, Urochloa mosambicensis, Digitaria* spp and *Sporobolus* spp. Shade grasses, in particular *Panicum maximum*, were sought out especially during the early dry season, when they tended to remain green longer than other grasses. During

Fig. 3.6 White rhino grazing in medium-tall *Themeda* grassland during the dry
season (Umfolozi, South Africa).

the dry season rhinos transferred their attention increasingly to medium-tall
climax grassland dominated by *Themeda triandra* (Fig. 3.6). *T. triandra* was
also favored during the wet season when kept short. This species provided
the greatest fraction (about 30%) of the food intake on a year-round basis.
The only grasses that were strongly rejected were *Cymbopogon* spp (which
are aromatic), *Aristida* spp (which are wirey), and *Tragus berterionanus*
(which is a very low-growing annual). However, another aromatic species,
Bothriochloa insculpta, was readily eaten, at least when short. Forbs made
up only 1% of the annual diet, and seemed mostly to be ingested acciden-
tally along with grass. No browsing was observed, apart from occasional
instances of chewing on woody stems.

White rhinos selected mainly for grassland type, rather than for particular grass species. During the wet season months they concentrated their
grazing on short grass grasslands (Fig. 3.7). As the dry season advanced,
they shifted their grazing to areas of medium-tall *Themeda* grassland,
though initially seeking out patches of short grass. The fringe of short grass
associated with termite mounds was especially favored. By the end of the

WHITE RHINO GRAZING DISTRIBUTION

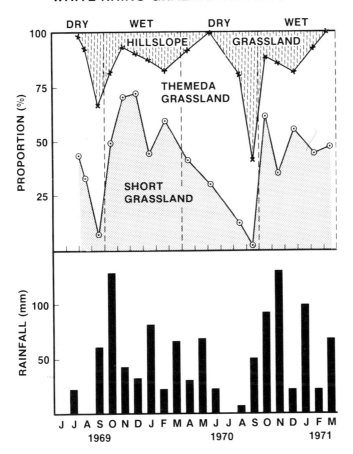

Fig. 3.7 Seasonal changes in the distribution of grazing white rhinos between different grassland types in the western section of Umfolozi Game Reserve.

dry season most of the *Themeda* grassland on gentle terrain had been grazed down, and animals then moved up onto hillslopes to graze remaining reserves of taller grassland.

While grazing, white rhinos swung the head in an arc to crop the grass that came within reach with each forward step. The mean height of the grass grazed increased from about 100 mm during the wet season to about 200 mm during the dry season. This was cropped down to a height of 25–60 mm except when tall dry grass was being eaten (Table 3.2). On short grass the feeding rate averaged 72 bites per minute.

Soil type also influenced grassland selection. White rhinos favored areas of grassland growing on soils derived from shale or dolerite, but were rarely seen grazing in areas underlaid by sandstone. The mean crude protein content in whole plant samples of *Themeda triandra* growing on sandy soils in Umfolozi was 5.9%, compared with 7.8% for plants of the same species collected from bottomland sites with clayey soils. *Panicum maximum* showed a similar difference (11.7% versus 14.9%; Downing 1979).

The prevalence of short grass grasslands at Umfolozi is largely due to the grazing effects of white rhinos. With sustained close cropping, erect tufted species like *Themeda* lose vigor and become replaced by short creeping species like *Panicum coloratum*, *Urochloa* spp, *Sporobolus smutsii* and *S. nitens* (Downing 1972).

In the Hluhluwe Reserve medium-tall grasslands predominate, mostly underlaid by sandstone or granite. White rhinos sought out short grass patches, the favored grass species in such habitats being *Dactyloctenium australe*. White rhinos introduced from Umfolozi into the Matopos Park in Zimbabwe, where granitic sands predominate, likewise favored mostly short grass species (Wilson 1969b). I made some brief observations on the white rhinos held in a 250 ha enclosure in the Pretorius Kop region of the Kruger Park in South Africa, also a sourveld region underlaid by granite. The predominant grass species eaten were *P. maximum*, *Setaria perennis*, *Cynodon dactylon*, *U. mosambicensis* and *Digitaria* spp, under wet season conditions. No use was made of the tall *Hyperthelia* and *Cymbopogon* predominating in parts of the enclosure.

White rhinos of the northern subspecies were recorded in Uganda feeding mainly on medium-height grasses, including *Hyparrhenia*, *P. maximum*, *Chloris gayana*, *Heteropogon contortis* and *Brachiaria brizantha*. The general height of the grass cropped was 250–300 mm, this being reduced to a level of about 50 mm (Foster 1967; van Gyseghem 1984).

I had one opportunity to weigh the stomach contents of a white rhino cow, which had died of acute peritonitis. The wet mass of the food material in the stomach was 72 kg, equivalent to about 4.5% of body mass. What proportion of the daily food intake this represents is unknown. The defecation rate of bulls varied betwen 4 and 6 times per 24 hours, based on monitoring the dungheaps used regularly by these individuals.

Summary

Both African and Asian elephants are mixed feeders, favoring grass when it is green, but becoming more dependent on woody browse in the dry season. Furthermore elephant diets vary flexibly between different regions depending on the relative availability of grasses and woody plants.

The dry season food intake may include a high proportion of fibrous plant tissues such as twigs, bark and roots. Nevertheless, high quality plant parts such as fruits are sought out when they are available. Elephants tend to reject shade-tolerant, late succession species in forest habitats. They also ignore some savanna species eaten readily by giraffe (e.g. *Melia volkensii* at Tsavo). The high defecation rate suggests a fast turnover of food in the digestive tract.

Indian rhinos are also mixed feeders, but generally favor grass. Both tall and short grass species are eaten. Indian rhinos are dependent upon flood plain areas where some green growth remains available year-round, although the proportion of browse in the diet increases in the dry season.

Both white rhino and hippo are strictly grazers, including no woody browse in their diets. They are generally area selective rather than species selective grazers. Both favor short grasses, but can switch to taller relatively fibrous grassland when short grass in unavailable. However grasslands growing on nutrient-poor soils tend to be avoided.

Black rhino, Javan rhino, Sumatran rhino and giraffe are browsers, including only small amounts of grass in the diet. Black rhinos favor herbaceous browse (forbs), but become more dependent on woody plants and succulents during the dry season. Giraffe and other browsing rhinos favor the new shoots of woody trees or shrubs. Fruits and pods are sought out when available.

Water and other habitat needs

The water requirement of herbivores may be largely met from the liquid content of plant tissues, plus water released during carbohydrate metabolism. Nevertheless, these sources may prove inadequate at certain times of the year, so that the liquid intake may need to be supplemented by drinking from standing surface water.

Body liquid balance becomes especially important in thermoregulation, when animals may become dependent upon evaporative cooling. Heat loads may also be avoided or reduced by behavioral responses, such as seeking shade or wallowing in water or mud. Animals may also need to restrict heat losses during cold or windy weather by seeking shelter in protected sites.

Certain minerals, in particular sodium, may be present in inadequate amounts in the diet, and may be sought out at sites where such minerals have become concentrated in soil.

Fig. 3.8 Elephants wallowing in a muddy pool (Mana Pools, Zimbabwe).

Elephants

African elephants drink one to three times daily when water is readily available. During the dry season they may go for periods of two or three days between waterhole visits, but remain dependent on regular access to surface water. In the Hwange Park in Zimbabwe, where soils are sandy and leached of minerals, pools with relatively high sodium contents are favored for drinking (Eltringham 1982; Laws 1970; Weir 1972; Wyatt & Eltringham 1974).

Surface water availability restricts the dry season distribution of elephants. At Tsavo in Kenya, elephants were restricted to a radius of about 15 km from water under conditions where they were dying from starvation. However at Hwange elephants undertook journeys of up to 24 km to and from water during a drought (Corfield 1973; Williamson 1975a).

Elephants lie and roll over in muddy pools (Fig. 3.8), and immerse themselves in deeper bodies of water. They use their trunks to spray water or mud over their bodies, and sometimes also blow dusty soil over themselves. After mud wallows they frequently rub themselves against trees (Hendrichs 1971; personal observations).

Elephants seek shade during the hot time of the day, with groups clustering under the remaining large trees where woodlands have been depleted. The enormous ear pinnae of African elephants facilitate cooling. Ear fanning rates are correlated with ambient temperature, and the temperature of the blood leaving the ear is cooler than that of the blood entering it (Buss & Estes 1971).

African elephants dig up and eat sodium-rich soil in saline depressions in Hwange. At one site in Kenya they penetrate into the deep recesses of a cave for the salt deposits there (Weir 1972).

Asian elephants usually drink one or more times daily. Males commonly lie and roll in muddy hollows, while females more often collect muddy liquid in their trunks and spray it over their bodies. Mud and dust is also collected in the curled trunk and thrown over the body, frequently after bathing in water. This may be followed by rubbing the skin against suitable trees or rocks (McKay 1973).

Hippopotamus

Hippos generally spend the daytime largely submerged under water, thereby avoiding heat stress. Sun basking on sandbanks may take place during the cooler winter period.

Giraffe

Giraffe can survive independently of surface water for long periods if adequate green foliage is available on trees, but drink regularly when water is readily available. Giraffe do not immerse themselves in water or mud. They may continue feeding through the midday period without seeking shade (J. T. du Toit personal communication 1986; Foster & Dagg 1972; Western 1975; personal observations).

Rhinoceroses

Indian rhinos drink daily, and ingest mineral-rich soil when it is available. They spend long periods lying in pools of water, especially during the hot summer months (Ullrich 1964; Laurie 1978; Fig. 3.9). Both Javan rhinos and Sumatran rhinos commonly wallow in muddy pools (Hubback 1939; Schenkel & Schenkel-Hulliger 1969b; Borner 1979).

At Hluhluwe, where water is widespread, black rhinos generally drink nightly. In Namibia, black rhinos drink every second night during the cool dry season months of June–July, but nightly during the hot dry month of October. However, in the Tsavo East Park in Kenya, black rhinos went for periods of 4–5 days without drinking.

Black rhinos commonly wallow in mud, and sometimes also in dust

Fig. 3.9 Indian rhino lying in a pool (Chitwan, Nepal, photo courtesy W. A. Laurie).

hollows. The favored time for mud wallowing is the late afternoon, but wallowing also occurs at night. Animals commonly rub against bushes or rocks following wallowing. Salt licks are commonly visited (Frame 1971; P. M. Hutchins personal communication; Joubert & Eloff 1971; Mukinya 1977; Schenkel & Schenkel–Hulliger 1969a).

White rhinoceros
    At Umfolozi water was abundantly available in numerous small depressions or pans during the wet season. White rhinos then drank daily, or even twice daily. During the dry season after these pools had dried up, animals were forced to make a journey to one of the longer-lasting pools or to one of the major rivers. To determine the frequency of drinking at this time of the year, I used three sets of evidence: (i) movements of those animals fitted with radio transmitters; (ii) identities of the animals drinking at particular waterholes on successive days; (iii) relationships between the interval between waterhole visits and the time spent drinking. Results showed that a drinking frequency of every 2–3 days was most usual during the dry season period, although some animals drank at four day intervals.
    Wallowing also took place at waterholes, and took two different forms. Animals lay down and rolled in muddy hollows to secure a thick coating of glutinous mud over the body. Following a mud wallow, animals rubbed their bodies against suitably inclined trees, stumps or rocks in the vicinity

Fig. 3.10 White rhino bull rubbing mud-plastered body against a stump (Umfolozi, South Africa).

(Fig. 3.10). Ticks were evident in the mud rubbed off. Animals also at times lay down in pools of water, for periods of up to several hours.

Mud-wallowing occurred most frequently in the early afternoon, but was recorded at all times of the day and even during the night. Lying in water was most common over midday, but on occasions took place even during the coolest part of the early morning (Fig. 3.11). Lying in water was much more common during the earlier part of the wet season than during the later part, while mud-wallowing was recorded equally frequently in both of these periods. Little wallowing took place during the dry season months, even when suitable mud was available.

During the heat of midday, white rhinos sought out suitably shady trees. Animals tended to congregate at favored sites, generally located on breezy ridgecrests. Under cool, windy conditions, white rhinos secluded themselves in lowlying areas of woodland or thicket.

No saltlicks were known to me in Umfolozi. However the water issuing from springs tended to be brackish. A few instances of animals licking termite mounds were noted.

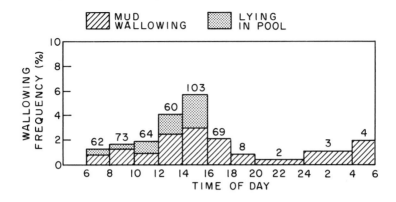

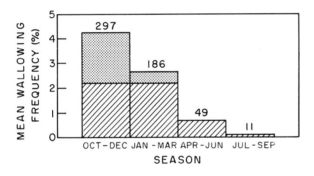

Fig. 3.11 Wallowing frequency of white rhinos in relation to time of day and season.

Summary

Hippos, Indian rhinos and elephants are strongly dependent on surface water availability, while white rhinos and black rhinos can go for periods of several days without water. Giraffe can be independent of water if green foliage remains available. Hippos are aquatic by day, while elephants and rhinos wallow both in water and mud. Giraffe do not wallow. All species except giraffe are shade-dependent over the midday period.

Comparisons with smaller ungulates

Megaherbivores show distinctions between grazers, browsers and mixed grazer-browsers as is typical among African bovids. White rhinos and hippos are as strict grass-feeders as African buffalo, wildebeest and topi. Black rhinos include no more grass in their diet than browsing antelopes like kudu, bushbuck and Grant's gazelle (although the term 'concentrate selector' can hardly be applied to black rhinos). Elephants and Indian rhinos switch between a primarily graminoid diet in the wet season, and increased representation of woody browse in the dry season, like impala, Thomson's gazelle and, to a lesser degree, eland; but, like impala in particular, elephants of both species show a wide regional variation in grass:browse proportions (Hofmann & Stewart 1972; Owen-Smith 1982).

Megaherbivores seek out the same high quality plant parts as those selected by smaller ungulates. White rhinos and hippos favor leafy short grasses, as do wildebeest, impala and warthog. Elephants, browsing rhinos and giraffe seek out seed pods and other fruits when available, as do most browsers from dikdik size upwards; while giraffe search for new unlignified shoots, as do duikers. Black rhinos favor legumes and other herbaceous browse, like kudus. However, megaherbivores include in their diets a higher proportion of stem and other fibrous tissues than do smaller ungulates, most particularly during the dry season period. In their consumption of bark and woody twigs megaherbivores are paralleled by small cecalids, such as African porcupine and hares (Bryant *et al.* 1985; Jarman 1974; van Jaarsveld & Knight-Eloff 1984).

Megaherbivores tend to favor the same species of plant as those sought out by smaller ungulates. However, African elephants make abundant use of *Colophospermum mopane*, which is eaten by giraffe and browsing antelopes such as eland and kudu only in the late dry season when little choice is available; while the *Brachystegia* species that elephants eat are also little used by other browsers. Hippos commonly graze *Sporobolus pyramidalis* in Uganda, a coarse tall grass which is generally avoided by other grazers. White rhinos in Zululand readily eat *Bothriochloa insculpta*, a strongly aromatic grass species, at least while it is young, although grazing antelope ignore this species unless hard pressed for food. Certain very large fruits, in particular the woody pods of *Kigelia pinnata*, are eaten only by giraffe and black rhinos.

All megaherbivores except giraffe are dependent upon the availability of surface water, like most antelope species except for many of the browsers. However white rhinos are able to go for longer periods between drinking than most grazing bovids and zebras under dry season conditions when no green grass is available (Western 1975).

Mud-wallowing and bathing in pools are habits shared by elephants, rhinos and hippos with other ungulates having relatively hairless skins, for example African buffalo, water buffalo and warthog. While some hairy ungulates also wallow in muddy hollows, for example American bison and red deer, this behavior is restricted to males during the rutting season

(Cumming 1975; Darling 1937; Eisenberg & Lockhart 1972; Lott 1974; Sinclair 1977; Struhsaker 1967).

Megaherbivores, apart from giraffe, seem similarly shade-dependent to most medium-sized ungulates, although some of the latter are able to spend long periods exposed to full sunlight due to their protective hair coats (Finch 1972).

4

Space–time patterns of habitat use

Introduction
In this chapter I consider how megaherbivores go about securing their habitat requirements in time and space. What times of day or night are favored for feeding or other maintenance activities, how much time per day is spent foraging, and how does this vary through the seasonal cycle? What size area do animals cover in seeking their food or water needs, and how does this change seasonally?

Temporal patterning of activities
Animals engage in a number of daily activities. These include feeding, travelling between feeding areas and perhaps to and from water, resting, other maintenance behaviors such as drinking, wallowing and grooming, and various forms of social interaction. These need to be scheduled optimally within the diel (day-night cycle), while ensuring that an appropriate amount of time is allocated to each. The animals need to accommodate for variations in temperature, cloud cover, wind and precipitation. Superimposed on these variations is the progression of the seasons, involving changes in the day-night ratio, prevailing temperatures and rainfall, and associated changes in food availability and reproductive physiology.

In the following account I will make a distinction between feeding and foraging. The former is synonymous with eating, i.e. gathering, chewing and swallowing, while the latter also includes movements made while searching for food.

Elephants
African elephants devote roughly equal proportions of the day and night to foraging. They tend to show three peaks in activity, occurring during the early morning, the later part of the afternoon, and around

Table 4.1. *Daily foraging time of African elephants in different regions*
Figures represent percentage of total time.

Area	Sex	Season	Daylight	24 hour	Reference
Uganda					
Queen Elizabeth	Female	Year	—	75	1
Tanzania					
Serengeti	Male	Year	—	75	2
Ruaha	Male	Wet	70–85	—	3
		Dry	56	—	
	Female	Wet	82	—	
		Dry	64	—	
Zimbabwe					
Sengwa	Combined	Wet	56.5[a]		4
		Dry	41.5[a]		

Note: [a] 8 hour day only.
Source references: 1 – Wyatt & Eltringham 1974; 2 – Hendrichs 1971;
3 – Barnes 1979; 4 – Guy 1976b.

midnight. Their main rest period is just before dawn when they may lie down for an hour or two. Most other resting is accomplished standing up. Most travelling takes place shortly after dusk.

The overall proportion of the 24 hour cycle devoted to foraging varies between 60% and 75% in different areas (assuming 80% of the night is devoted to foraging; Table 4.1). More time is spent foraging in the wet season than in the dry season. Elephants spend more of their time foraging in grassland areas than in woodlands. Males and females spend very similar proportions of their time foraging. Resting, whether standing or lying, occupies only 2–3 hours during the day and 1–3 hours at night; whilst travelling (i.e. walking without feeding) takes up about 2–3 hours in total (Barnes 1979; Guy 1976b; Hendrichs 1971; Wyatt & Eltringham 1974).

The peak drinking time of African elephants is during the early evening, with the majority of animals arriving at water between 18.30 and 21.00 (du Preez & Grobler 1977; Weir & Davison 1965).

Asian elephants forage for 75% of the daylight hours, or an estimated 17–19 hours per 24 (Vancuylenberg 1977).

Hippopotamus

Hippos spend the entire day resting in water, and emerge on land to forage only at night. Animals generally leave the water after nightfall and return during the early pre-dawn hours. Adult males are the last to leave, emerging from the water as late as 20.00–21.00, though somewhat earlier in dull weather. Animals begin drifting back to their pools between 02.00 and 03.00, with the majority of hippos back in the water by 04.00. This leaves no more than 7–8 hours for foraging, though after reaching their grazing areas animals feed with little pause (Verheyen 1954; personal observations).

Giraffe
Giraffe forage for most of the day, but devote much less time to foraging at night. There is also a sex difference in foraging time. At Serengeti cows foraged for 72% of the day versus 55% for bulls. The mean proportion of the night spent foraging by both sexes combined was 34% on moonlit nights and 22% on dark nights. More time was spent foraging during the dry season than in the wet season, but the time spent actually eating may be less. The main feeding periods are during the three hours post-dawn and again pre-dusk. At Kyle in Zimbabwe animals moved into patches of woodland at night, possibly because of the high concentration of browse in a small area. However, ruminating is the dominant nocturnal activity, and time is also spent lying down. Sleep periods are brief, lasting no more than 5.5 min on average. Adult males spent 5% more time walking per day than adult females (Lightfoot, 1978; Pellew, 1984c).

Giraffe have no special drinking time, and may appear at water at any time of the day or night (du Preez & Grobler, 1977; Weir & Davison 1965).

Rhinoceroses
An Indian rhino cow, watched for 24 hours in January (mid-winter), spent 57% of its time foraging and 40% lying. Another cow watched for 22 hours in April (towards the end of the dry pre-monsoon period) spent 65% of its time foraging. Less time was spent foraging during the rainy monsoon season (Laurie 1978).

Black rhino females at Hluhluwe were active for 47% of the day and 95% of the night, and males for 33% of the day and 93% of the night, as indicated by radio-telemetric monitoring. In East Africa black rhinos spent 30% of the daylight hours eating and 20% walking, including movements between feeding stations. Their peak drinking time is in the early evening between 18.00 and 21.00 (du Preez & Grobler 1977; Goddard 1967; Hitchins 1971; Mukinya 1977).

White rhinoceros
I recorded the activities of all white rhinos when encountered, and maintained a continuous record of the activities of animals kept under

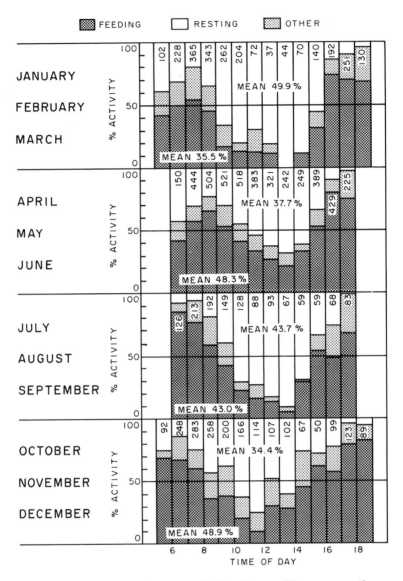

Fig. 4.1 Diurnal activity pattern of white rhinos at different seasons (for sunny days only). Figures indicate total sample sizes, in 15-min rhino activity units.

observation for extended periods. From continuous watches, the prevailing activity over each 15 min period was extracted. Brief records were assigned to the 15 min period in which they were observed. Thus records were analyzed in terms of 15-min rhino activity units.

White rhinos showed a bimodal activity pattern (Figs. 4.1 and 4.2). The

a

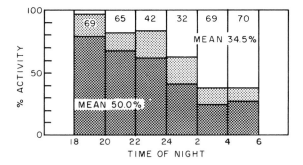

TIME OF NIGHT

b

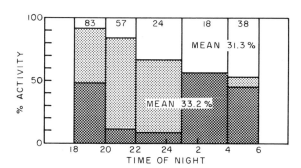

TIME OF NIGHT

Fig. 4.2 Nocturnal activity pattern of white rhinos, for two times of the year. Figures indicate total sample sizes, in 15-min rhino activity units. a. *Wet season conditions* (data mostly March–June). b. *Dry season conditions* (data derived from June 1966 and July 1969, and biased towards neighborhood of waterholes).

main active periods were the early part of the morning, and the late afternoon, extending into the evening. A long rest was taken through the middle part of the day, during which period animals slept for up to eight hours under hot summer conditions. During the late summer period through January to March, the majority of rhinos were inactive between 09.00 and 16.00. The length of the inactive period was shorter during the cooler months of the early dry season of April–June, and if the weather was

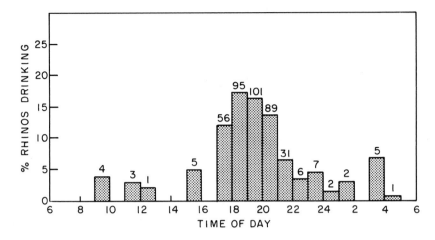

Fig.4.3 Drinking schedule of white rhinos at waterholes during the late dry season. Figures indicate total sample of rhinos observed drinking in each time interval (analyzed from composite records).

mild a feeding spell sometimes occurred over midday. Rhinos were less active over midday during the late dry season months of July–September, although mean temperatures were only slightly higher than those prevailing over the April–June period. Another rest period occurred through dawn, when animals commonly lay in sandy patches. The dawn rest disappeared in the late dry season. At this time of the year much of the evening was taken up by travelling to and from waterholes.

During the early wet season months of October–December, rhinos were more active during the day than they were during the later wet season, even though prevailing temperatures differed little between the two periods. Animals frequently interrupted their midday rest to revert to feeding for short spells, and much wallowing in pools or mud took place.

During cloudy weather rhinos were much more active during the day than they were under sunny conditions, with little seasonal variation. Short spells of feeding and resting alternated throughout the day, with relatively more time spent standing, neither feeding nor resting. On sunny days with a midday temperature of 24 °C the mean level of midday activity was 43%, compared with 81% on cloudy days with the same maximum temperature. Midday activity was depressed to 10% on sunny days with temperatures reaching 34 °C.

The main drinking period was in the early evening between 17.00 and 21.00 (Fig. 4.3).

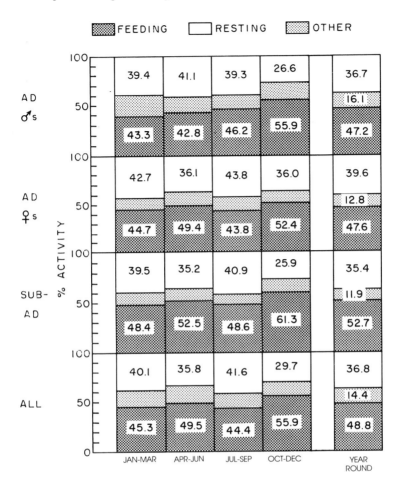

Fig. 4.4 Daytime activity budgets of white rhinos for different seasons and age/ sex categories. Figures represent percentages.

As a year-round average for all age/sex classes, foraging occupied 49% of the daylight hours (Fig. 4.4) and about 50% of the night (Fig. 4.2). Adult males and adult females spent similar proportions of their time foraging over the year. However, females spent more time foraging than males during the early dry season, which followed the peak calving months of March and April. Cows that were either heavily pregnant or in early lactation spent more time foraging than other females at this time of the year. Subadults of both sexes devoted slightly more time to foraging than adults at all times of the year.

The daily foraging time for all age/sex classes peaked in November, while

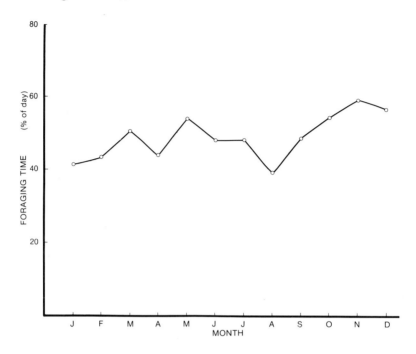

Fig. 4.5 Monthly variations in the daily foraging time of white rhinos (daytime only, all age/sex classes combined).

lowest foraging times were in January and in August (Fig 4.5). The increase in feeding time from January through to May is associated with decreasing mean temperature. After a minimum in August daily feeding time rose, despite rising temperatures after July.

Summary

Elephants of both species devote about three-quarters of their time to foraging, and spend relatively little time inactive. White rhinos, Indian rhinos, black rhinos and giraffes forage for 50–60% of the 24 hours; while for hippos foraging occupies no more than one third of the time. African elephants and white rhinos spend less time foraging in the dry season than in the wet season, whereas Indian rhinos and giraffes spend more time foraging in the dry season than in the wet season.

Giraffe are more active diurnally than nocturnally, and black rhinos more active nocturnally than during the day. Hippos accomplish all their feeding at night. Elephants, white rhinos and Indian rhinos are equally active during the day and night periods. However, all species except giraffe undertake most of their travelling, including journeys to and from water, shortly after nightfall.

Utilization of space

Animals tend to restrict their movements to a particular segment of the available habitat. The area traversed during routine activities is generally termed the home range (Jewell 1966); but what is routine is left vague. Walther (1972a) suggests that the home range is the area familiar to the animal, within which it establishes some space–time pattern of movement. However, there are occasions when animals wander beyond the limits of their usual home area. The term 'lifetime range' has been suggested for the total area covered by an animal, including temporarily used migration routes and exploratory sallies (Jewell 1966). However, unless observations are extended over a period comparable with the lifespan of the animal, the term 'annual range' seems preferable. This allows for the possibility that animals may move seasonally between different home ranges. Within the home range, certain sections may be used more frequently than others, these being referred to as 'core areas'. The term territory is applied only to areas that are defended, or at least monopolized by one particular individual or group to the exclusion of others.

Daily distances travelled may be measured along the path of movement, or simply as the straight-line distance between locations on successive days. The latter distance I will refer to as the 'daily translocation'.

Elephants

African elephants exhibit an exceptionally wide variability in home range extent. The home ranges of family units in the ground water forest habitat of Lake Manyara Park, Tanzania, covered only 14–52 km². In contrast, in the semi-arid steppe of Tsavo East National Park, Kenya, female elephants were radio-tracked over areas of up to 3750 km², with a mean of 1800 km². Under the wetter and topographically more diverse conditions of the adjoining Tsavo West Park, female ranges averaged 408 km² (based on two individuals). Male ranges had a mean extent of 1180 km² in Tsavo East and 840 km² in Tsavo West. The extensive movements in Tsavo East were monitored following a drought when much vegetation destruction and elephant mortality had occurred, and may thus have been exceptional. Preliminary data from the Kruger National Park in South Africa indicate range sizes of 240–720 km² for cows, and 140–1140 km² for bulls (Douglas-Hamilton 1972; Hall-Martin 1984; Leuthold 1977c).

Dry season home ranges cover only about 10% of the area of wet season ranges, and are based around permanent water sources. Distant areas are used opportunistically in response to local rainstorms and resultant green growth, areas having a dense grass cover being especially favored. Wet season dispersal movements away from permanent water were under 50 km in the Zambezi and Luangwa Valleys. However, movements over distances of 140 km or more between the Chobe and Hwange National Parks in Botswana and Zimbabwe respectively have recently been documented (G. Calef personal communication; Caughley & Goddard 1975; Dunham 1986; Leuthold 1977c; Melton 1985; Rodgers & Elder 1977).

Daily distances travelled average 4–6 km during daylight, and about 12 km over 24 hours. Walking speed is about 5–7 km h^{-1}. Mean daily translocation was 2.4 km at Manyara, 6 km at Tarangire and over 20 km at Serengeti (all in Tanzania), with wide day to day variability (Douglas-Hamilton 1971; Guy 1976b; Hendrichs 1971; Merz 1986b; Wyatt & Eltringham 1974).

For Asian elephants in the Ruhunu Park in Sri Lanka, home range sizes were estimated to be about 40 km^2, but surrounding settlements inhibited wider movements. Daily distances travelled during daylight varied between 1.0 and 8.5 km (McKay 1973).

Hippopotamus

Hippos in the Queen Elizabeth and Murchison Falls parks in Uganda had an area of heavy grazing extending up to 5 km from the lakeshore or river margin, with a mean extent of about 3 km. Some animals travelled as far as 10 km from water to graze at the end of the dry season. Along the Mara River in Kenya hippo paths generally extended 1.0–1.4 km from the river, with a maximum distance of 2.5 km. Along the Lundi River in Zimbabwe, the mean nightly distance travelled by hippos away from water varied from 0.4 km (maximum 1.2 km) during the wet season, to 0.7 km (maximum 2.1 km) during the dry season; but under drought conditions these distances increased to a maximum of 2.8 km in the wet season and 10 km in the dry season (Field 1970; Laws, Parker & Johnstone 1975; Lock 1972; Mackie 1973; O'Connor & Campbell 1986; Olivier & Laurie 1974).

Giraffe

Home ranges of female giraffe typically cover 80–120 km^2, but extended up to 480 km^2 in Tsavo East National Park. Seventy-five percent of sightings fell within a core area covering 30% of the total range. Home ranges of mature males tend to be slightly smaller than those of cows, while young adult males wander over a much wider area.

Daytime distances travelled averaged 6 km for males and 3 km for females in the Timbavati Reserve in South Africa; while daily translocation averaged 1.3 km (range 0.1–7.6 km) at Serengeti (Foster & Dagg 1972;

Langman 1973; Berry 1978; Leuthold & Leuthold 1978; Pellew 1981, 1984a).

Rhinoceroses

Indian rhinos at Chitwan occupied long narrow home ranges bordering the river. Individual cows were recorded over areas of up to 19.5 km², but most sightings fell within core areas covering 2–4 km². Male home ranges were similar in size (Laurie 1978, 1982). For Sumatran rhinos, home range extents of up to 50 km² were estimated, on the basis of the recognizable tracks of particular individuals (Borner 1979).

For black rhino cows home extent varied from a minimum of less than 2.6 km² in the Lerai Forest of the Ngorongoro Crater, Tanzania, to 99 km² on the Serengeti plains (Table 4.2). Typical home range sizes were about 7–35 km². However, desert-dwelling black rhinos in Namibia move over areas of up to 500 km². Home range sizes for adult males tended to be a little smaller than those of females, though the largest home range at Serengeti covered 133 km². Since these estimates include all sightings of particular animals, they represent annual ranges.

Animals moved over a wider area during the wet season than during the dry season, due to the availability of leguminous forbs in grassland areas at this time of the year (Goddard 1967; Hitchins 1971; Joubert & Eloff 1971; Loutit, Louw & Seely 1987; Mukinya 1973; Frame 1980).

White rhinoceros

I recorded the movements of white rhinos both by radio telemetry and by chance sightings of recognizable individuals. Six adult females, one young adult male, and three subadults were fitted with functioning radio transmitters. Transmitter life varied between 2.5 and 12.5 months. Methods are described elsewhere (Owen-Smith 1971b, 1974b).

The size of the area covered by white rhino cows varied depending on the prevailing habitat conditions. When both green grass and water were plentifully available, animals restricted their movements to a fairly small area, which may be termed the core area. During periods when the grass was drying out, movements were extended over a wider area. Such conditions were generally associated with the late wet season and early dry season months, but could occur temporarily during the wet season. The total area including such feeding movements will be termed the home range. Over the late dry season, when water sources became restricted to only a few points, movements further afield in the direction of one of the long-lasting water points were recorded. Between excursions to water, cows reappeared within

Table 4.2. *Home range extents of black rhinos in different areas*

Area	Age/sex	Range extent (km²)		Reference
		Mean	Min.–max.	
Tanzania				
Ngorongoro	Adult female	15	2.6–26	1
	Adult female	16	2.6–44	
	Imm. female	28	14–58	
	Imm. male	36	14–58	
Olduvai	Adult female	36	3.6–91	
	Adult male	22	5.5–52	
	Imm. female	22	7.5–36	
	Imm. male	38	29–47	
Serengeti	Adult female	76	43–99	2
	Adult male	92	59–133	
Kenya				
Mara	Adult female	14	5.6–23	3
	Adult male	13	7.1–19	
South Africa				
Hluhluwe	Adult female	6.7	5.8–7.7	4
	Adult male	4.1	3.7–4.7	
Umfolozi	Adult female	10	—	5

Note: Imm. = Immature.
Source references: 1 – Goddard 1967; 2 – Frame 1980; 3 – Mukinya 1973; 4 – Hitchins 1971; 5 – personal observations.

their usual home ranges. Animals with a permanent water point within their home range did not exhibit these excursions. Thus the water corridors are regarded as temporary additions to the home range (Fig. 4.6). Home ranges of cows covered between 9 and 15 km². Core areas encompassed 5–10 km², and annual ranges including water corridors covered an area of 8.9–20.5 km² (Table 4.3). Some cows moved beyond radio-tracking range, so that the latter figures must be regarded as a minimum estimate of the annual range.

Home ranges of adult males covered 0.7–2.6 km²; but since these were mutually exclusive, they represent territories and will be discussed in Chapter 7 under social organization.

Though defined on the basis of habitat conditions, the core areas were the most favored sections of the home ranges of cows at all times of the year. For the two best-known cows, 72% of all points of location over the year fell within the core area. Movements beyond the core area seemed to represent random probes in search of better grazing conditions. Drying conditions

HOME RANGE DISPERSIONS FOR TWO COWS
MADLOZI STUDY AREA, 1 DEC. 1968 – 4 SEPT. 1971

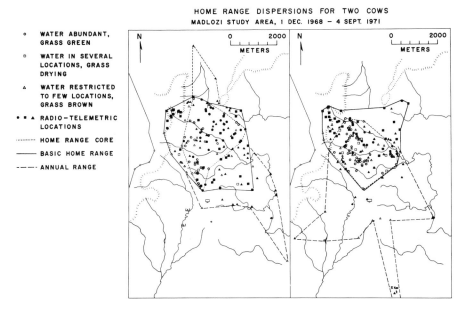

◦	WATER ABUNDANT, GRASS GREEN
▫	WATER IN SEVERAL LOCATIONS, GRASS DRYING
△	WATER RESTRICTED TO FEW LOCATIONS, GRASS BROWN
• ▪ ▲	RADIO–TELEMETRIC LOCATIONS
-------	HOME RANGE CORE
———	BASIC HOME RANGE
—–—·	ANNUAL RANGE

Fig. 4.6 Home ranges of two white rhino cows at Umfolozi, showing delineation of the core area and the dry season extensions in relation to grass condition and water availability. From Owen-Smith 1975.

prompted a general shuffling around of individuals, as also did the arrival of rains breaking a dry spell. When grass was green and water plentiful, cows remained grazing in the same area of about 1–2 km² for several consecutive days.

The observed distance covered by white rhinos during a 24 hour period was 4–5 km under wet season conditions. Daily translocation distances varied little through the year, apart from journeys to and from water in the late dry season (Table 4.4). Some cows travelled routinely a distance of 6–8 km to a waterhole and back every 3–4 days towards the end of the dry season. The furthest distance that a white rhino bull had to travel from his territory to water and back was nearly 10 km, involving a total time of 4–5 hours out of his territory on each occasion. Longer journeys were necessary after waterholes away from the two Umfolozi rivers had dried up. One cow moved overnight a distance of 11 km to the Black Umfolozi River, but was back in her usual home range a day later following a thunderstorm which broke the winter drought. The walking speed of white rhinos travelling to water was about 3 km h⁻¹.

Occasional wandering movements over greater distances may be made. One cow with a distinctive horn shape that occasionally visited my study

Table 4.3. *Home range sizes of white rhinos at Umfolozi*
Areas measured planimetrically after connecting outlying points. Upper figure
represents the mean area, lower figures (in brackets) the range in area.

Age/sex class	No. of individuals	Area (km²)		
		Core area	Home range	Annual range
Adult female	8	6.7 (5.3–9.6)	11.6 (8.9–14.7)	16.2 (8.9–20.5)
Adult male	27	1.7 (0.8–2.6)	—	—
Imm. female	2	5.0 —	6.0 (5.0–7.0)	7.3 —
Imm. male	4	4.6 (4.3–5.0)	4.6 (2.2–7.0)	6.7 (6.1–7.3)

Note: Imm. = Immature.

Table 4.4. *Daily translocations by white rhino cows*
Based on straight-line distances between early morning locations on successsive
days.

Conditions	Translocation distance (km)		N
	Mean	Min.–max.	(days)
Late wet season	1.20	0.1–2.3	12
Early dry season	1.47	0.05–3.1	16
Late dry season	1.66	0.2–5.1	25
	1.16[a]	0.2–2.8[a]	18
End of dry season	1.69	0.15–4.0	37

Note: [a] Excluding waterhole movements.

area was seen by a park ranger 13 km away. Another cow with a broken off anterior horn was seen grazing the green flush following a fire on the hillslopes to the west of my study area in November, 1968, when conditions were generally very dry. She was not seen again in the study area, but a broken-horned cow, which was almost certainly the same individual, was seen from the road some 28 km to the east.

Some subadult white rhinos appeared resident in the study area, while others wandered in, remained for a period, then disappeared again. Home range extents of resident subadults varied between 2.1 km² and 7.3 km². Two subadult females marked with eartags in the study area were subsequently seen 15 km and 25 km away respectively. Thus a proportion of subadults had evidently not established a fixed home range.

In the Kyle Game Park in Zimbabwe, white rhino cows were recorded moving over home ranges of up to 20 km². In the Murchison Falls National Park in Uganda, female white rhinos of the northern race (which had been moved there from west of the Nile River) covered annual ranges of between 50 km² and 97 km². Similarly large home ranges have been observed for animals of the southern race introduced into the Kruger Park in South Africa (Condy 1973; A. J. Hall-Martin personal communication; van Gyseghem 1984).

Summary

African elephants move over large annual ranges typically covering several hundred square kilometres, while giraffe also move over large areas a hundred or more square kilometers in extent. Rhinos of various species occupy somewhat more moderate-sized ranges typically covering 10–100 km², excluding movements to water. Asian elephants occupy restricted home ranges in sanctuaries, but may have moved over somewhat larger areas in the past.

African elephants move seasonally between a restricted dry season range near permanent water, and a wet season expansion area, which they use opportunistically in response to rainfall and resultant vegetation growth. A similar pulsating pattern is shown by black rhinos, at least in relatively open grassland habitats, and giraffe. In contrast, white rhinos restrict their movements during the wet season, and move over a larger feeding area during the dry season.

Comparisons with smaller ungulates

Medium-sized ungulates typically devote 35–60% of the daytime to foraging (Duncan 1975; Jarman & Jarman 1973a; Owen–Smith 1979; Stanley-Price 1977; Spinage 1968). This is not very different from the proportion of the daylight hours spent foraging by megaherbivores (50–75%, except for hippos). However, megaherbivores (excluding giraffe) are equally active, or even more active, at night as during daylight; while medium-sized ungulates (from gazelle size upwards) spend less time feeding and more time ruminating at night than in the daytime. Thus over a 24 hour cycle smaller ungulates feed for less time than do megaherbivores, excepting hippos.

Many grazing ungulates show decreased feeding times during the late dry season. White rhinos and African elephants do likewise. This effect may be largely due to the constraint on digestive throughput rates imposed by the slow digesting forage such as dry grass or woody twigs and bark. However, another factor to be considered, at least in the case of white rhinos, is the increased height of the grass eaten during the dry season compared with the wet season, so that more herbage is ingested per bite. Notably white rhinos spent the greatest proportion of their time feeding during the early wet season when the grass was shortest.

While the ranges covered by African elephants are large, they are exceeded in extent by the area covered by migratory antelope such as the Serengeti wildebeest, which may encompass up to 20 000 km² annually (Pennycuick 1975). Migratory zebra move over ranges covering several hundred square kilometers (Klingel 1967). Other wide-ranging species of bovid move over home ranges equal or larger in size to those covered by rhinos and giraffe, for example African buffalo (Leuthold 1972; Sinclair 1977) and both roan and sable antelope (Joubert 1974; Estes & Estes 1974).

5

Body size and nutritional physiology

Introduction

An animal's food choice is constrained by its metabolic requirements and by the functional anatomy and physiology of its digestive tract.
In this chapter I consider how these constraints operate, and how their
effects vary with body size. For example, larger animals generally eat more
food per day than smaller animals. However the more critical sorts of
questions that I will ask are these:

1. Does a 5000 kg elephant bull eat 1000 times as much food per day
   as a 5 kg dikdik?
2. What allometric relation best predicts the trend in food intake with
   increasing body mass?
3. Do particular species deviate notably from the overall trend, in
   particular those of very large body size?

In this chapter and others of its kind I will introduce each section with a
deductive proposition as to how the particular attribute being considered
ought to vary in relation to body mass. I will then test whether the published
data on large herbivores support or refute this starting hypothesis. The
statistical technique to be used is that of least squares regression. The reader
must first be forewarned of potential pitfalls in this method, as discussed by
Peters (1983).

1. Standard regression techniques assume that the X-variate (i.e.
   body mass in our case) is measured without statistical error.
   Generally I will use the mean body mass for the age/sex category
   being considered (see Appendix I), except in those few cases where
   more precise figures are available for the particular animals observed. Some error may be introduced here, but it should be fairly
   minor on a log scale.
2. The major influence on the slope of the regression line is exerted by
   the extreme points; in our case the values reported for very small

and very large animals. To allow for a possible biasing influence by megaherbivores, I will report regression coefficients for the subset of data excluding megaherbivores as well as for the complete set.

3. Values for the coefficient of determination R^2 are influenced by the total range of variation in the data, since this parameter represents the reduction in variance accounted for by the regression line. Thus, data for a sample of animals varying widely in body size will tend to yield a higher value for R^2 than any subset of the same data spanning a narrower range in body mass.

Potential biological pitfalls in using published data for comparative purposes were discussed by Clutton-Brock & Harvey (1977a):

1. Reported differences between species could be due to differences in the methods used in the particular studies. I have alluded to some of these in the preceding chapters, e.g. the distinction between foraging time and feeding time, and between annual and seasonal ranges.

2. The available data may be unrepresentative of particular species, especially where ecological features vary seasonally or between different regional populations or habitat types. Wide intraspecific variability in home range sizes was clearly evident in Chapter 4 both for African elephant and black rhino. For most ungulates, quantitative data are restricted to a single, supposedly definitive, study.

3. The data may be weighted heavily towards particular taxonomic groups, and thus not be typical of large herbivores in general. However, as will be evident in the graphical plots that follow, published studies now span a wide range of ungulate species. Thus, for most ecological features biases of this kind should be less than what they might have been one or two decades ago.

Nevertheless, the empirically derived relationships presented should be regarded as suggestive rather than as conclusive, awaiting more critical testing by matched comparisons or directed experiments.

Metabolic requirements

The basal or maintenance energy requirements of a wide range of animals, measured in calories or joules per day, vary in relation to $M^{0.75}$, where M represents body mass (Kleiber 1961; Peters 1983; Schmidt-Nielsen 1984). Linstedt & Calder (1981) suggest that this follows from underlying physiological mechanisms. The time periods for a wide range of processes, from muscle twitch times to potential lifespans, vary as a function of body mass raised to the power about one quarter. Hence metabolic rate, which represents a volume factor (the animal's size) divided by a time factor, is scaled in relation to $M^{1.0}/M^{0.25} = M^{0.75}$. By similar basic algebra, the specific metabolic rate per unit of body mass decreases with body size in relation to $M^{-0.25}$ (i.e. $M^{0.75}/M^{1.0}$). The factor $M^{0.75}$ is referred to as the metabolic mass equivalent. In terms of this relation, the maintenance energy requirements of a 5000 kg elephant should be only 180 times those of a 5 kg dikdik, rather than a thousand times as great.

The above relations refer strictly to maintenance requirements, with no allowance made for the extra costs for activity or growth or for temperature regulation. Daily metabolic expenditures measured for 23 species of eutherian mammals under natural conditions (using doubly-labelled water) varied according to a scaling coefficient of 0.81. However, three of the four large mammal species for which data were available were pinnipeds, which tend to exhibit higher metabolic rates than other mammals. For herbivores alone, field metabolic rate scaled in relation to $M^{0.73}$, although the only large herbivore included was white-tailed deer (Nagy 1987). In the absence of further information, I will provisionally assume that the daily metabolic requirements of large herbivores vary similarly according to a scaling coefficient of 0.73.

Protein requirements are derived from two components: (i) endogenous urinary nitrogen excretion, which is related to protein turnover and hence to metabolic rate; (ii) fecal nitrogen losses, including not only undigested protein but also enzymes added during digestion and cells sloughed off from the digestive tract lining. Fecal nitrogen losses are related to the quantity of food processed rather than directly to metabolic rate. Protein requirements vary even more widely than energy needs in relation to growth increments and reproduction.

Gut anatomy

As a volume factor, the capacity of the digestive tract should increase in direct proportion to body mass, in the absence of any adaptive trends. Despite contrasting digestive anatomies, the capacity of the fermentation chamber (as measured by the wet mass of its contents) forms a similar fraction of body mass in both foregut and hindgut fermenters, as also does the total digestive capacity of the gastro-intestinal tract. However the mass of gut contents increases in relation to $M^{1.08}$, rather than $M^{1.0}$ (Paraa 1978).

Demment (1982) compared the ruminoreticulum capacity of a variety of African ungulates, as measured by liquid filling, against the capacity as indicated by mass of contents. He found that, while the mass of ruminoreticulum contents increased in relation to $M^{1.11}$, liquid fill capacity varied in relation to $M^{0.94}$. It is likely that contents mass underestimates the ruminoreticulum capacity of small ungulates, because their rumens are often not filled to capacity; while liquid fill overestimates capacity for small species, with rumen walls that are thinner and thus more prone to stretching. The best estimate for the power coefficient of variation between gut capacity and body mass is hence the mean of the above two estimates, i.e. 1.03. Thus overall gut capacity, as well as the capacity of the fermentation chambers (whether ruminoreticulum or cecum plus colon), effectively increases in direct proportion to body size.

Food intake and digestion

Jarman (1968, 1974) and Bell (1969, 1971) focussed attention on patterns of food selection among African ungulates of varying body size. They noted that, while specific metabolic rate decreases with increasing body mass, gut capacity remains a constant fraction of body mass. Hence larger ungulates should be able to tolerate a lower minimum dietary quality than smaller species. Geist (1974a) labelled this concept the 'Jarman–Bell Principle'.

Bell (1971) noted further that the turnover rate of rumen contents decreases with increasing fiber content of the diet, due to the fact that particles have to reach a certain degree of comminution before they can pass out of the ruminoreticulum through the narrow passage connecting it to the rest of the gut. This restriction does not apply to non-ruminants like zebras, which hence show a faster passage rate of material through the gut than ruminants. Thus a non-ruminant should be able to tolerate a diet of higher fiber content, and thus lower nutritional quality, than a ruminant of similar body size (see also Janis 1976).

More specifically, if nutrient requirements increase in proportion to metabolic expenditures, and food intake is restricted directly by gut capacity, nutrient concentrations in the diet should vary as a function of $M^{-0.30}$ ($M^{0.73}/M^{1.03}$). This means that if a 5 kg dikdik required a diet that was 80% digestible, a 5000 kg elephant would be able to accept a diet that was only 10% digestible. However, this assumes that: (i) the turnover rate of the digestive tract contents remains constant; and (ii) digestive efficiency is the same. In fact, more nutritious herbage ferments faster than more fibrous material, and hence passes through the digestive tract more rapidly. Digestive efficiency is influenced by passage rate. Material that passes through the fermentation chamber faster is fermentated less completely than material that is retained for longer. Thus non-ruminants like horses show a lower digestive efficiency, in terms of cell wall breakdown, than ruminants.

In compensation the more rapid rate of food passage allows non-ruminants to eat more food per day than ruminants. On high fiber diets non-ruminants may assimilate more nutrients per unit time than ruminants, despite the superiority of the latter in extent of digestion.

Demment & Van Soest (1985) noted that: (i) the retention time of digesta in the gut tends to get longer as absolute gut capacity increases with increasing body mass, and (ii) the extent of digestion of cell wall is a function of retention time. Hence they suggest that at some body mass hindgut fermenters should be able to achieve virtually complete digestion of potentially digestible cell wall components, despite the absence of the selective delaying structures promoting efficient digestion in the rumen of ruminants. From model calculations, they suggest that this should happen at a body mass of 600 kg for rapidly fermenting forage (such as dicotyledonous foliage), or 1200 kg for slowly fermenting forage (such as grass leaves). Interestingly, these values lie close to the size criterion for megaherbivores adopted in this book.

Foose (1982) investigated the comparative digestive efficiencies of foregut and hindgut fermenters. The 36 species of large herbivore that he studied span a wide range in body size, so that his data also reveal the effects of body mass on digestive processes.

Foose's measurements were carried out on captive animals held in zoos, and are thus subject to the limitations that these conditions impose, not least of which is the limited sample of animals of each species that could be studied. Two fairly standardized diets were fed: (i) a high fiber grass hay, varying between 4.5% and 7.4% crude protein and 65–70% cell wall content (assayed as neutral detergent fiber); (ii) a moderate fiber content legume (alfalfa) hay, containing 17.4–22.2% crude protein and 31–56% cell wall constituents. The measures that we are interested in are (i) rate of food intake, (ii) rate of digestive passage, (iii) efficiency of cell wall digestion and (iv) overall nutritional balance.

Daily food intake

Foose's data show a decline in organic matter intake, expressed as a proportion of body mass, with increasing body mass (Fig. 5.1). For the grass hay the correlation is only marginally significant, and is due only to the equids, which as a group exhibit food intakes per unit of body mass about twice those of other ungulates of comparable body mass. For the legume hay, on the other hand, the correlation is highly significant. (Giraffe and pigmy hippos, which were reluctant to eat the hay diets, have been omitted from the regressions).

Evidently the food intake of foregut fermenters is restricted on the low

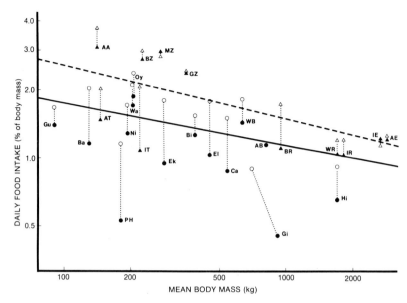

Fig. 5.1 Daily food intake, expressed as organic matter, in relation to body mass, from the work of Foose (1982) using standardized diets. Solid = grass hay, open = legume hay, circles = foregut fermenters, triangles = hindgut fermenters. For key to species labels, see Appendix I.
Regressions:

    (i) grass hay, all species except pigmy hippo and giraffe (solid line): Daily Food Intake (*DFI*) (organic matter as % of body mass) = $4.04M^{-0.184}$ (SE(b) = 0.074, R^2 = 0.217, N = 24, $P < 0.05$).

    (ii) grass hay, hindgut fermenters only: $DFI = 6.95M^{-0.236}$ (SE(b) = 0.101, R^2 = 0.376, N = 11, $P = <0.05$).

    (iii) legume hay, all species except pigmy hippo and giraffe (dashed line): $DFI = 7.31M^{-0.231}$ (SE(b) = 0.049, R^2 = 0.500, N = 24, $P < 0.0001$).

    (iv) legume hay, hindgut fermenters only: $DFI = 13.8M^{-0.315}$ (SE(b) = 0.049, R^2 = 0.819, N = 11, $P < 0.001$).

quality grass hay, but not with the faster digesting legume hay. For hindgut fermenters on the legume hay diet, the slope of the regression (0.31) is almost identical to the value needed to compensate for metabolic rate variations (0.30). Equids in particular compensated for their higher specific metabolic rates relative to larger hindgut fermenters by means of increased food intakes. However, this does not represent a facultative increase in intake to compensate for poor digestibility (as implied by Janis 1976), but is rather a constitutive anatomic ability to process more food per day than similar sized ruminants. In Foose's experiments the equids generally ate less of the grass hay than of the legume hay, indicating that their food intake was

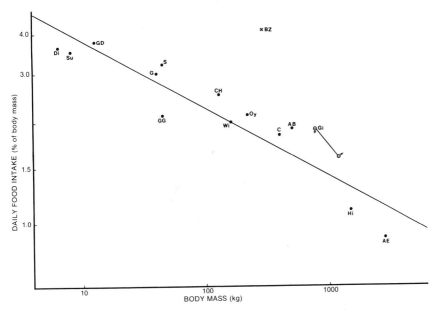

Fig. 5.2 Daily food intake in relation to body mass, from various published
sources. For key to species labels, see Appendix I. Data from Arman & Field
(1973), Arman & Hopcraft (1975), Hoppe (1977a), Laws *et al.* (1975), Nge'the
(1976), Pellew (1984c), Sinclair (1977) and Taylor & Maloiy (1967).
Regression: Daily Food Intake (dry mass as % of livemass) = $6.0M^{-0.191}$
(SE(b) = 0.039, R^2 = 0.647, N = 15, P < 0.01).

constrained on poor quality forage, but to a much lesser degree than was the
case for ruminants (see also Hintz *et al.* 1978).

Other data reported in the literature using various diets show a significant
tendency for food intake per unit of body mass to decrease with increasing
body mass (Fig. 5.2). In particular, the daily food intake of the smallest
antelope represents 3.5–4.0% of body mass, compared with only about 1%
of body mass for hippos and elephants. For cattle and other medium sized
ruminants, daily food intake ranges between 1.5% and 3.0% of body mass.
The majority of the available data refer again to captive animals fed
standardized rations, except in the case of megaherbivores (see Chapter 3).

Thus large herbivores do tend to eat less food per day, as a proportion of
their body mass, than small herbivores. For hindgut fermenters fed standardized diets, the decline in mass-specific food intake with increasing body
mass parallels the corresponding decrease in metabolic requirements; while
for ruminants the compensation is only partial, particularly for the low
quality grass forage. The mass-specific food intakes of megaherbivores are

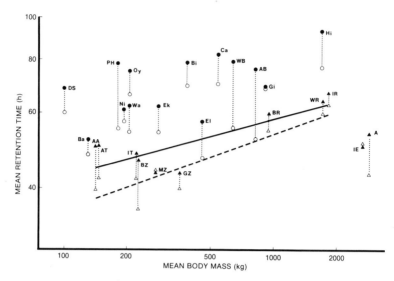

Fig. 5.3 Mean retention time of digesta in relation to body mass, from the work of Foose (1982) using standardized diets. Solid = grass hay, open = legume hay, circles = foregut fermenters, triangles = hindgut fermenters. For key to species labels, see Appendix I.

Regressions:

(i) grass hay, all species: Retention Time RT (in hours) = $46.1M^{0.048}$ (SE(b) = 0.044, R^2 = 0.046, N = 26, P not significant).

(ii) grass hay, hindgut fermenters only: RT = $32.0M^{0.075}$ (SE(b) = 0.033, R^2 = 0.358, N = 11, P < 0.05).

(iii) grass hay, perissodactyls only (solid line): RT = $22.8M^{0.135}$ (SE(b) = 0.035, R^2 = 0.673, N = 9, P < 0.01).

(iv) legume hay, all species: RT = $36.6M^{0.061}$ (SE(b) = 0.040, R^2 = 0.088, N = 26, P not significant).

(v) legume hay, hindgut fermenters only: RT = $23.4M^{0.106}$ (SE(b) = 0.035, R^2 = 0.504, N = 11, P < 0.05).

(vi) legume hay, perissodactyls only (dashed line): RT = $12.8M^{0.177}$ (SE(b) = 0.28, R^2 = 0.849, N = 9, P < 0.01).

typically about half those of medium-sized ruminants and about one quarter those of equids.

Retention time

In Foose's experiments the rates of digestive passage were assessed by mixing a proportion of dye-marked hay with the forage and recording the time of appearance of increasing fractions of the dye in the feces. From these data the mean retention times of food residues in the gut (the inverse of passage rates) were calculated.

In the overall data set, combining both foregut fermenters and hindgut

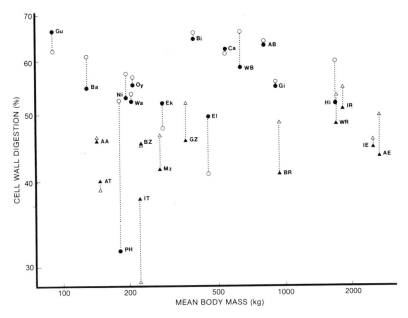

Fig. 5.4 The extent of cell wall digestion achieved in relation to body mass, from the work of Foose (1982) using standardized diets. Sold = grass hay, open = legume hay, circles = foregut fermenters, triangles = hindgut fermenters. For key to species labels, see Appendix I.

fermenters, there is no correlation between mean retention time and body mass (Fig. 5.3). However, for perissodactyls retention times do show a significant tendency to increase from tapirs and equids through to rhinos. Moreover, hindgut fermenters show shorter retention times (or faster passage rates) than foregut fermenters of similar body mass. Elephants exhibit retention times intermediate between those of equids and rhinos; while hippos exhibit an exceptionally long retention time, amounting to almost four days for the grass hay diet. The retention times shown by the two largest rhino species are comparable with those of some of the medium sized ruminants, such as waterbuck and American elk, though not as great as those of bovines or camels.

Digestive efficiency

The extent of digestion of grass cell wall achieved by the two grazing rhinos was somewhat less than that shown by grazing ruminants (Fig. 5.4) (apart from eland, a species which despite its mixed diet has the rumen anatomy typical of a browser; Hofmann 1973). Hippos, despite a long retention time, achieved a cell wall digestion only marginally greater

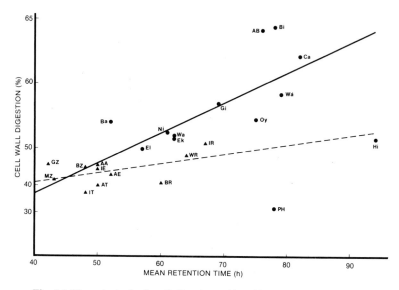

Fig. 5.5 The extent of cell wall digestion achieved in relation to mean retention time for grass hay diets, from the work of Foose (1982). The Y-axis has been scaled in terms of the log of the residual amount of digestible cell wall remaining, based on Waldo *et al.* (1972). It is assumed that the potential cell wall digestibility is 70%.
Regressions:

    (i) all species, except for hippos: Log_e (residual digestible cell wall *RDCW*, as %) = 4.95 − 0.034 (mean Retention Time *RT*, in hours) ($R^2 = 0.721$, $N = 23$, $P < 0.001$).

    (ii) foregut fermenters, excluding hippos (solid line): Log_e (*RDCW*) = 4.75 − 0.032 *RT* ($R^2 = 0.588$, $N = 12$, $P = < 0.05$).

    (iii) hindgut fermenters (dashed line): Log_e (*RDCW*) = 3.77 − 0.010 *RT* ($R^2 = 0.292$, $N = 11$, P not significant).

than that of the grazing rhinos. However, the two largest rhinos attained a higher cell wall digestion than equids despite their similar digestive anatomy, and also digested the grass hay somewhat better than did the elephants despite the body size advantage of the latter.

However, to compare digestive efficiencies among different categories of herbivore, the effects of varying retention time need to be separated out. For domestic ruminants, the rate of cell wall digestion has been found to follow linear kinetics, if the indigestible fraction is subtracted from the total cell wall pool (Waldo *et al.* 1972). This means that the log of the residual amount of digestible cell wall remaining should be linearly related to time. Thus a plot of this function against time reveals differences between species in their rates of cell wall digestion.

The measurements made by Foose (1982) indicate that hindgut fermenters achieve a lower extent of cell wall digestion than ruminants for the same retention time (Fig. 5.5). Moreover, the extent of cell wall digestion increases more rapidly with time in ruminants than among hindgut fermenters. At the short retention times typical of equids the two regression lines cross, suggesting that there is not much difference in cell wall digestion between foregut fermenters and hindgut fermenters for retention times of the order of two days. Elephants appear no different from equids in their rate efficiency of digestion. Hippos exhibit a rate efficiency of cell wall digestion that fits the regression line derived for hindgut fermenters, with pigmy hippos notably inefficient.

Nutritional balance

It was hypothesized by Bell (1971), Janis (1976) and Foose (1982) that nonruminants should outperform ruminants on forages of high fiber content, since their food passage rate is not slowed down much by indigestible material. Furthermore, on account of their very large body size and hence low specific metabolic rate, elephants and rhinos should be the most successful among extant species at utilizing low quality forage.

Foose (1982) estimated rates of energy extraction from the extent of digestion of the organic matter content of the diets supplied, and standardized values for the energy content of plant tissues. Energy gains were then related to energy requirements, estimated to be about 1.5 times basal metabolic rate calculated on the basis of $M^{0.75}$. His results indicate that equids, rather than rhinos or elephants, were best able to extract energy from the low quality grass hays (Fig. 5.6). However, elephants and rhinos were as successful as most of the bovids; while hippos, despite their large body size, were unable to meet their energy requirements from the grass hay diet. Giraffes and pigmy hippos were notably unsuccessful on both diets. Black rhinos performed poorly on the grass diet, but did very much better on the legume hay, due largely to the much higher rates of food intake that they showed for the latter.

Foose (1982) made similar calculations for protein balance. This appeared to be less restrictive than energy balance, although no allowance was made for protein needed during pregnancy and lactation. However, the nutritional value of the temperate grass hays used in Foose's experiments was considerably higher than that typical of tropical (C_4) grass species during the African dry season (see Owen-Smith 1982). Thus Foose was unable to identify the lowest tolerance limits in terms of dietary quality for the large herbivore species that he investigated.

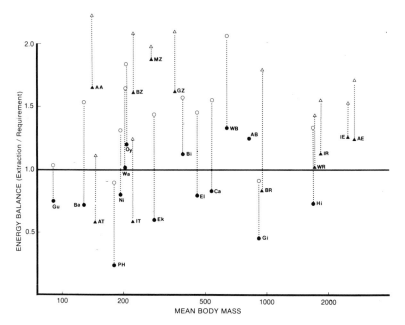

Fig. 5.6 Nutritional balance for energy calculated from Foose's (1982) measurements, for standardized diets. Solid = grass hay, open = legume hay, circles = foregut fermenters, triangles = hindgut fermenters. For key to species labels, see Appendix I.

Summary

In summary, different megaherbivores employ varying strategies to cope with low quality forage:

1. Elephants show a moderate food intake and fairly rapid turnover rate of digesta relative to their body mass. Because of their large size they achieve similar digestive efficiencies to equids, but with a lower specific food intake. Notably, both surviving species of elephant are mixed feeders, adapted anatomically (at least in their dentition) to handle both grass and woody browse.

2. Rhinos gain an advantage from their large body size relative to other perissodactyls in the form of increased retention times. This allows them to achieve a greater extent of cell wall digestion than equids and tapirs, though less complete than that attained by medium-sized ruminants. The black rhino, a browser, appears to be less efficient at cell wall digestion than the two grazing rhino species, though this may be partly a consequence of its smaller body size.

3. Hippos rely on an extremely long retention time to eventually achieve fairly complete digestion of cell wall components, but at the expense of a restricted rate of food intake. Foose (1982) suggested that, due to their semi-aquatic habits and consequently low thermoregulatory demands, hippos may have a lower metabolic requirement per unit of body mass than other ungulates.

4. Giraffe performed poorly on the diets supplied, due either to their difficulties in ingesting the forage supplied, or to an inability to digest the high fiber content

6

Body size and feeding ecology

Introduction

It is evident from Chapters 3 and 4 that megaherbivores select high quality green herbage or fruit when these are available, but switch to more fibrous forage during the dry season when choice is restricted. In superficial terms this pattern is not very different from those displayed by smaller ungulates, except for the amount of woody material eaten by elephants at times. Megaherbivores tend to forage for longer over the 24 hour day than do smaller ruminants; but daily foraging times are similar to those of medium-sized non-ruminants like zebras. The home range sizes of megaherbivores are no larger than those of many medium-sized ungulates, again with elephants being a clear exception. To discern body size influences, quantitative data for a range of species of widely varying body size need to be examined.

Diet quality

The nutritional value of food ingested must be adequate to satisfy metabolic demands, otherwise survival chances will be reduced. From the results reported in Chapter 5, total daily metabolic requirements (for maintenance plus activity) are predicted to vary with body size as a function of $M^{0.73}$. The assimilation rate of nutrients depends both on the capacity of the digestive tract and on the passage rate of its contents. Larger animals can support their lower specific metabolic requirements either by eating less food per day, or by accepting food with lower nutrient concentrations, or some combination of both. The data reported in Chapter 5, based largely on captive animals, suggest that daily food intake declines with increasing body mass approximately in relation to $M^{-0.2}$ (Figs. 5.1 and 5.2), while gut capacity is proportional to $M^{1.03}$. This implies that diet quality should vary as a function of $M^{-0.10}$ $\{(M^{0.73}/M^{1.03})/M^{-0.2}\}$. In other words, larger animals should eat diets of lower nutritional value than those selected by smaller herbivores, but the difference should be less marked than predicted simply on the basis of the body size–metabolic rate relationship. The question I will now consider is the extent to which this relation, based largely on artificial diets fed to captive animals, holds for free-ranging wild herbivores.

The first problem is to assess dietary quality. Ideally this should be expressed in terms of the concentration of available, i.e. both digestible and metabolizable, energy. However, digestibility is influenced by a number of factors, including plant species and part, stage of maturity, microbial activity in the fermentation chamber of the herbivore and rate of passage through it.

Protein concentrations are more readily measured than energy. Commonly dietary protein content is analyzed from samples of the food taken from the stomach or rumen. However, these results underestimate the protein concentrations in the ingested material, since more nutritious components pass out of the rumen more rapidly than slowly digesting fibrous material. For example, for topi the crude protein concentrations within food samples collected *via* an esophageal fistula averaged 2.7 percentage points higher than those analyzed in samples of rumen contents (Duncan 1975).

Nevertheless, the only comparative data available for a sufficiently wide range of species are those based on samples of rumen or stomach contents. The published data for crude protein concentrations in stomach contents show considerable variability, but much of this is related to whether or not the material was washed free of rumen liquor and associated microbes before analysis. If only washed samples are considered, there is a significantly negative correlation between crude protein contents and body mass for ruminants up to the size of African buffalo; but the value of the regression coefficient (-0.23 ± 0.07) is somewhat higher than that predicted allowing for variations in daily food intake (Fig. 6.1).

Megaherbivores exhibit protein levels in their stomach contents comparable to those of much smaller ruminants such as gazelles. For giraffe and black rhino this can be related to the predominance of dicotyledonous foliage in the diet, while elephants also include a high proportion of woody plant leaves in their diet during most seasons. Nevertheless the two grazing megaherbivores, white rhinos and hippos, apparently select diets no different in their protein concentrations from those eaten by most grazing bovids.

Some indication of digestible energy concentrations in the diet can be obtained from the rate of fermentation of the contents of the rumen, cecum or colon, assessed *in vitro* in terms of rates of gas production using samples

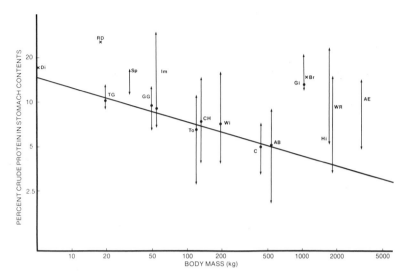

Fig. 6.1 Crude protein concentrations in the stomach contents of various large herbivores during the dry season in relation to body size. For key to species labels, see Appendix I. Data from Blankenship & Qvortrup (1974), Clemens & Maloiy (1982), Duncan (1975), Field & Blankenship (1973), Hall-Martin & Basson (1975), Hoppe *et al.* (1977a and b), Hoppe *et al.* (1981), Malpas (1977), Robinson (1979), Sinclair (1977) and Stanley-Price (1977); for white rhinos based on the crude protein contents of the grasses eaten, from Downing (1979) and Du Toit *et al.* (1940).

Regression line based on means for washed samples for ruminants only (as indicated by dots), excluding giraffe: CP (%) $= 21.6M^{-0.23}$ (SE(b) $= 0.028$, $R^2 = 0.92$, $N = 9$, $P = 0.0002$).

from freshly killed animals. Hoppe (1977a) reported results obtained from 72 specimens of 11 species of wild ruminant collected during the African dry season. His findings show that rates of gas production decline as a function of $M^{-0.22}$, a very similar relation to that found for the variation in crude protein concentrations (Fig. 6.2). If this trend is extrapolated into the megaherbivore range, it is evident that African elephants exhibit high fermentation rates for their size; while hippos show exceptionally low fermentation rates. For elephants the high fermentation rate can be explained in terms of the high proportion of browse in their diet, since woody plant leaves tend to digest more rapidly than grass leaves (Owen-Smith 1982). No comparable data are available for any of the rhino species; but based on their rates of digestive turnover, rhinos would be expected to be intermediate between elephants and hippos, and hence to fall not very far from the projected regression line.

An alternative indication of dietary quality is in terms of the proportions of plant parts ingested. Bell (1971) and Jarman (1974) suggested that small

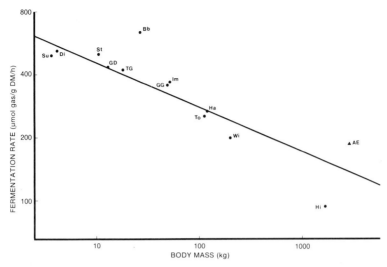

Fig. 6.2 Fermentation rates, expressed in terms of gas production, of rumen, cecal or colonic contents in relation to body mass. Circles = foregut fermenters, triangles = hindgut fermenters. Data from Hoppe (1977a), van Hoven (1977), van Hoven *et al.* (1981).

Regression, for ruminants only, as given by Hoppe (1977a): Digestive fermentation rate (μmole gas g $(DM)^{-1}h^{-1}$) $= 945M^{-0.22}$ ($SE(b) = 0.047$, $R^2 = 0.695$, $N = 11$, $P < 0.05$).

ungulates are highly selective for the more nutritious plant parts, such as new leaves, flowers and fruits; while large ungulates are tolerant of a relatively high proportion of fibrous stems in their diets, although they seek out high quality plant parts when these are available. It is under dry season conditions that the metabolic tolerance of larger animals should be most clearly expressed. Hoppe (1977a and b) and others have reported the proportions of leaves, stems and other plant parts in the rumen or stomach contents of various large herbivores collected during the dry season. The fraction formed by non-stem material (i.e. leaves, leaf sheaths, flowers, fruits, whether from grasses or woody plants) declines significantly with increasing body mass (Fig. 6.3). Small antelopes like dikdik and grey duiker managed even in the dry season to secure diets containing 86–96% leaf material, 70–100% of which consisted of green leaf. In contrast, at this time of the year the rumens of medium-sized grazers, such as topi, hartebeest and wildebeest, contained only 28–46% grass leaf, only half of which was green. During the dry season of a drought year in Tanzania, elephant diets included about 55% wood, bark and roots, and only 38% leaf (Barnes 1982a).

The data reported above show that dietary quality does decline with

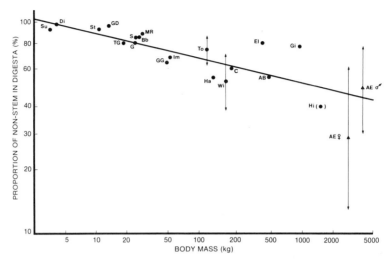

Fig. 6.3 Proportion of non-stem material (i.e. leaves, leaf sheaths, fruits, etc.) in the rumen or stomach contents of various large herbivores in relation to body mass (except for African elephants based on replicated food intake, and for hippos (bracketed) derived from fecal samples).
Circles = foregut fermenters, triangles = hindgut fermenters. Data from Barnes (1982a), Bell (1970), Berry (1980), Hall-Martin (1974), Hillman (1979), Hoppe (1977a and b), Irby (1977), Jarman & Sinclair (1979), Owaga (1975) and Scotcher *et al.* (1978).
Regression: % of non-stem material = $116M^{-0.118}$ (SE(b) = 0.018, R^2 = 0.721, N = 21, P < 0.01).

increasing body mass, despite some compensation in daily food intake. However, the wet season diets eaten by megaherbivores may differ little from those selected by medium-sized ruminants, at least as measured in terms of protein content. It is particularly during the dry season period of restricted herbage availability that the ability of very large herbivores to tolerate a low quality diet, characterized in particular by a high content of indigestible fiber, becomes evident.

On account of the metabolic rate–body size relation, larger animals lose condition more slowly on a submaintenance diet than do smaller animals (Bell 1971). Furthermore, Lindstedt and Boyce (1985) show that stored fat reserves become a greater fraction of body mass as size increases. Hence increased body mass could be an adaptation to compensate for extreme seasonal fluctuations in food availability. Notably, white rhinos may build up deposits of subcutaneous fat to aid their survival through the dry season. Selous (1899) commented that 'towards the end of the rainy season, in February and March, white rhinos used to become excessively fat, and would keep in good condition until late in the dry season. I have seen them so fat that between the skin and the flesh over the greater part of the body there was a layer of fat over one inch in thickness, whilst the whole belly was covered in fat to two inches thick.' The existence of similar fat deposits in the northern subspecies of white rhino was confirmed by Cave & Allbrook (1958). Hippos were renowned among early hunters for their fat, and Ledger's (1968) data show carcass fat contents for hippos of 7–11%, more than twice the mean value for African wild ruminants.

Foraging time

Bell (1971) suggested that, because larger herbivores have a greater absolute food requirement than smaller ones, they need to spend a some- what larger proportion of the day eating than the latter. However, eating time depends on the rate of food intake obtained relative to body mass. If this rate varies in direct proportion to body mass then, since megaherbivores eat less food per day per unit of body mass than smaller ungulates, very large animals should be able to satisfy their nutritional needs within a shorter feeding time than that required by smaller species. Nevertheless, it was apparent in Chapter 4 that megaherbivores tend to forage for a greater proportion of their time than smaller ungulates. Thus we need to consider more critically the factors controlling food ingestion rate.

Food ingestion rate

Rates of food ingestion while feeding are a function of bite size, biting rates and the time spent apprehending or manipulating food items. Bite size is controlled by the width of the mouth parts used for plucking food and the extent of the gape, and by the depth of the bite taken. Biting rates are influenced by bite size, with large mouthfuls of food requiring more time to be chewed and swallowed than smaller ones. For giraffe, Pellew (1984c) found that these two factors tended to compensate, so that eating rates varied little despite a wide variation in the bite sizes obtained from different plant species. However, this does not hold for other browsing ungulates (Cooper & Owen-Smith 1986).

The proportion of time spent searching during foraging spells is related to the spatial dispersion of acceptable food items in the vegetation. If food items are fairly continuously distributed, so that some morsel can be located with nearly every advancing step, little time is diverted to searching; chewing and swallowing can be synchronized with stepping. If food items are patchily distributed, time must be spent walking from one patch to another, during which no eating occurs.

Bite dimensions should vary as a volume factor (width × gape extent ×

depth), provided that bite depth is not restricted by plant structure. However, a grazing ungulate cannot pluck more than the height of grass leaves available above ground level, whatever its potential bite depth. While browsers can remove complete leaves plus petioles from woody plants, a further increase in bite depth yields only the woody stem to which the leaves are attached. Thus plant structure can impose an upper limit on the bite depth that can be ingested – at least without incurring an abrupt increase in the fibrous content of the ingested material.

If the constraint imposed by plant structure on bite depth operated equally for large and small ungulates, bite dimensions would vary as an area factor. Rates of food ingestion per unit of body mass should then increase in relation to $M^{\frac{2}{3}}$, i.e. larger herbivores would need to feed for much longer than smaller ones in order to ingest the same proportion of their body mass. On the other hand, if bite depth was not a constraint (i.e. larger herbivores took proportionately deeper bites than smaller herbivores), then elephants and dikdiks could consume the same fraction of their body mass in the same eating time. However, dikdiks would have to forage for longer than elephants, because of the additional time they would need to search out the discontinuously distributed plant parts upon which they are dependent.

Rates of food ingestion are ultimately restricted by rates of digestive throughput. In the simplest physical model, the ruminoreticulum (or cecum, or stomach) can be envisaged as a container being emptied by a pipe (represented by the intestine). If the linear dimensions of the system were doubled, the volume of the container would be increased eight times, while the cross-sectional area of the pipe would be increased four times. If there was no compensating increase in the linear rate of flow of material through the pipe, it would then take twice as long to empty the container (ignoring frictional drag).

However, material can pass out of the gut either by passage along it (the slow route) or by absorption through the gut wall (the fast route). The fraction of material following either the one or the other pathway is dependent upon the digestibility of the diet. If dietary fiber contents increase with increasing body mass, rates of digestive throughput should decrease. Hence constraints of digestive passage should operate more severely on very large herbivores than on small herbivores.

Earlier it was found that megaherbivores ingested a lower fraction of their body mass per day than smaller ungulates. This could be the result either of a constraint on digestive throughput rates, due to lower dietary quality; or of their lower metabolic requirements, assuming similar quality diets. In either case, megaherbivores should be able to satisfy their quantita-

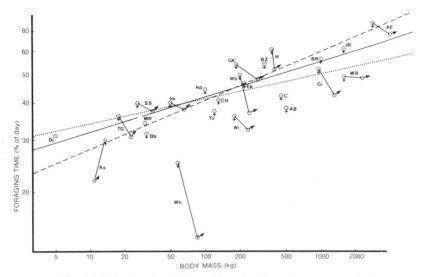

Fig. 6.4 Daily foraging budgets of large herbivores in relation to body mass.
For key to species labels see Appendix I. Data from Clough & Hassam (1970),
Craighead *et al.* (1973), Duncan (1975, 1980), Gogan (1973), Grimsdell & Field
(1976), Grubb & Jewell (1974), Hendrichs & Hendrichs (1971), Hitchins (1971),
Irby (1981), Jarman & Jarman (1973a), Laurie (1978), Low *et al.* (1981),
Norton (1981), Novellie *et al.* (1984), Owen-Smith (1979), Pellew (1984b),
Spinage (1969), Stanley-Price (1977), Thomson (1973), Walther (1973), Waser
(1975) and Wyatt & Eltringham (1974).
Regression line for all species (solid): $FB(\%) = 24.2M^{0.12}$ (SE(b) $= 0.029$, R^2
$= 0.47$, $N = 21$, $P = 0.0006$); for ruminants only (dotted): $FB(\%) =$
$27.9M^{0.08}$ (SE(b) $= 0.026$, $R^2 = 0.46$, $N = 12$, $P = 0.0076$); for non-
ruminants only (dashed): $FB(\%) = 19.0M^{0.17}$ (SE(b) $= 0.11$, $R^2 = 0.32$, $N =$
7, $P = 0.184$).

tive food requirements within a shorter daily feeding time than smaller
species if plant structure did not restrict ingestion rate.

Foraging time budgets over the 24 hour diet cycle show a significant
tendency to increase with increasing body mass (Fig. 6.4). Thus larger
herbivores forage for longer per day than smaller species, despite lower food
requirements and less time diverted to searching. Notably, white rhinos
exhibit foraging times just as long as those of zebras, even though their
specific food requirements are only one quarter as great as those of zebras.

Thus it is evident that plant structure does impose a restriction on food
ingestion rate. However, the increase in daily feeding time with increased
body mass is less steep than it would be if this constraint affected very large
herbivores and small herbivores equally. We thus turn next to a consideration of possible adaptations in the mouth or other feeding structures that
might compensate for a bite depth restriction. If bite depth is limited say by

grass height, we need to examine the scaling of bite width in relation to body mass.

Mouth parts and other feeding structures

As an index of the scaling of bite width to body size, the ratio of effective mouth width, in mm, to the cube root of body mass, in kg, will be used. Among ruminants the effective mouth width is controlled by the breadth of the incisor tooth row, while for rhinos and hippos it is related to the width of the lips used for cropping grass. For a relatively broad-muzzled grazer like wildebeest, this index works out at 11.4, versus 7.0 for a narrow-muzzled browser like greater kudu (based on measurements kindly supplied by N. Caithness). For a white rhino with a bite width of 200 mm, the bite width index comes to 17.1; while for hippos with a bite width of nearly 0.5 m (Laws 1981a) the index works out at about 35. The 50% increase in the effective bite width of white rhinos over wildebeest would be sufficient to reduce the foraging time needed by white rhinos to ingest a similar fraction of their body mass by about one third. If both animal species were equally constrained by grass height in their bite depth (i.e. eating rates were a function of $M^{2/3}$), a 1600 kg white rhino would need to feed for 1.3 times as long as a 180 kg wildebeest to ingest the same fraction of its body mass – instead of twice as long, as would be the case without the compensating expansion of bite width. A hippo with its even wider mouth would need to feed for only two-thirds as long as a wildebeest for the same specific food intake. In fact wildebeest feed for about 8 hours per day (Berry *et al.* 1982), as also do hippos, compared with 12 hours per day for white rhinos. This suggests that effective bite gape must also be constrained, taking into account the lower quantitative food requirements of the megaherbivores relative to wildebeest.

An elephant has a very different feeding adaptation in the form of its mobile trunk, which enables it to pluck large amounts of herbage at a time. For African elephants the mean weight of food gathered per trunk-load was 75 g (wet mass) in a Zimbabwean woodland (Guy 1975), while for Asian elephants in Sri Lanka an average weight of 150 g per trunk-load was calculated (Vancuylenberg 1977). Converted to dry mass this represents 10 $\times$ 10^{-6} to 20 $\times$ 10^{-6} of body mass. For cattle the mean bite mass of 0.33 g (dry weight) (Chacon & Stobbs 1976; Zimmerman 1978) represents 0.8 $\times$ 10^{-6} of body mass, i.e. only 4–8% as much.

However, the plucking rate of an elephant in terms of trunk-loads per minute is very much less than the biting rates achieved by cattle and other ungulates. The feeding rate of African elephants in Zimbabwe averaged 2.4

trunk-loads per minute (Guy 1975). Feeding rates of 5.9 trunk-loads per minute were recorded in Uganda (Wyatt and Eltringham, 1974), while in Tanzania feeding rates decreased from 8.4 trunk-loads per minute during the wet season to 3.1 trunk-loads per minute in the dry season (Barnes, 1979). For Asian elephants feeding rates of 2.1 trunk-loads per minute and 0.9 trunk-loads per minute were found by McKay (1973) and Vancuylenberg (1977) respectively, in the same park in Sri Lanka. In Malaya a feeding rate of 0.9 trunk-loads per minute was recorded (Olivier 1978, cited by Eltringham, 1982).

Based on Guy's data for a predominantly woody browse diet, the mean food ingestion rate of African elephants is about 72 g dry mass per minute, equivalent to 24×10^{-6} of body mass. For cattle with a biting rate of 55 per minute (Chacon & Stobbs 1976; Zimmerman 1978), the food ingestion rate amounts to 45×10^{-6} of body mass. Thus the large amount of herbage plucked per trunk-load by elephants is not quite adequate to compensate for the slower plucking rate.

To achieve the same mass-specific food intake as cattle, elephants would need to feed for nearly twice as long. However, the daily food requirement of an elephant per unit of body mass is only about 55% of the food intake of cattle. This reduced food intake rate almost exactly compensates for the reduced daily food requirement, so that elephants should be able to satisfy their appetite within about the same feeding time as cattle. However, the daily foraging times recorded for African elephants by Wyatt & Eltringham (1974) in the Queen Elizabeth National Park in Uganda were about 50–80% longer than those typically shown by cattle. This suggests that these elephants, occupying a largely grassland area, were in less favorable habitat than those studied by Guy (1975) in mopane woodlands in Zimbabwe.

Grazers and browsers

Notably most of the species falling above the regression line of daily foraging time on body mass (Fig. 6.4) are browsers or mixed feeders, while those falling below the line are grazers (with the exception of bushbuck). This pattern can be explained on the basis of the patchy distribution of woody plants compared with the even spread of grasses. For example, black rhinos spend about 20% of their day walking (Goddard 1970a), compared with only 7% for white rhinos (my study). Black rhinos feed for about 30% of the daylight hours, so that their food ingestion rate is reduced by about one third relative to white rhinos due to this effect. Likewise browsing kudus spend about 70% of their foraging time actually feeding (Owen-Smith & Novellie 1982), compared with 85–90% for largely grazing impalas (Dunham 1982).

Males and females

For most ungulates for which separate data for the two sexes are available, females forage for longer than males (Fig. 6.4). A similar pattern exists among primates (Clutton-Brock & Harvey 1977b). This difference follows from the additional nutritional demands imposed on females by pregnancy and lactation. Thus females generally need either to eat relatively more food per day than males, or to seek out higher quality food than that acceptable to males. Lactating females show higher stomach fills than non-lactating females both for hippos (Laws 1968b) and for African elephants (Laws *et al*. 1975). However, no data seem to be available to indicate whether females usually ingest more protein-rich food than males.

For white rhinos there is no sex difference in foraging budgets. White rhino males need just as long to satisfy their food requirements as females. This may be an outcome of the constraint imposed by grass height, i.e. white rhino males cannot increase their bite depth in proportion to their increased body size relative to females.

Diurnal and nocturnal feeding

The trend towards increasing foraging times with increasing body mass is apparent only in data based on the complete 24 hour cycle. Foraging times for the daylight hours alone show no obvious trend. The ratio of night-time to daytime feeding, for species from the size of Thomson's gazelle upwards, shows a downward trend with increasing body mass (with giraffe anomalous) (Fig. 6.5). Extreme examples, which cannot be denoted in the graph, include (i) warthogs, which feed for nearly half the day but spend the entire night hidden away in burrows (Cumming 1975); (ii) hippos, which remain inactive in pools during the day, and carry out all of their foraging at night. Moreover, megaherbivores generally travel to and from water after nightfall, while most medium-sized ungulates drink during the day.

However, very small antelope, like grysbok, grey duiker and suni, are more active at night than during the day (Novellie *et al*. 1984; T. Allan–Rowlandson personal communication; R. H. V. Bell personal communication), as also are both mountain and common reedbuck (Irby 1981, and

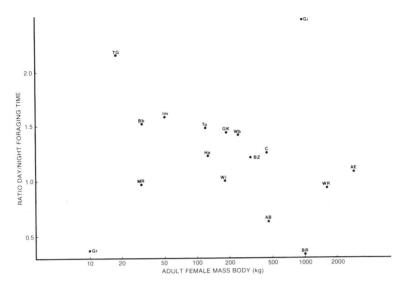

Fig. 6.5 Ratio of daytime to night-time feeding in relation to body mass. Data from references given in caption of Fig. 6.4 (excluding giraffe, $N = 14$, $R^2 = 0.442$, $P < 0.1$).

personal observations). These are all cryptic species relying on concealment for predator evasion. Medium-sized antelope such as impala feed for nearly one third of the night, but move around much less while doing so than during the day, and tend to congregate in open areas during the hours of darkness (Jarman & Jarman 1973a). Megaherbivores, being relatively invulnerable to predators, can afford to move around more at night than can smaller species. African buffalo, which are also more nocturnal than diurnal, probably gain some advantage in terms of predator protection from their very large, compact herd structure even at night.

Thermoregulatory constraints

Another factor promoting increased nocturnal activity by megaherbivores is the problem of maintaining thermal equilibrium. Maintenance metabolic rates are adjusted nearly in proportion to the surface area available for dissipating body heat. But besides body maintenance, the additional heat generated during muscular activity also needs to be taken into account.

Locomotion costs less energy per unit distance covered for big animals than for small animals (Taylor 1973). This is only because small animals need to take more steps to cover the same distance. A wide variety of data on the energy expended in running, flying and swimming support a hypothesis that all animals require the same quantity of energy in joules per unit of body mass to move one step (Gold, 1973). However, because the mass-specific resting metabolic rate decreases with increasing body mass, the costs of movement relative to maintenance metabolism are greater in larger animals than in smaller animals (Garland 1983; Peters 1983).

White rhinos averaged 35 steps per minute while walking, while 50 steps per minute were recorded for greater kudu females weighing 180 kg (Owen-Smith & Novellie 1982). Thus a white rhino should be expending only about two thirds as much energy per unit of walking time, relative to its body mass, as a kudu. However, relative to its basal metabolic rate the white rhino would be expending three times as much energy. Assuming both animal species to be constructed as simple cylinders, the ratio of surface area to volume for a 1600 kg white rhino is only about half that for a 180 kg kudu. Thus white rhinos have a smaller surface area available to dissipate the excess body heat produced by this activity than do kudus. Hence the problem of getting rid of the excess body heat produced during periods of muscular activity is more acute for megaherbivores than for smaller ungulates.

Accordingly, it is advantageous for megaherbivores to be more active at night, when there is less environmental heat load, than during the day. In fact, white rhinos and elephants might tend to perform all their feeding during the night, were it not for the constraints imposed by food ingestion and digestion rates. Black rhinos are much more nocturnal than white rhinos, probably as a consequence of the more rapid fermentation of browse leaves compared with grasses. Hippos are able to do all of their feeding at night, on account of the enhanced food ingestion rate conferred by their very large mouths.

The lack of hair on extant megaherbivores is also an aid to heat dissipation. Any liquid applied to the bare skin, whether in the form of water or mud, assists the cooling process. Observations that rhinos wallow at night as well as during the day support the hypothesis that it is the extra heat load generated by muscular activity that creates special problems for megaherbivores. Notably, white rhinos also possess unusually large sweat glands distributed over the whole body surface, evidently an adaptation for the rapid release of liquid for emergency cooling (Cave & Allbrook 1958). Elephants have a special adaptation in their enormous ear pinnae, which serve as heat radiating surfaces. Blood leaving the ear is notably cooler than that entering it (Buss & Estes 1971). It is this feature that allows African elephants to forage for as long as 9 hours during the daytime even in tropical African climates.

Summary

Daily foraging time shows a tendency to increase with increasing body mass, suggesting that bite size and hence feeding rate is constrained by features of plant structure. Megaherbivores exhibit compensatory adaptations to counter this restriction, either in terms of exceptionally wide mouths, or in the form of a special feeding apparatus such as the elephant's trunk. Foraging time budgets are also influenced by: (i) food type – browsers spend more time walking between feeding stations than grazers of similar body mass, and hence need to forage for longer; (ii) digestive passage rates – these constrain feeding times less in browsers than in grazers, and less in equids compared with other ungulates; (iii) reproductive status – pregnant or lactating females forage for longer than males or non-pregnant females; (iv) predation risks – smaller species in open habitats forage less at night than during the day; (v) heat balance related to muscular exertion – megaherbivores require special adaptations for dissipating body heat to allow them to forage for a large part of the day.

Home range extent

Megaherbivores tend to cover larger home ranges than smaller ungulates. McNab (1963) proposed that home range extent was directly dependant on bioenergetic requirements, i.e. it should vary in relation to $M^{0.75}$. However, Harestad & Bunnell (1979), using data drawn mostly from North American mammals, found that the power coefficient b relating home range area to body mass was significantly greater than 0.75; for herbivores (including squirrels and other granivores) they found $b = 1.02$, while for carnivores $b = 1.36$. For primate home ranges, Milton & May (1976) and Clutton-Brock & Harvey (1977a) found values for b of 0.83–0.99 for frugivores and 1.06–1.15 for folivores. Harestad & Bunnell (1979) explained the discrepancy with the suggestion that mean resource density within the home range decreased with increasing body size, because larger home ranges tended to include more lacunae or patches of low productivity.

Clutton-Brock & Harvey (1977b) pointed out that home range size should be related to total energy requirements, including expenditures for activities, rather than to resting metabolic rate. However, data for primates (Harvey & Clutton-Brock 1981), and for combined samples of birds and rodents (Mace & Harvey 1983), show that home range size increases in

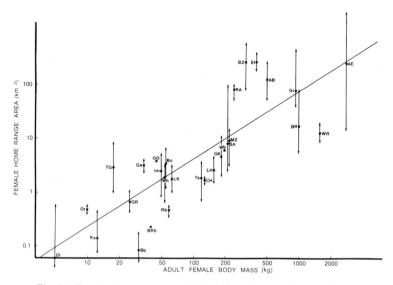

Fig. 6.6 Female home range extents for African large herbivores in relation to body mass. Key to species labels in Appendix I. Data from Cumming (1975), David (1973), Douglas-Hamilton (1972), Dowsett (1966), Dunbar & Dunbar (1974), Estes & Estes (1974), Goddard (1967), Gosling (1974), Grobler (1974), Hendrichs & Hendrichs (1971), Hillman (1979), Hitchins (1971), Jarman & Jarman (1973a), E. Joubert (1972), S. Joubert (1972, 1974), Jungius (1971b), Leuthold (1970, 1971, 1972 and 1974), Leuthold & Sale (1973), Leuthold & Leuthold (1978), Langman (1973), Murray (1980), Owen-Smith (1979), Pellew (1981), Simpson (1974), Sinclair (1977), Spinage (1969), Walther (1964, 1965), Waser (1974). Regression based on the geometric means of the figures reported from different localities (indicated by dots). Lines indicate the complete range of published data.

Regression line: $HR(ha) = 1.35M^{1.25}$ (SE(b) = 0.16, $R^2 = 0.70$, $N = 29$, $P = 0.0001$).

relation to daily metabolic expenditure raised to the power 2.2, rather than 1.0. In effect, larger animals require disproportionately larger home ranges than can be explained simply on the basis of energy requirements.

Damuth (1981a) suggested an alternative hypothesis. He pointed out that the extent of home range overlap increases with increasing body size, and that this effect could account for the observed departure from the metabolic rate–body size relation.

For most African ungulates only female home ranges can be related directly to nutritional requirements, because male home ranges are restricted by social pressures. The data show that the total extent of the home ranges covered by adult female ungulates increases more steeply with body mass than was found for North American herbivores, with an estimate for the power coefficient *b* of 1.25 ± 0.32 (Fig. 6.6). Notably the species falling above the regression line tend to be those forming large herds, for example Thomson's gazelle, African buffalo and eland; while solitary species like white rhino fall below the line. Sociality is more prevalent among larger ungulates than among smaller species (Jarman 1974); and the need to move over a relatively larger area can be viewed as a cost of sharing the home range with companions.

In terms of resource requirements it is more appropriate to divide the home range area by the size of the social group, to obtain the home range extent available per individual. For this purpose the social group represents not the number of animals associated together at any one time, but rather those individuals sharing a common home range. For example, impala females form clans of 50–100 animals occupying fairly discrete areas covering 0.8–1.8 km² (Murray 1982b). African elephants likewise form clans of 100 or more individuals sharing a common home range of about 200–700 km², with little overlap between different clan areas (Martin 1978 and personal communication; Hall-Martin 1984). For white rhinos no such clan organization exists, so that it is impossible to draw a direct comparison with other species on this basis. If a typical group size of three is assumed, then the home range area per individual white rhino is about 4 km². However, the population density of 5.3 per km² prevailing in the Madlozi study area indicates that the actual extent of habitat available to each individual white rhino was only about 0.2 km².

The regression of home range extent per individual against body mass yields an estimate for *b* of 0.83 ± 0.43, which is not significantly different from 0.75 (Fig. 6.7). However, this analysis does not allow for range overlap between social units. Home ranges tend to be fairly discrete among the smallest antelope species, among which females frequently share the territories of individual males. Among the larger ungulates a pattern of overlapping home ranges between different female groups is usual.

Home range size thus represents a compromise between individual metabolic requirements and social pressures, so that no simple functional relationship with body size should be expected (Damuth 1981a). Metabolic demands increase with body mass, but so does group size, and also the degree of range overlap between different individuals or groups. Bioenergetic influences are better revealed by the relationship between prevailing population densities and body mass. This approach will be taken up in Chapter 14 when body size relationships at the demographic level are considered.

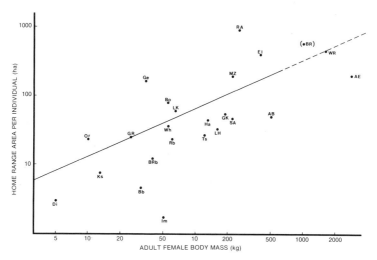

Fig. 6.7 Home range extent per individual for African large herbivores in relation to body mass. Based on dividing the home range extent by the size of the social group. Key to species labels in Appendix I. References as in Fig. 6.6. Regression line (excluding megaherbivores): HRI(ha) = $1.07M^{0.83}$ (SE(b) = 0.21, R^2 = 0.44, N = 21, P = 0.001).

Trophic ecology of megaherbivores: summary

Let me now summarise the general findings of both this chapter and the previous one concerning the trophic adaptations of megaherbivores. Large body size automatically confers an advantage in terms of ability to digest structural cellulose through bacterial fermentation, due to a prolonged retention time of food in the gut. It thus represents an alternative mechanism for enhancing cell wall digestion to the development of delaying compartments in regions of the gut as shown by the ruminant artiodactyls. However, the digestive efficiency achieved through large size does not quite match that attained by certain grazing ruminants. This is because megaherbivores lack the ability to remasticate the ingested forage and thereby increase the surface area for bacterial action, which is a major advantage of the ruminant digestive system.

The increase in fermentative degradation of cell wall that occurs with increasing body size is probably the basic explanation for the trends towards large size shown repeatedly among lineages of hindgut fermenters. The two grazing rhinos epitomize this solution to the problem of being a herbivore, while hippos show that similar results can be achieved using a foregut fermentative chamber.

An alternative adaptive route open to hindgut fermenters is to compensate for digestive inefficiency through high food intake. What is quickly digestible is digested, and the remainder is passed out of the gut to make way for more material. This is the solution adopted by equids, and by small hindgut fermenters such as lagomorphs and certain rodents. Such species select 'duplex' food sources consisting of a suitable mix of quickly fermentable carbohydrates with completely indigestible fiber. For example, zebras consume a mixture of green grass leaves diluted by fibrous stem, while similar-sized ruminants filter out some of the structural fiber using the selective apprehension ability conferred by their lower incisor-palatal bite. Compared to modern rhinos, both extant species of elephant tend towards the equid pattern. When feeding on grasses in the dry season, they tend to select underground parts where carbohydrates are stored, rather than trying to digest the fibrous dry leaves, as white rhinos do. This follows from the fact that both African and Asian elephants are adapted as mixed grazer–browsers, and woody plant branchlets offer a suitable mix of rapidly digesting leaves and indigestible stems. In fact, African elephants frequently discard poor quality leaves in favor of bark containing some fraction of soluble carbohydrates. The extinct mammoths and grazing gomphotheres such as *Stegomastodon*, which from their dentition were adapted for a diet of fine grass leaves, most probably resembled the grazing rhinos in having a relatively efficient hindgut fermentation.

Megaherbivores following the rhino digestive solution are restricted in their food intake by the inherently slow rate of passage of digesta through their gut. They are thus dependent upon a high enough concentration of fermentable fiber in their diet to achieve both an adequate throughput rate and a sufficient level of digestible energy. The result of the interaction between these two factors is that, despite their much lower metabolic demands per unit of body mass, such species are dependent on a dietary quality not that much lower than that needed by much smaller ruminants, to satisfy their maintenance requirements. Thus, during the dry season white rhinos occupy regions where the soil nutrient–rainfall combination causes grasses to build up only moderate levels of indigestible fiber in their leaves (so-called 'sweetveld' conditions, as prevalent at Umfolozi). In 'sourveld' regions, where grasses are mostly highly fibrous and lack nutrients in above-ground parts during the dry season, white rhinos feed mostly in those areas of the landscape where soil nutrients accumulate, for example around termitaria and along the margins of drainage sump grasslands (which was a feature of much of their former distribution in Zimbabwe). The northern steppe-tundra favored by the extinct woolly mammoths was characterized by a combination of low precipitation and fertile loess soils, and so was a boreal counterpart of southern African sweetveld. Nevertheless, under submaintenance conditions during the dormant season for plant growth, all megaherbivores can survive somewhat longer on a starvation diet than would be the case for smaller animals with a higher metabolic rate.

Relatively nutritious vegetation components tend to be available at a somewhat low standing biomass. Thus megaherbivores dependent upon such vegetation encounter problems in achieving an adequate rate of food ingestion relative to their needs. Their ability to compensate through increased foraging time is restricted by problems of thermoregulation, exacerbated by the heat generated through muscular activity relative to their restricted surface area to muscular mass ratio. Thus they need adaptive modifications of food gathering structures to compensate for the limitations imposed by plant growth patterns. In grazing rhinos and hippos, this is achieved by expansions of the lips. Interestingly, mammoths exhibited flanges on the sides of the trunktip, which probably served to increase the amount of grass that could be lifted per trunk-load (Guthrie 1982).

7

Social organization and behavior

Introduction

Patterns of social organization reflect the cooperative and competitive interactions occurring among animals within local populations relating to survival and reproduction. Generally different age/sex classes differ in their spatial dispersion, i.e. group membership and the spatial relationships both within and between groups. Other social relationships may be evident from the patterns of behavior displayed in encounters, for example those signifying dominant/subordinate relations. Dominance is particularly a feature of adult males, which are inevitably competitors for reproductive opportunities. Females in turn may exert some selection over the sire of their offspring. Anti-predator responses are also appropriately considered in this chapter, since the affiliative relationships established among adult females serve largely to reduce the risks of predation, not only for self but also for progeny.

Group structure

The term group refers to a close spatial association between individuals. However, socially the temporal cohesion of the group is relatively more important than short-term spatial proximity. Groups may vary in size, and in the age/sex classes of animals composing them. Different groups may either move independently of one another, or tend to associate together, or space themselves out with respect to other groups. Grouping patterns may furthermore change seasonally, particularly in relation to variations in reproductive activity.

Elephants

Among African elephants, the nuclear group comprises an adult female together with 1–3 immature progeny of varying ages. However, generally one finds 2–4 mothers plus young associated together to form

Fig. 7.1 African elephant family unit (photo courtesy N. Leader-Williams).

family units typically numbering about 4 to 12 animals (Fig. 7.1). From an analysis of the ages of the animals (determined from culling operations), many of these units evidently consist of an old female or matriarch (aged 38–60 years) together with her mature daughters and their offspring. Other family units include two mature females of similar age, which may be sisters. While members of family units are usually found close together, they may separate into nuclear mother–young units under dry season or drought conditions. When alarmed family unit members bunch around the matriarch, and if she is shot they mill around in confusion. The few solitary females encountered are almost invariably senile individuals over 50 years in age. Groups recorded for forest elephants are somewhat smaller than those typical of savanna elephants. In the Tai National Park, Ivory Coast, the mean group size was 3.4 (maximum 9), and solitary females were not unusual.

Different family units are commonly found in close proximity, within a kilometer or less, and tend to move in a coordinated fashion along parallel lines. Large aggregations of elephants numbering 100–1300 animals, including a number of males, may also be observed during the wet season, and these may maintain cohesion over several weeks. Radio-telemetric monitoring of movements in the Sengwa Research Area in Zimbabwe indicated the existence of distinct clans of females plus young, including a hundred or more animals. While the ranges of neighboring clans overlap peripherally, members of different clans are seldom in the overlap area at the same time. The very large herds of a thousand or more elephants observed in the Murchison Falls Park in Uganda during the 1960s could have been a result of the compression of clans by human settlements occupying part of their former range (Barnes 1982b; Buss & Savidge 1966; Douglas-Hamilton 1972; Hendrichs 1971; Laws 1969b, 1974; Laws, Parker & Johnstone 1975; Leuthold 1976b; Martin 1978; Merz 1986b; Moss 1983; Western & Lindsay 1984; Wing & Buss 1970).

Males over 16 years of age associate only transiently with the family units. Bulls join together in groups averaging 2–4 animals (maximum 35, although there is one report of a group of 144 bulls from the Tsavo region of Kenya). Associations between particular individuals are temporary. The longest recorded attachment lasted 51 days, but most bulls remain together only a day or two. Between 13% and 60% of adult males are solitary, the proportion tending to be highest during the wet season. Adult males favor so-called bull areas distinct from the home ranges of female clans, but wander widely at certain times (Barnes 1982; Croze 1974a; Hall-Martin 1984; Hendrichs 1971; Laws 1969; Laws, Parker & Johnstone 1975; Martin 1978; Moss 1983).

Asian elephants show a grouping pattern similar to that of African elephants, although population levels are too low for large aggregations to form. In Sri Lanka the modal group size of female–young units is 2–5, with the largest recorded groups numbering 39 animals at Gal Oya and 71 at Amparai (but there is a report of a herd of 150 from elsewhere in Sri Lanka). Large herds generally split into smaller foraging groups, which move in a coordinated manner. Infrasonic calls may help maintain coordination between subgroups separated by distances of several kilometers (and have recently been confirmed also for African elephants). So-called herds comprising between 20 and 150 animals occupied fairly discrete home ranges.

About 70–80% of Asian elephant bulls are solitary, although groups of up to 7 males form occasionally. Bulls may join family units, but remain in their company for only a few days before departing (Eisenberg, McKay & Jainudeen 1971; Ishwaran 1981; Kurt 1974; McKay 1973; Payne, Langbauer & Thomas 1986; Santiapillai, Chambers & Ishwaran 1984).

Hippopotamus

Hippos congregate in their daytime refuge pools in discrete groups typically numbering 10–30 animals. These generally include a large dominant bull, several females and young, and from two to six peripheral males.

Such groups appear to remain fairly constant in their membership, at least over periods of a month or two. However, when rivers dry up animals may be forced together in aggregations of up to 150 animals in remaining pools. Solitary males commonly occupy outlying small pools, although sometimes all-male groups may share a pool. Hippos travel solitarily while feeding on land at night, except for females accompanied by one or more offspring (Marshall & Sayer 1976; Olivier & Laurie 1974; Verheyen 1954; Viljoen 1980).

Giraffe

Giraffe females and young form groups averaging between 3 and 17 animals in different areas. A maximum group size of 239 was recorded in the Serengeti, but no group larger than 35 has been reported elsewhere. However, observers vary in the spatial limits they use in identifying groups, some regarding all animals within less than a kilometer as being part of the same group (which is reasonable considering the wide visual field that a giraffe commands from its height). Groupings are not cohesive, but change in composition from day to day. Female home ranges overlap extensively. Individual females return repeatedly to the same localities to give birth, even when their home ranges have shifted away from these areas.

Adult male giraffe are most commonly solitary. However they may join female groups temporarily, or associate in loose all-male groups. Male home ranges also show much overlap (Foster & Dagg 1972; Hall-Martin 1975; Langman 1977; B. M. Leuthold 1979; Pellew 1983a).

Rhinoceroses

For Indian rhinos the cow-calf unit is the only enduring association. The largest recorded groups comprised 3 animals, but such associations rarely persisted longer than 2–3 days. Adult males are solitary, apart from temporary associations with females. Sixty-three percent of subadults were recorded alone, the remainder being associated with another subadult. Home ranges show much overlap, both between bulls and cows as well as between different individuals of each sex (Laurie 1982). Javan rhinos and Sumatran rhinos are likewise mostly solitary in their habits (Borner 1979; Groves & Kurt 1972; Schenkel & Schenkel-Hulliger 1969b).

Black rhino females were solitary or accompanied only by a calf in 75% of sightings at Ngorongoro and Olduvai in Tanzania, while 80% of the bulls seen were solitary. Subadults were solitary in about 40% of sightings at Ngorongoro and 68% of sightings at Olduvai. Only 1.7% of black rhino groups at Hluhluwe consisted of more than 3 animals, the largest association numbering 7 animals. The largest group recorded at Ngorongoro consisted of 5 animals – 2 cows, a calf, a subadult female and a bull – but these animals remained together only four days. One temporary aggregation of 13 animals disbanded after two hours. The largest group recorded at Amboseli, Kenya, consisted of four animals: an old cow, a young cow, a subadult female and a bull (Goddard 1967; P. M. Hitchins personal communication; Klingel & Klingel 1966; Schenkel & Schenkel-Hulliger 1969a).

Radio-telemetric tracking of black rhinos at Hluhluwe showed that female home ranges overlapped; while adult males occupied mutually exclusive home ranges, except for a case in which two bulls shared the same range. In Namibia no two bulls shared the same home range. Elsewhere extensive overlap between the home ranges of both males and females has been reported, but the possibility of dominant and subordinate bulls sharing home ranges was not considered, nor was allowance made for the possibility of bulls leaving their home area to travel to and from water (Goddard 1967; Hitchins 1971; Joubert & Eloff 1971; Schenkel 1966).

White rhinoceros

My observations on white rhinos were based on animals that were individually identifiable, from variations in horn shapes and other features. At Umfolozi most white rhino cows were accompanied only by a single offspring; while white rhino bulls were most often solitary (Fig. 7.2). Subadults tended to be associated in pairs, either of the same or opposite sex. Groups of three generally consisted of either a subadult attached to a cow–calf pair, or an adult male accompanying a cow plus calf. A few groups comprised three or more subadults, including in some instances a cow lacking a calf. The largest cohesive group in the study area numbered seven animals, a cow and six young subadults. One group of nine subadults was seen outside the study area. Most of the larger groups represented temporary associations, and broke up into units of two or three within a few days.

A few groups included two adult females, both without calves. In most of these cases one of the females was a young adult that may have been an offspring of the older cow. In one case both females were in early prime and of closely similar age, so that a mother–offspring relation was precluded. These two cows remained together, along with 2–3 adolescent companions, for four months, until one of them gave birth. The bond between the two cows was quite amicable, with no dominance or leadership difference expressed.

Cow–calf pairs as well as pairs of subadults tended to keep close together, generally within one body length (about 4 m). A subadult or second adult

(a) Cows

(i) All groups

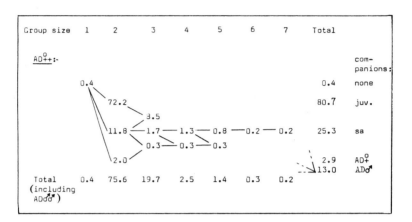

```
Group size   1     2     3     4     5     6     7     Total

AD♀♀:-                                                          com-
                                                               panions:
           0.4                                       0.4       none
                72.2                                  80.7      juv.
                      3.5
               11.8—1.7 —1.3—0.8 —0.2—0.2            25.3      sa
                    0.3—0.3—0.3
               2.0                                   2.9       AD♀
                                                     13.0      AD♂
Total      0.4   75.6  19.7   2.5   1.4   0.3   0.2
(including
AD♂♂)
```

(ii) Stable associations

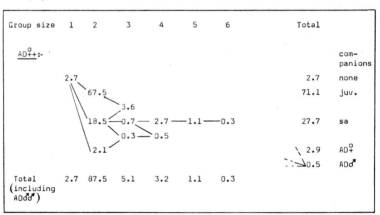

```
Group size   1     2     3     4     5     6     Total

AD♀♀:-                                                    com-
                                                          panions
           2.7                                   2.7      none
                67.5                              71.1     juv.
                      3.6
               18.5—0.7 —2.7 —1.1—0.3            27.7     sa
                    0.3—0.5
               2.1                               2.9      AD♀
                                                 0.5      AD♂
Total      2.7   87.5   5.1   3.2   1.1   0.3
(including
AD♂♂)
```

(b) Adult ♂♂

(i) All groups

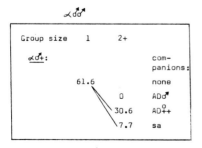

α♂♂

```
Group size   1     2+

α♂:                            com-
                               panions:
           61.6                none
                   0           AD♂
                   30.6        AD♀♀
                   7.7         sa
```

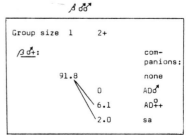

β♂♂

```
Group size   1     2+

β♂:                            com-
                               panions:
           91.8                none
                   0           AD♂
                   6.1         AD♀♀
                   2.0         sa
```

(ii) Stable associations

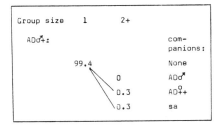

(c) Subadults

(i) All groups

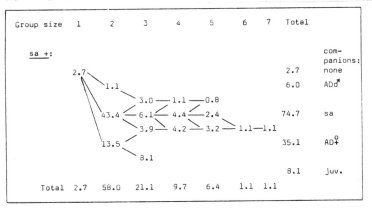

(ii) Stable associations

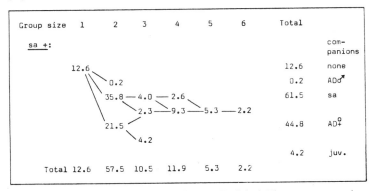

Fig. 7.2 Grouping patterns of white rhinos at Umfolozi. Figures represent the percentage of individuals of each age/sex category that were associated with each combination of group size and age/sex class of companions. Lines clarify successive additions in building up larger groups. Results are analyzed separately for stable associations, i.e. groupings persisting for longer than a month. For all groups, $N = 1432$ records; for stable associations, $N = 935$ records, a♂♂ = territorial adult males, β♂♂ = subordinate adult males. From Owen-Smith 1975.

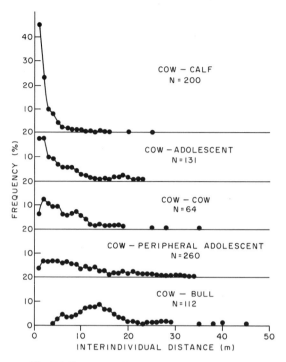

Fig. 7.3 Frequency distribution of interindividual distances in different white rhino groupings.

female attached to a cow maintained somewhat greater spacing, while a bull accompanying a cow tended to lag about 15 m behind her (Fig. 7.3).

White rhino cows generally responded neutrally when they met. They grazed in the same areas, and lay close together at restplaces, without antagonism. Sometimes cows approached one another to make nasonasal contact, and about a quarter of these meetings developed into playful horn sparring. In one instance a prolonged but playful horn wrestling and chasing game developed between two cows, with the subadult companions of both cows also joining in. In a small proportion of meetings cows exchanged threatening snorts or roars, usually in circumstances where one was accompanied by a young calf. Cows also exchanged aggressive roars and snorts at waterholes, where animals from different areas came into close proximity.

The home ranges of white rhino cows were individually distinctive but overlapped extensively. There was no indication of clans sharing a common home range. Nevertheless, there appeared to be some subdivision of the population with regard to the use of particular waterholes. None of the study area cows was recorded south of a spring located at the southern limit of the study area, while few of the cows seen to the south of the spring appeared in the study area.

Summary

Elephants of both species are highly social, showing a hierarchical pattern of groupings including mother–offspring associations, extended family units and coordinated aggregations of a hundred or more animals. Distinct clans of females plus young occupy fairly discrete home ranges. Adult males may be either solitary, associated in small all-male groups, or attached to female units. Hippos associate together in groups, including several females plus young and one or more males, in their refuge pools; but animals forage solitarily on land at night. Giraffe females form small groups which change in composition from day to day; while males are solitary or in small all-male groups. Rhinoceroses of all species are generally solitary. Cohesive groups consist mostly of mother–offspring associations, or of pairs formed between subadults. Among Indian rhinos and black rhinos subadults are found alone more often than not. In contrast white rhino subadults generally join together in pairs or larger groups, while cows without calves may join up with another cow or with one or more subadults.

Male dominance relations

Dominance relationships may be evident either from differences in the behavioral actions displayed in encounters, or from patterns of avoidance. They may be expressed within a spatial context, as in territoriality, or be dependent upon individual identity independently of location. Where dominance is most strongly established, overt aggression may be rare. An initial contest may be needed to establish relative dominance, but in many cases it is assessed on the basis of auditory, olfactory or visual displays advertizing potential fighting ability.

Elephants

Asian elephant bulls exhibit periods of heightened sexual and aggressive activity, termed musth. Musth is associated with a copious secretion from temporal glands, and by continual dribbling of small amounts of smelly urine. Bulls in musth rub the temporal gland secretion on trees, and wander from one herd to another inspecting females for signs of estrus. A bull in musth manifests dominance over all other bulls in his vicinity, and non-musth bulls give way without contest. Bulls tend to show non-overlapping musth periods, so that it is rare for two males to be in musth in the same area. When this happens, one male either moves away, or comes out of musth. Musth periods generally last 2–3 months, and usually recur once annually, most often during the rainy season (Eisenberg, McKay & Jainudeen 1971; Jainudeen, McKay & Eisenberg 1972; Kurt 1974).

African elephant bulls exhibit a similar condition, revealed by a green exudate from the penis. However, temporal gland secretions are not restricted to males in musth, being shown by adult females at times. Scent marking of the environment has not been recorded, although twigs are sometimes inserted into the temporal gland orifice. Bulls in musth wander widely, and readily displace equally large bulls that are not in musth. It is rare for two bulls in the same area to be in musth at the same time. Mature males over 35 years of age typically show one musth period lasting 2–3 months each year. Among bulls that are not in musth a dominance ranking is evident. A more dominant animal holds his head higher, with ears out but relaxed, and may sniff the end of the trunk or the temporal region of the other bull. A subordinate bull places its trunk in the mouth or under the chin of the dominant bull. Several males not in musth may gather in the vicinity of a female in estrus, and fights may break out among them (Moss 1983; Poole & Moss, 1981).

Hippopotamus

Among hippos up to six adult or subadult males may share the same pool, but one bull is clearly dominant. Other bulls remain around the periphery of the pool, and when challenged demonstrate their submission by defecating with tail wagging in front of the dominant bull. Dominant bulls have lower testis testosterone levels than subordinate bulls, although there is no difference in sperm production.

Some pools or sections of lakeshore may be occupied by solitary hippo bulls without females, or by all-male groups. Since no more than one dominant bull occupies the same pool or region of lakeshore, these areas in effect represent exclusive territories. Neighboring dominant bulls meeting at a boundary stare, then present their rear ends and defecate, scattering their dung with their tails before returning to their own territories. Violent fights may break out when one dominantly behaving bull intrudes into another's territory, and lethal injuries can be inflicted with the lower canine tusks. However, males moving on land at night tend to ignore one another, and there is some overlap in the use of particular foraging areas. Both dominant and subordinate males scatter their dung with tail wagging movements, and commonly also urinate at particular sites alongside trails (H. Klingel personal communication; Laws 1984; Verheyen 1954).

Giraffe

Dominant bull giraffe demonstrate their status by standing tall upon encountering another bull. Young bulls engage in ritualized contests, striking one another with swinging blows of the head. A superior male may attempt to mount the other male at the end of the contest. This may serve to establish relative dominance without the need for a fight. No spatial exclusion among males in evident. Dominant bulls spend much of their time patrolling their home range core areas, reinforcing their dominance over males encountered, and investigating females. Fighting, involving powerful blows with the head, may develop when a strange male is encountered in an area (Pellew 1984a).

Rhinoceroses

Among Indian rhinos dominant males perform squirt-urination and foot-dragging displays. They are commonly found associated with females, but never with one another, although home ranges overlap. Males occupying adjoining home ranges rarely interact aggressively. However, violent fights, sometimes with fatal results, develop when strange squirt-urinating males intrude into an area. Submissive males do not squirt-urinate, with rare exceptions. They share the ranges of dominant males, and are sometimes tolerated close to the latter, while on other occasions they are chased. Some of the submissive males are young adults, while others are old individuals. Subadult males are also sometimes attacked and killed. Dung is deposited in piles bordering trails and feeding areas. Both dominant and subordinate males as well as females add to these accumulations, and the dung is not scattered (Laurie 1982; Ullrich 1964).

Sumatran rhino males commonly urinate in a spray directed over a bush, and may couple this with backwards scraping of the hindlegs and horning of the bush. Dominance relationships remain unknown. Both Sumatran rhinos and Javan rhinos accumulate their dung at particular sites (Borner 1979; Schenkel & Schenkel-Hulliger 1969b).

Black rhino bulls at Hluhluwe in Zululand occupy mutually exclusive territories, with one documented case of an apparently subordinate bull sharing a territory. A fight between two bulls was observed to take place at the boundary between their adjoining territories. In Namibia black rhino bulls occupy mutually exclusive home ranges, although other manifestations of territoriality have not been recorded. In East Africa male home ranges are much larger than observed in Zululand and overlap extensively. Bulls meeting one another respond in a variety of ways. Sometimes they stare at one another, but decline to meet. Sometimes a 'complex bull ceremony' takes place, involving stiff-legged scraping, 'imposing' postures, and charges. In some meetings the resident bull emits 'screaming groans', while the intruder is silent. Dominance rankings are thus not evident; but behavioral observations in East Africa have probably not been sufficiently intensive to discern them, considering the possibility that dominant and subordinate bulls may share the same home areas.

Black rhino bulls commonly eject their urine in a spray without any preceding actions. On other occasions they horn the ground, drag the feet (sometimes for distances of up to 10 m), and spray-urinate. Black rhinos commonly defecate on dung-heaps, and both sexes scatter their droppings with backwardly directed kicks (Goddard 1967; P. M. Hitchins personal communication; Joubert & Eloff 1971; G. Owen-Smith personal communication; Schenkel & Schenkel-Hulliger 1969a)

White rhinoceros

White rhino bulls occupy nonoverlapping home ranges, which are appropriately termed territories (Owen-Smith 1972 and 1975). Bulls leave their territories only to proceed to and from water, apart from occasional exploratory sallies. These territories covered areas of between 0.75 and 2.6 km², and were thus considerably smaller than the home ranges covered by white rhino cows. In the Kyle Game Park in Zimbabwe, somewhat larger territories were recorded, varying between 5 km² to 11 km² in extent (Condy 1973). A single adult male introduced into the Murchison Falls Park in Uganda moved over an annual range of 30 km², although he restricted his movements to an area of 6 km² during the wet season (van Gyseghem 1984).

In some instances two, or even three, white rhino bulls shared the same home territory, but in these cases one bull was clearly dominant and the other subordinate. Subordinate bulls stood with ears back giving loud roars when confronted by the dominant bull (Fig. 7.4). Most such confrontations were brief, lasting less than a minute, and ended when the dominant bull moved away. Dominant bulls ejected their urine in powerful sprays, while subordinate bulls and cows urinated in a conventional stream. Spray-urination was commonly preceded by a ritualized wiping of the anterior horn over a low bush or the ground, followed by scraping the legs over the site. Dominant bulls scattered their dung after defecating, while subordinate bulls only occasionally made a few ineffectual kicking movements. Females and subadults did not scatter their dung. Subordinate bulls sharing

Fig. 7.4 A white rhino subordinate bull stands giving the snarl display when approached by the resident territorial bull. From Owen-Smith 1975.

the same home territory appeared neutral in their relations with one another.

If two dominant bulls occupying adjoining territories met at a common border, they engaged in a ritualized confrontation (Fig. 7.5). The animals stared silently horn against horn, then backed away to wipe the anterior horn on the ground. These actions were repeated for periods varying from a few minutes to over an hour. Eventually the two bulls moved apart back into their own home areas. A dominant bull intruding into a neighbor's territory, generally on his way to or from water, displayed the same actions, but backed away steadily during the course of the confrontation, until he reached the border of his own territory.

Dominant bulls crossing more distant territories behaved like subordinate bulls. A wandering bull did not spray-urinate while outside his own territory, although he still scattered his dung. If accosted by another bull, the intruder stood with ears back making loud roars and shrieks. Generally such confrontations were brief, with the resident territory occupant moving

Fig. 7.5 Two white rhino territory holders stare horn against horn during a territory border confrontation. From Owen-Smith 1975.

off allowing the intruder to continue on his journey. However, some encounters with intruding bulls were more prolonged, and led to attacks or even fights. Subadult males in pairs were generally confronted briefly, while solitary subadult males were subjected to prolonged confrontations and sometimes attacked.

Dominant bulls spent more time engaged in 'other activities' than subordinate bulls, cows or subadults (Table 7.1). Part of this time was taken up by social interactions (mainly with females), and the remainder by walking, related to the patrolling of territory boundaries. Spray-urination sites were most densely concentrated along border regions between adjoining territories, and also along trails frequently used by rhinos. Dungheaps in border regions showed a characteristic central hollow, resulting from repeated visits by one or other of the neighboring territory holders every couple of days. Subordinate bulls spent more of their time standing looking around than did rhinos of other social classes.

Eleven instances of changes in dominance status were observed during my study. In three cases a strange bull moved into the study area displacing

Table 7.1. *Relative apportionment of time among 'other activities', by different social categories of white rhino*
Upper figure = percent of 'other activity' time. Lower figure (bracketed) = percent of total daily activity.

Age/sex class	Walking	Standing alert	Wallowing + rubbing	Drinking	Social interactions	Total	N days	N mins
Territorial bull	53 (9.7)	$17\frac{1}{2}$ (3.2)	8 (1.4)	$3\frac{1}{2}$ (0.6)	18 (3.3)	100 (18.2)	43	3174
Subordinate bull	$47\frac{1}{2}$ (6.5)	44 (6.0)	2 (0.3)	3 (0.4)	$3\frac{1}{2}$ (0.5)	100 (13.7)	14	851
Cow	42 (5.4)	20 (2.6)	12 (1.6)	$3\frac{1}{2}$ (0.5)	22 (2.8)	100 (12.9)	34	2503
Subadult	$48\frac{1}{2}$ (5.8)	20 (2.4)	25 (3.0)	4 (0.5)	2 (0.2)	100 (11.9)	20	1239

one of the resident bulls. In three cases previously subordinate bulls assumed dominance in a nearby territory (though never in the territory that they had occupied while a subordinate). One chain displacement occurred: a bull displaced by a newcomer remained in his home territory as a subordinate for three months, then assumed dominance in a territory 2 km away. The bull that he displaced remained one week, then took over dominance from an ageing male in the adjoining territory. The latter bull was still present as a subordinate in the same territory at the conclusion of my study 15 months later. Another deposed dominant bull shifted to the neighboring territory where he assumed subordinate status. One subadult male moved into the study area and took up residence as a subordinate bull.

In some cases a serious fight clinched the change in relative status between the bulls concerned, but in other instances the two bulls showed no more than minor gashes after the transition. Newly instated dominant bulls immediately set about placing their spray-urination marks all over the territory. Deposed bulls behaved like subordinate males: they ceased spray-urinating, stopped scattering their dung, and stood defensively with roars when confronted by the new territorial dominant. The latter treated them like any other subordinate bull, and did not drive them out of the area.

Summary

Male dominance relations are organized territorially in white rhinos, hippos, and in at least some populations of black rhinos. Elephants of both species and giraffes show a rank dominance system. Elephants furthermore show a temporally restricted period of heightened aggressive and sexual behavior. Indian rhinos show no spatial exclusion, but males appear tolerant of neighbours while strangers are likely to be attacked.

Courtship and mating

There are changes in female behavior and physiology associated with estrus. These elicit courtship responses from males, leading to copulation. Matings and resulting births may vary in the degree to which they are restricted seasonally. Prior dominance relations may govern mate access by males more or less strictly. Females can exert some choice over the sire of their offspring by avoiding less acceptable suitors.

Elephants

Among African elephants most conceptions occur during the rainy season. In the Luangwa Valley in Zambia, 77% of estimated conception dates fell during the 4 month period January–April. In the Kruger Park in

South Africa, 70% of conception dates occurred during the 6 wet season months of November–April. In the Hwange and Gonarezhou Parks in Zimbabwe, 88% of conception dates fell during November–April, except in one year with an extended rainy season when conceptions continued through May–June. Following a gestation period of 22 months, the birth peak usually occurs over October–December at the beginning of the wet season. In Uganda, where there is a bimodal rainfall distribution, there are two peaks in conceptions. At Amboseli in Kenya, which is on the equator, there is no strong seasonal pattern in reproductive activity (Hanks 1969b; Laws, Parker & Johnstone 1970; Sherry 1975; Smuts 1975c; Western & Lindsay 1984).

However, all of the above estimates of conception times incorporate an error in the ageing of small fetuses, causing these to be assigned birth dates about two months too late. The result is an underestimation of the degree of seasonality in births. Using a corrected formula, it was found that 93% of conception dates for elephants culled in the Chirisa and Chizarira Game Reserves in Zimbabwe fell during the rainy season months of November–April, with no conceptions occuring during the dry season months of June–August. The peak birth months in this area were December–February (Craig 1984).

During the wet season African elephant bulls roam widely, contacting and checking any females encountered. Searching behavior is not restricted to bulls in musth. However, searching may be largely a feature of low density populations where female units are dispersed. In other populations family units aggregate into large herds in the wet season, and may have a number of adult males associated with them. At Amboseli large males older than 35 years spent less time with female groups than younger males did, and were usually in musth when associated with a female group. Large dominant males tend to have their musth periods during the late wet season, with less dominant males coming into musth during the dry season.

Females in the early stages of estrus behave in such a way as to draw attention to themselves. Large numbers of bulls gather in the vicinity of such females – 10 on average at Amboseli, with a maximum of 67 recorded. The female actively remains close to a large musth bull if one is present. Other bulls keep their distance from the musth bull, but approach and pursue the female if the large bull moves away. Females try to avoid such overtures from younger males, and her companions may assist by attracting the attention of larger bulls with loud noises if she is caught and mounted by a medium-sized bull. Copulations are brief, lasting only 40 s on average. Estrus persists for 2–6 days, during which time a female may be mounted by several different males (Barnes, 1982b; Martin 1978; Moss 1983; Short 1966; Western & Lindsay 1984).

Asian elephant bulls wander from one female herd to another while in musth. Mating is not restricted to musth bulls. However, since bulls in musth are able to dominate other bulls, they monopolize matings when present. It is not uncommon for females to be mated by several different males in succession. Mountings last about half a minute, but intromission persists for only 8 s. In Sri Lanka there is a tendency for most births to occur over August–October, the transition period between the dry season and the start of the wet season, but the pattern varies somewhat between years (Eisenberg, McKay & Jainudeen 1971; Kurt 1974; Santiapillai, Chambers & Ishwaran 1984).

Hippopotamus

Among hippos in the Kruger Park in South Africa, 70% of births occur during the wet season months of October–March, with a peak in January–February. Since the gestation period is 8 months, the peak conception period is May–June in the early dry season. In Uganda conception peaks occur in February and August towards the end of each dry season, with corresponding birth peaks in October and April during the early rains (Laws & Clough 1966; Smuts & Whyte 1981).

Dominant hippo bulls periodically investigate the females clustered in their pools. An estrous female is pursued until she turns round and the pair clash jaws. This leads to a pushing contest until the female lies down in the water allowing the male to mount. Peripheral males keep away. Copulations are lengthy (Kingdon 1979; Verheyen 1954).

Giraffe

Among giraffe in the Timbavati Reserve in South Africa, 60% of conceptions occur during the late wet season months of December–March. With a gestation period of 15 months, the peak in births falls over March–July in the early dry season. In the Nairobi Park in Kenya, there is likewise a birth peak during the dry season, the peak months being August–September. However, in the Serengeti in Tanzania 45% of births occur over May–August, thus extending from the later part of the rainy season into the beginning of the dry season, and there is a lesser birth peak in December–January at the end of the short rains (Foster & Dagg 1972; Hall-Martin, Skinner & Van Dyk 1975; Pellew 1983a).

Dominant giraffe bulls investigate all female groups encountered, sampling the urine of each female in turn to test for estrus status. If no female is in heat, the bull departs. If the bull locates a female in estrus, he follows her, displacing any other male in attendance. The consort period lasts up to 3 days. A female may actively avoid the advances of other males. Fights between bulls can develop in the presence of an estrous female, but generally a dominant bull is able to displace other males simply by walking towards them. Copulations are brief. The dominant bull mates with most of the cows that come into estrus within his home range core area (Kingdon 1979; B. M. Leuthold 1979; Pellew 1984a).

Rhinoceroses

Indian rhinos at Chitwan in Nepal show a weak peak in the number of females in estrus over January–June, spanning the late winter and pre-monsoon period. Most births occur in July–August during the monsoon. The gestation period is 16 months (Laurie 1978).

Indian rhinos form pre-mating consort associations lasting a few days. Prolonged chases over distances exceeding 800 m are sometimes a feature of courtship. During these chases females make loud honking noises, while the male makes squeak-pants. Both sexes urinate frequently. These sounds and smells may serve to attract other males. The male remains with the female for a day or two after mating. Three dominant bulls occupying overlapping ranges at Chitwan each consorted with cows on different occasions, but submissive bulls were rarely seen with females. Matings average 60 min in duration under zoo conditions, and apparently endure for similar periods in the wild. Estrus recurs at intervals of 36–58 days if mating is unsuccessful (Lang 1961, 1967; Laurie 1978, 1982; Ripley 1952).

At Hluhluwe 65% of black rhino matings take place over October–December during the early wet season. Birth peaks occur in January–February (19% of observed cases) and over June–August (41% of observed cases). Allowing for a gestation period of 15 months, the latter period indicates a second peak in conceptions over March–May. In East Africa no seasonal variations in reproduction are evident (Hitchins & Anderson 1983; Goddard 1967).

Black rhino bulls remain in attendance for a period of 6–7 days prior to mating. Estrus lasts one day only. At Hluhluwe females may be mated several times; but in East Africa a single mating is the rule, although multiple mountings may occur. However, in East Africa there are cases of females being mounted by two or three different bulls in succession, while at Hluhluwe generally only a single bull is present. Attacks by the female on the male are sometimes a feature of courtship. Copulation durations vary between 20 min and 43 min. Intervals between successive estrus periods

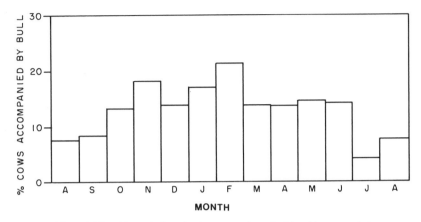

Fig. 7.6 Monthly variations in the proportion of white rhino cows being accompanied by bulls at Umfolozi (1968–71).

average 35 days (range 28–46 days) (Goddard 1966; Guggisberg 1966; Hitchins & Anderson 1983).

White rhinoceros

Among white rhinos the proportion of cows being accompanied by bulls (an indication of estrus) remains high through the wet season and early dry season months, dropping thereafter to low levels. Peaks are evident in November and in February (Fig. 7.6). Most of the associations between cows and bulls during the dry season are transient.

My records from Umfolozi indicated a gestation period of 16 months (based on the interval between the last occurrence of estrus, as revealed by an accompanying bull, and the birth date of the subsequent calf, from its estimated age when first seen). More exact estimates are available from captive animals: gestation periods of 484 days were recorded in the Krugersdorp Game Park in South Africa (Schaurte 1969) and 476 days in the Pretorius Kop enclosure in the Kruger Park in South Africa (M. C. Mostert personal communication). Zoo records show a modal gestation period of 490–500 days (Lindemann 1982).

Correspondingly calving peaks at Umfolozi occurred in March and in July (Fig. 7.7). The bimodal pattern could have been influenced by the midsummer droughts that occurred during the study period. Detailed observations show that the number of cows being accompanied by bulls tended to rise 2–4 weeks after good falls of rain, sufficient to induce a flush of green grass. Conversely, few cows were associated with bulls during dry periods. In the Kyle Game Park in Zimbabwe a similar bimodal pattern in births is evident, with peaks in April and in July (Condy 1973).

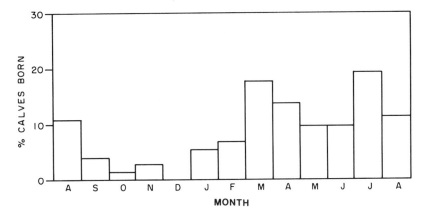

Fig. 7.7 Monthly variations in white rhino births at Umfolozi over 1968–71. (N = 74 calves).

Dominant white rhino bulls investigated cows encountered within their territories, while cows responded with threatening snorts or roars. Usually bulls moved away after perhaps a minute. If a bull remained with a cow for more than a day, this was a sign that the cow was coming into estrus.

During the pre-estrus consort period, the bull followed behind the cow and her companions. However, if the cow neared a territory boundary, the bull moved in front squealing to block her progress (Fig. 7.8). A confrontation sometimes ensued, with roars from the cow and squeals from the bull, occasionally lasting several hours before the cow turned back. Neighboring territorial bulls did not interfere, even though the sounds of such boundary blocking interactions were clearly audible over some distance. However, if a cow was able to evade the bull and cross into the next territory, the bull did not follow and she was joined by the neighboring bull.

The pre-estrus consort period typically lasted 1–2 weeks (Table 7.2). The onset of estrus was indicated by the commencement of repeated approaches by the bull, accompanied by a hic-throbbing sound. Initially the cow chased the bull away with a snort or roar. Eventually the cow tolerated the bull resting his chin on her rump. Several mounts usually occurred before intromission was achieved. The time interval between the first hic-throbbing advances and mating was about 24 hours. Copulations lasted 15–28 mins, with ejaculations repeated every few minutes. In one case a second copulation was observed after a bull had been forced off the cow by an overhanging branch at his first attempt (Fig. 7.9), but in other cases bulls made no further advances after a successful mating. However zoo records show instances in which two or three copulations occurred over 2–3 consecutive days (Lindemann 1982).

Fig. 7.8 A white rhino territorial bull moves round to block a cow and calf nearing a territory boundary.

Bulls usually remained associated with cows for 2–6 days after mating. Estrus recurred about 30 days later if mating was unsuccessful, but was not manifested if dry conditions prevailed. Zoo records show an estrous cycle period of 27–44 days, with estrus lasting 1–3 days (Lindemann 1982).

Subordinate bulls sharing the territory sometimes remained near a courting pair, but did not interfere. One case was observed in which a subordinate bull (who had been the dominant bull in the adjoining territory a year previously) attached himself to a cow and accompanied her for two days, at a time when the dominant bull in the territory was consorting with another cow. However the dominant bull then displaced him and courted and mated with this cow, while the subordinate bull circled around the pair displaying great agitation. After completing copulation, the dominant bull clashed horns briefly with the subordinate bull, then wandered off. The subordinate bull then attempted to court the cow, but was warded off vigorously.

One mating by a subordinate bull was observed, in unusual circumstances. There was no preceding consort period, and the dominant bull in the area showed no interest in the cow until the following month. The

Table 7.2. *Data on estrus and mating behavior by white rhinos*

	Mean	Range	N
Duration of consort period (days)			
(a) Pre-estrus	13	0–20	8
(b) Post-estrus	4	0–6	10
(c) Total	15	1–30	9
Duration of courtship (hours)	20	17–25	6
Number of mountings before copulation	3	1–7	6
Duration of copulation (mins)			
(a) Mounted	26	23–30	5
(b) Intromission	22	15–28	5

Fig. 7.9 A copulating white rhino bull being gradually forced off the cow by an overhanging branch.

copulation took place in an area that had formed part of the home area of the subordinate bull, but which had three weeks previously been incorporated into the neighboring territory following its takeover by a new bull. The subordinate bull did not remain with the cow after mating. Two and a half months later this subordinate bull moved away to assume dominance in a small territory elsewhere.

Summary

Among white rhinos, hippos, elephants and giraffes mate access is governed largely by prior dominance relations established among males. Among African elephants subdominant (i.e. non-musth) males may contest matings in the absence of a large dominant bull. Among black rhinos and Indian rhinos the relation between prior dominance and mating remains unclear. Chases, fights or loud vocalizations, which may serve to attract other males to the scene and thus allow some degree of mate choice to be exerted, are shown by elephants and the various rhino species. Copulations are prolonged in rhinos and hippos, brief in elephants and giraffes. Post-mating guarding is evident in white rhinos and hippos. Gestation lasts over a year in all species except hippo. Seasonal variations in reproduction are evident in all species. In elephants, white rhino, black rhino and giraffe, conception peaks occur during the wet season, while in Indian rhino and hippo mating peaks fall within the dry season. Birth peaks fall during the dry season in white rhino, black rhino and giraffe, while in elephants the birth peak occurs in the early part of the rainy season.

Responses to predators

Animals may respond to the threat posed by potential predators either by fleeing, attacking or standing defensively. Alarm displays may be given to draw the attention of companions to the threat, or to signal to the predator that it has been detected. Of special interest are the actions of mothers to protect their vulnerable offspring. Potential predators include not only lions and other carnivores, but importantly also humans.

Elephants

African elephants have no audible alarm call, and the cessation of the low growling rumbles exchanged among group companions supposedly signals a threat. The recent discovery that elephants make low frequency sounds below the range of human hearing (Payne, Langbauer & Thomas 1986) may change this interpretation. Members of family units bunch tightly together at threatened danger, whether from lions or humans, with

Fig. 7.10 A family unit of African elephants bunches together when threatened by a nearby vehicle.

young animals in the center (Fig. 7.10). Family units approaching water chase aside any lions encountered. Adult males take little notice of lions (Kingdon 1979).

For human observers the risk of following family units is much greater than that of following elephant bulls. This is due not only to the problem of avoiding a greater number of animals, but also because cows more often display aggression on detecting a nearby person than do bulls. However, most commonly family members bunch together and make off in haste. In the Kruger Park in South Africa, attacks on vehicles are most often by adult females, and commonly occur when a car drives between a mother and her calf. I was once chased in a vehicle over a distance of about one kilometer by a family group of elephants, the matriarch in the group leading the charge. At Lake Manyara in Tanzania, a family group of elephants made a sudden and silent coordinated attack, involving three females, on a vehicle parked close by. Bulls commonly demonstrate at a threatening person or vehicle by flapping their ears, trumpeting and making advancing rushes. Fatal attacks by both bulls and cows on humans are not uncommon where humans on foot frequently contact elephants (Barnes 1979; Douglas-Hamilton 1972; Guy 1976a; personal observations).

Asian elephants used to convey tourists in the Kaziranga Sanctuary in

Assam display nervousness in the vicinity of tigers. They are regarded as less aggressive towards humans than African elephants (Eltringham 1982).

Hippopotamus
Adult hippos encountering lions on land may charge open-mouthed. However, at night hippos amble past lions as if they did not exist. There are records of females killing lions in defense of their young. Lions may occasionally kill fully-grown hippos by rolling them onto their backs and biting the chest and throat.

Hippos surprised on land by humans generally take evasive action by rushing towards water along the nearest path. However, any human in the way may be chomped by the huge jaws with potentially fatal results. In the Virunga National Park in Zaire, hippos are responsible for more human fatalities than any other large mammal. Hippos feeding on land at night appear to be less of a danger. Alarm calls are undescribed (Bourliere & Verschuren 1960; Guggisberg 1961; Kingdon 1979; Schaller 1972).

Giraffe
Adult giraffe may stand watching lions from ranges as close as 20 m, but run off if the lions appear to be stalking. In one recorded instance a giraffe reared up and slashed at an attacking lioness with both forelegs, before fleeing. Giraffe mothers may successfully defend their calves against lions, standing over the calf directing powerful kicks at any approaching lion. Giraffe are inoffensive so far as humans are concerned. Giraffe have no audible alarm call (Pellew 1974a; Schaller 1972).

Rhinoceroses
Indian rhino mothers generally run off behind their calves when alarmed. When disturbed by humans, Indian rhinos sometimes respond by rushing towards the intruder making snorting sounds, although usually they can be turned aside by shouts. Mothers with small infants frequently charge. Fatal attacks on local villagers occur, injuries being inflicted with the lower incisor tusks. Indian rhinos give a snort when disturbed by the sounds of an approaching animal, including another rhino (Laurie 1978; Ullrich 1964).

Adult black rhinos commonly pay little attention to lions. At waterholes they may advance on lions forcing the latter to give way. In one incident at Ngorongoro in Tanzania, a subadult male lion attacked and pursued an 11 month old black rhino calf, separating it from its mother. The mother followed, and when the lion diverted its attack to her, she killed it with several horn thrusts. Hyenas have been observed attempting to pull down calves up to 4 months in age by grabbing them from behind. In these incidents the mother charged the hyenas when the attacked calf squealed, forcing them to retreat. Black rhino calves generally walk behind the mother; but when disturbed the calf may flee in front of the mother for the first 50–100 m, after which the mother takes the lead (Goddard 1967; Schenkel & Schenkel-Hulliger 1969a).

Black rhinos are renowned for their aggressive charges both towards vehicles, and humans on foot. Many of their advances are merely curious approaches, the animal wheeling and making off once the human is clearly identified. Others seem to be directed charges, accompanied by steam-engine-like puffs, especially if an animal is suddenly disturbed at close quarters. Attacking blows are struck with the horn if the person or vehicle is contacted, but if evasive action is taken the animal continues on its way. At Tsavo in Kenya black rhinos commonly reacted to human scent with a rapid advance over a distance of 5–20 m, then turned and made off. However, if human scent was combined with visual or auditory stimuli, their responses were less predictable. Often they reacted with immediate flight. On occasions inquisitive approaches and short rushes occurred. Black rhinos in a group sometimes stopped in a fan formation, after having fled for a distance. In one incident a black rhino bull charged from 40–50 m range upon detecting a human silhouette visible against the skyline (having been alerted by oxpeckers). The animal chased the person around a tree, and inflicted a horn wound when the person fell out of the tree. The rhino then ran off. In the Ngulia area of Kenya where black rhinos had been hunted from early times by people armed with poison arrows, rhinos were extremely truculent and ready to charge instantly (Guggisberg 1966; Ritchie 1963; Schenkel & Schenkel-Hulliger 1969a).

White rhinoceros

I observed four encounters between white rhinos and lions, which had recently colonized the western section of Umfolozi where the study area was situated. In one instance an adult male white rhino lay down to sleep while the lions settled 30 m away. In another case a white rhino cow with a year-old calf ignored a group of four young lions that lay watching 15–20 m away, while the lions made no move to attack. At a waterhole a group of seven lions was present while numerous white rhinos moved past to drink. One cow accompanied by a subadult advanced towards the lions, which bounded out of her way at 12 m range. Another cow with a two-year-old calf stood peering at two lions lying about 5 m in front of her. The lions made no

Fig. 7.11 A white rhino calf runs off in front of its mother.

attempt to stalk the white rhinos, though they advanced towards an approaching group of zebra.

White rhino cows with newborn infants did not charge even when I approached quite close. They displayed great agitation, but remained standing over the calf. If alarmed by detection of a human intruder, a cow immediately rushed beside her infant with loud pants, and *vice versa*. The pair then ran off, the calf galloping in front of the mother (Fig. 7.11). The horns of white rhino cows tend to be longer and more slender than those of bulls, and in some females the horns project forwards rather than upwards from the snout (Fig. 7.12), a condition not recorded among males. Female horns thus seem more effectively designed for warding off attacking carnivores, while the stouter horns of males are better able to resist horn-to-horn blows during intraspecific combat.

At distances of up to 800 m, white rhinos responded to human scent by standing peering about uneasily. At closer ranges their immediate response was to flee, usually downwind. If the breeze eddied, they shuffled about agitatedly facing in different directions, until able to detect an auditory or visual clue, then they ran off. In one incident, a white rhino subadult that

Fig. 7.12 A white rhino cow with a forward-pointing horn.

had detected signs of my presence had difficulty in communicating its alarm to its group companions. It shuffled about agitatedly making intention movements towards running, but as the other rhinos detected nothing they resumed feeding. The panting sounds made by white rhinos are not alarm calls, but serve as contact sounds in a variety of contexts.

Red-billed oxpeckers (*Buphagus erythrorynchus*) generally reacted to a nearby human by giving loud churring calls. This caused the rhinos they were sitting on to react by looking around apprehensively. If the birds continued calling, the rhinos shuffled around nervously searching for a sign of the cause of the birds' alarm. At any slight noise or movement, the rhinos immediately ran off.

White rhinos were unable to identify me visually as a human intruder at ranges greater than 20–30 m, provided that I remained immobile. On occasions they appeared confused and uncertain how to respond, shuffling around agitatedly, perhaps starting to run off, but then standing their ground again. Approached on foot from a vehicle near one of the tourist roads, white rhinos were less likely to run off than elsewhere in Umfolozi, provided that the approach was made openly in full view. Under these

Fig. 7.13 A group of four white rhinos stands in a defensive rump-against-rump formation.

conditions a group of white rhinos commonly adopted a rump-to-rump formation, standing facing outwards in different directions (Fig. 7.13).

Occasionally white rhinos rushed towards me when alarmed, but they were usually turned aside by noises. However, on two occasions I experienced a directed charge. In the one instance a white rhino bull suddenly charged when I stepped aside from the bush behind which I had been standing at 5 m range. When I slipped on wet grass the rhino dodged me, but tossed my assistant to one side with its head while he was trying to run away. In another case a white rhino bull accompanying a cow suddenly charged when I clicked my camera from 5 m range, but turned aside at a few meters range when I shouted. Steele (1968) describes an incident in which he was chased by a white rhino bull over some distance, and eluded it only by diving into an antbear hole. There is a record of a woman being killed by a white rhino just outside Umfolozi Game Reserve. In the Sudan Lang (1920) experienced only one charge while hunting white rhinos, and that was from an animal that had been wounded.

Summary

Adult megaherbivores generally respond indifferently or aggressively towards lions and other predators. Carnivores attempting to catch young animals may be killed by defending mothers. Responses towards humans vary between fleeing and attacking, depending on circumstances. White rhinos and giraffes are generally inoffensive, while elephants, Indian rhinos, black rhinos and hippos are quite commonly aggressive. Alarm calls are lacking in all species, except possibly elephants.

Comparisons with smaller ungulates

Small species of antelope are either solitary or occur in small loose groups; while females plus young of larger species form stable herds typically numbering 10 to 30 or more animals. African buffalo occur in large herds of 100–2000 animals including adults of both sexes. In their lack of sociality rhinos, and to a lesser degree hippo and giraffe, resemble small bovids. Elephants differ from other species in the hierarchical structure of their groupings. Female–young units occupying discrete home ranges have been identified in impala, and may occur among other antelope species (Jarman 1974; Leuthold 1977a; Murray 1982b; Owen-Smith 1977).

In most African ungulates male dominance relations are organized in the form of the spatially restricted dominance of territoriality. Exceptions include species forming large mixed-sex herds like African buffalo; semi-nomadic species like plains zebra; and browsing antelope species such as kudu. White rhino and hippo, and in some areas black rhino, conform to the territorial dominance pattern; while giraffe resemble other browsers in their loose rank dominance system. Elephants are unique in having the temporally restricted dominance of musth superimposed on a rank dominance pattern. Satellite males sharing territories, as occur in white rhino, hippo and perhaps black rhino, are found also in waterbuck and Grevy's zebra (Owen-Smith 1977).

Most ungulate species have narrowly restricted breeding seasons, with birth peaks occurring during the wet season when food availability is optimal. Breeding is less seasonally concentrated in certain species occupying equatorial regions. A few ungulates, for example waterbuck, roan antelope and African buffalo, show a fairly broad birth peak, like that typical of megaherbivores, even in southern Africa. The longest gestation periods among African bovids and equids are 12 months, in the case of zebra, and 11 months in the case of African buffalo; while all megaherbivores except hippo have gestation periods exceeding one year.

All medium-sized ungulate species have alarm calls, and in some cases also visual displays, which they give on detecting a nearby predator. These lead to coordinated flight if the predator is within attack range. Megaherbivores lack specific alarm signals; while small antelope like grey duiker have sniffing calls that they give while running off from a disturbance. All bovids and equids flee from human intruders, except for African buffalo bulls. Megaherbivores flee only from humans, and may on occasions attack humans as well as threatening carnivores. Among bovids where females possess horns, the horns of females tend to be designed as stabbing weapons, while male horns are stouter at the base and so better able to resist the forces of horn-to-horn combat (Packer 1983). A similar pattern is manifested by the two African rhino species, though not among other megaherbivores.

8

Life history

Introduction

As animals grow and age they pass through different functional stages in terms of their social relations and contribution to reproduction. These stages may be subdivided as follows (i) infancy and juvenilehood – the period of complete or partial dependence on the mother for sustenance and protection; (ii) adolescence and subadulthood – the early period of independence from the mother, through attainment of physiological sexual maturity; (iii) adulthood – the period following attainment of full social and sexual maturity. Interest lies in the timing of these stages, and in the changing behavioral patterns of animals as they pass through each stage.

Infancy and juvenilehood

This period commences with birth. During the early neonatal period the offspring is completely dependent upon its mother for sustenance in the form of milk. During later infancy the offspring starts supplementing its milk intake with vegetation, but it is some time before nursing ceases and weaning is complete. By use of the term infancy I imply the period during which the young animal could not survive if separated from its mother. Juvenilehood refers to the period of partial dependency on the mother for perhaps some food supplementation, or at least protection from predation. In most species the juvenile period ends when the young animal is driven away by the mother around the time of birth of the next progeny. However, in some species older offspring may remain associated with the mother and her companions through adolescence.

Elephants

African elephant cows give birth in the company of their group companions. Other females in the group gather around the newborn infant, the matriarch assisting the mother in removing the fetal membranes. A newborn elephant is able to stand, although unsteadily, about 20 min after birth, and nurses for the first time about 30 min after being born. About an hour after birth it is able to follow its mother and her companions, although requiring some assistance (Leuthold & Leuthold 1975).

African elephant calves begin feeding on vegetation at 3 months of age, and by 24 months spend a similar proportion of their time feeding to adults. Calves under 24 months in age do not survive if orphaned. Suckling takes place frequently, on average every 37 min for male calves and every 50 min for female calves. The mean duration of suckling bouts is about 90 s for both sexes. Nursing generally continues until the birth of the next offspring, i.e. typically for about 4–5 years. In a few instances calves stopped suckling after 4 years of age despite a longer interval to the birth of the next offspring, while a small proportion of calves continue suckling after the birth alongside the younger sibling. The oldest recorded age at weaning is 8 years. Instances of females nursing calves that are not their own offspring are rare.

Juvenile elephants spend most of their time within less than 5 m of their mothers (median distance 2 m) until 8 years of age (Fig. 8.1). Calves seek one another out and engage in much challenging, wrestling and mutual pushing play while the family is resting or drinking. Juvenile and adolescent females comfort, assist and protect calves in their family units which are not necessarily their siblings (Douglas-Hamilton 1972; Lee & Moss 1986; Lee 1987; Sikes 1971).

Hippopotamus

Hippo females move to a secluded area on the river or lakeshore, or to shallow water, to give birth. The neonate remains in the exclusive company of its mother for the first few days after birth. The young nurse under water. Some hippo groups contain fewer calves than lactating females, suggesting that calves may nurse from more than one female. Infants commence feeding on grass between 6 and 8 weeks of age. Under captive conditions calves continue nursing until about 14 months old (Laws & Clough 1966; Smuts & Whyte 1981; Verheyen 1954).

Giraffe

Giraffe mothers return to specific areas to give birth in seclusion. The newborn infant remains isolated from other giraffes for 1–3 weeks, with the mother returning to suckle it 2–4 times per day (night behavior is unrecorded). Infants start nibbling on plants at 2 weeks of age. For the first 3–4 months calves spend most of their time lying out alone, or in the company of other young giraffes. Mothers move as far as 3 km away from

Fig. 8.1 African elephant infant close behind its mother.

their offspring. Sometimes calf pools are formed, including one or two adult females in the company of several small calves. Juveniles nurse 2–3 times during daylight hours, for about a minute at a time. Male calves initiate 70% of their nursing bouts, but female calves only 50%. Youngsters may attempt to nurse from females besides their mothers, but are rarely successful. Calves engage in playful running and jumping, especially when in groups. By five months of age young giraffe move with a group for part of the day, and feed for more than half the day. By 6–9 months the activities of the calf are similar to those of its mother. Cows continue lactating for 12–13 months. Young giraffe drift away from their mothers between 12 and 18 months of age (Hall-Martin, Skinner & Smith 1977; Langman 1977; Leuthold & Leuthold 1978; B. M. Leuthold 1979; Pratt & Anderson 1979; Pellew 1984a).

Rhinoceroses

Indian rhino females seek seclusion in thick grassland or forest to give birth, and are particularly aggressive towards other rhinos while the calf is young. Up to 6 months of age the offspring may be left lying alone for periods of over an hour while the mother forages up to 800 m away. Calves aged up to one year nurse four or more times during daylight, older calves

Fig. 8.2 White rhino cow standing over her day-old calf in thick bush.

once or twice daily. Nursing bouts typically last 3–4 min, with suckling continuing until calves are about two years old. Young calves inquisitively approach other rhinos to engage in nasonasal nuzzling, which sometimes develops into playful sparring; but older calves tend to respond aggressively when approached by another rhino. Juveniles are driven away by the mother a week or more before the birth of the next offspring. In some cases periodic reunions between the older calf and its mother occur over the first few months following separation, especially if the offspring is a male (Laurie 1978).

Black rhino cows may leave a newborn infant hidden for the first week following birth. Nursing bouts last about four minutes, but frequency of nursing has not been documented. Infants start nibbling on bushes when only a few weeks old. Calves generally keep within 25 m or less of their mothers, except during locomotory play. Mothers separated from their calves make breathing calls. Suckling may continue until the calf is as old as 19 months. Youngsters become independent of their mother around the time of the birth of the next offspring, when aged about 2.2–3.3 years. In some situations the older offspring rejoins its mother after the birth of the

Table 8.1. *Frequency and duration of nursing by white rhino calves*

Age range (months)	Total no. of records	Nursing frequency		Duration of nursing (min)		
		Mean per hour	h. of observ.	Mean	Range	N
0–2	14	1/ 1.0	13	3.5	1.0–6.0	12
2–6	5	1/ 2.3	7	2.9	2.3–3.5	3
6–12	30	1/ 2.4	38	2.9	1.3–5.0	16
12–18	43	1/ 4.8	97	3.6	2.8–7.0	18
18–24	5	1/15.0	30	3.8	—	1

new calf (Frame & Goddard 1970; Hall-Martin & Penzhorn 1977; Schenkel & Schenkel-Hulliger 1969a).

White rhinoceros

I encountered white rhino females with newborn infants either in patches of dense thicket (Fig. 8.2), or in one case on a hillslope secluded from other rhinos. Mothers favored densely wooded areas until calves were about two months old. Infants less than two months old nursed hourly, while older calves nursed at intervals of about 2.5 hours. Nursing bouts typically lasted about three minutes (Table 8.1). Suckling frequency declined between 12 and 18 months of age, but one 24 month old calf was recorded still being suckled. It seems that cows ceased lactating in their third or fourth month of pregnancy, so that suckling continued longer if the birth interval was prolonged.

Infants started nibbling at grass at about 2 months, but not until they were over a year old did they spend as much time feeding as the mother (Table 8.2). Mother and offspring kept close together, generally within half a body length (2 m) of one another, with separation distances tending to increase with increasing age (Table 8.3). Cows that had lost their calves wandered around making repeated panting calls, while a calf separated from its mother made high-pitched squeaks.

Infant calves engaged in locomotory play, running back and forth over distances of 5–15 m at a stretch. Calves displayed great curiosity upon encountering other rhinos, and commonly approached to sniff. Meetings between calves frequently led to playful chasing and head to head tussles. White rhino calves remained associated with their mothers until the time of the birth of the next offspring. The older calf was then persistently driven away by the mother. Only in rare cases did older calves rejoin their mothers.

Table 8.2. *Proportionate time spent grazing by white rhino calves relative to the mother at different ages*

Activities recorded at 5 minute intervals during observations sessions.

Age range (months)	No. of records with cow grazing	No. of records calf also grazing	% time spent gr. rel. to cow
0–2	19	0	0
2–4	52	14	27
4–8	43	27	63
8–12	41	37	90
12–24	44	44	100

Table 8.3. *Separation distances from the mother maintained by white rhino calves of different ages*

Age of calf	Separation distances (m)						
	Mean	Range	Distribution (%)				N
			0–4	5–9	10–19	20+	
<2 months	1.4	0–15	100	0	0	0	21
2–4 months	2.5	0–15	83	13	4	0	52
4–12 months	2.2	0–15	91	7	2	0	84
12–24 months	3.7	0–25	81	11	4	4	47

Summary

Rhino and hippo calves are left lying out for a period of up to a week, while giraffe calves remain lying out for several months. Elephant calves move with their mothers' family units from the day of birth. In all species calves nurse several times daily, and mothers continue suckling for a year or longer. Despite the solitary habits of rhinos, social play is a feature of all species. Rhino mothers drive away the older offspring when the next calf is born, while young giraffe drift away of their own accord. Elephant and hippo females tolerate continuing associations by older offspring.

Adolescence and puberty

Following severance of the mother–offspring bond, immature animals may either wander alone or attach themselves to other companions. The onset of puberty is indicated by the beginning of spermatogenesis in males, and by the occurrence of the first ovulatory cycles in females. However, full fertility may not be attained until some time later. Males do not achieve social maturity until they have attained mature weight, and thus become able to compete successfully with older males. Females attain adult status following the birth of their first calves. I use the term adolescence for the prepubertal period following independence from the mother. The term subadult is applied generally to the complete period from breaking of the mother–offspring bond to attainment of social maturity.

Elephants

Young African elephants of both sexes remain with their maternal family units after weaning. Females typically undergo their first ovulation between 11 and 14 years of age. The earliest recorded age at first conception is 7 years, while under conditions of malnutrition ovulation may be retarded until as late as 20 years of age. The mean age at first parturition varies between 13 and 18 years in different populations, with a minimum age of 8 years recorded. Growth by females levels off between 15 and 20 years (Douglas–Hamilton 1972; Hanks 1972a; Jachmann 1986; Lang 1980; Laws 1969b; Laws, Parker & Johnstone 1975; Lee & Moss 1986; Sherry 1975; Smuts 1975c).

Male African elephants begin spermatogenesis between 7 and 15 years of age, but full sperm production is not reached until 10–17 years. Males leave family units between 10 and 16 years of age, in some cases in response to aggressive behavior by the adult females. They then join up with other males of similar age and older. The first musth periods start at about this time. Males seem to show a growth surge in height and weight between 20 and 30 years of age, although measurements of hindfoot length failed to confirm this. Males do not attain full weight and hence competitive ability until 30–35 years of age (Douglas–Hamilton 1972; Hanks 1973; Laws, Parker & Johnstone 1975; Lee & Moss 1986; Moss 1983; Sherry 1975).

The above ages were generally estimated using Laws' (1966) dental ageing criteria. However, there is evidence that this method may overestimate true ages by 2–4 years in the age range 10–25 years. If this is confirmed, some quoted ages might need to be revised accordingly (Croze, Hillman & Lang 1981; Jachmann 1985; Lang 1980; Lark 1984).

Asian elephant females may reach sexual maturity between 7 and 8 years of age, and produce their first calves at 9–10 years. Males tend to become peripheral to their maternal units at 6 years, suggesting that puberty is reached between 7 and 10 years. Musth periods usually commence around 19–20 years, but some males do not show musth until 30 years of age. Subadult males become semi-nomadic, remaining in an area for several days then disappearing (Eisenberg, McKay & Jainudeen 1971; Jainudeen, McKay & Eisenberg 1972; Kurt 1974; McKay 1973).

Hippopotamus

The age of first ovulation among hippo females varies between extremes of 3 and 20 years, with about 9–11 years most typical in wild populations. Zoo records indicate ages at first parturition between 3 and 8.5 years. Males first start producing sperm at 2 years of age, with peak sperm production occurring by 6 years. A captive male was sexually potent at 3.2 years. However, asymptotic weight is only reached at 25–30 years of age, and dominant bulls are generally aged at least 26 years. Young females tend to remain associated with their maternal group (Dittrich 1976; Laws 1968a; Sayer & Rahka 1974; Skinner, Scorer & Millar 1975; Smuts & Whyte 1981).

Giraffe

Female giraffe show a mean age at first conception of 4–5 years, with a minimum of 3.8 years in the wild. Captive giraffe females attain puberty at a mean age of 3.9 years, with the earliest record being 2.8 years. Males become sexually potent at about 3.5 years, but do not reach full weight and hence social maturity until aged 8 years or more. Subadult male giraffe wander widely (Hall-Martin & Skinner 1978; Pellew 1984a).

Rhinoceroses

In Indian rhinos, adolescent males tend to join other young males, while adolescent females usually attach themselves to adult females. However such associations generally last only a few days. Adolescents in groups are less likely to be attacked by adult males than lone adolescent males. Adolescent females sometimes remain in the maternal home range, while adolescent males tend to move out of high density areas. At Chitwan, females produce their first offspring between 6 and 8 years of age. In zoos, females achieve sexual maturity as early as 3 years of age, while males became sexually potent at 7 years (Lang 1967; Laurie 1978).

Black rhino adolescents sometimes join up with other adolescents or with cows, but the majority are solitary. Some subadults wander widely. Nasonasal contacts occur when adolescents encounter other subadults or cows, and playful horn sparring may develop between subadults (Goddard 1967; Schenkel & Schenkel-Hulliger 1969a).

In East Africa, black rhino females first mate at an age of 4.5 years, with first parturition occurring at about 6 years of age. At Hluhluwe, the earliest recorded matings occurred at 7–8 years of age, but earliest parturition did not occur until 12 years. In the adjoining Umfolozi-Corridor region, there were two records of females giving birth at ages of 6.5 years and 8.5 years respectively. Black rhinos introduced from Kenya into the Addo Park in the Cape showed a mean age at first parturition of 6.3 years (after the initial unsettled period), with a minimum of 5.1 years. Among black rhinos held in zoos, the mean age at first conception is 6.4 years, with a minimum of 5.0 years. Males at Hluhluwe do not commence spermatogenesis until 8 years of age, but a captive male reportedly sired a calf at an age of 3.5 years. There is suggestive evidence that one male mated at 6 years of age at Addo. Males do not become territorial until at least 9 years old at Hluhluwe (Goddard 1967; Hall-Martin & Penzhorn 1977; Hall-Martin 1986; Hitchins & Anderson 1983; Lindemann 1982).

White rhinoceros

White rhino adolescents became separated from their mothers when aged between 2 and 3.5 years. Thereafter they wandered about attaching themselves temporarily to cows or to other adolescents, until after a month or two a more stable bond was formed. In the Umfolozi study area two-thirds of adolescents joined other adolescents, while one third formed a persistent attachment with a cow lacking a calf. However, the proportion of subadults joining cows may have been atypically high, due to the number of cows that had had their calves removed by the rhino capture team operating on the periphery of the study area. Cows without calves readily accepted several subadults as companions. Only two cases were recorded in which a calf rejoined its mother and her new offspring.

In some instances particular pairs of adolescents remained together for periods of several years (Table 8.4). Bonds were formed equally readily with a companion of the same or opposite sex. However, bonds formed between pairs of young males tended to last longer than those involving females, because of the delayed maturity of males relative to females. Attachments with cows were of limited duration, since the subadult companion was driven away when the cow produced her next calf.

While some subadult pairs or trios were resident in the study area, other subadults appeared to be semi-nomadic. New individuals appeared in the study area, were seen there several times over the course of a few days, then disappeared again. Two subadult females marked with eartags in the study area were subsequently seen 15 km and 25 km away respectively.

Subadults of all ages displayed great interest in meeting other rhinos. Nearly half of their encounters with other subadults, and one third of their encounters with cows, led to nasonasal contacts (Fig. 8.3). Forty percent of such contacts developed into playful horn sparring engagements.

Solitary tendencies started to develop in some young males at an age of

Table 8.4. *Social nature and duration of stable bonds formed by white rhino subadults*

A bond enduring one month or longer is regarded as stable.

Age/sex categories	No. of stable bonds recorded	Av. min. bond duration[a] (months)	Max. recorded duration (months)
A. Subadult pairs			
(i) Male–male:			
Adolescents < 6 yrs	6	8.7	20.5
Subadults > 6 yrs	7	13.6	26
All ages combined	13	11.3	
(ii) Female–female:			
Adolescents < 6 yrs	7	3.7	9.5
Subadults > 6 yrs	7	3.7	5
All ages combined	14	3.7	
(iii) Male–female:			
Adolescents < 6 yrs	8	4.1	15.5
Subadults > 6 yrs	5	4.2	8.5
All ages combined	13	4.2	
B. Cow–subadult pairs			
(i) Cow–subadult male:			
Adolescent < 6 yrs	7	7.4	22
Subadult > 6 yrs	1	21	21
All ages combined	8	8.1	
(ii) Cow–subadult female:			
Adolescent < 6 yrs	10	10.2	26
Subadult > 6 yrs	2	11.2	12
All ages combined	12	10.3	

Note: [a] The period between first and last sightings of the same two individuals still together.

about 8 years, while other males remained associated with a subadult companion until 11–12 years. One young male was monitored regularly through the transition period from subadulthood to subordinate bull status. Between the ages of 8 and 9 years, he attached himself temporarily to various cows or subadults, being associated with such companions in 54% of sightings ($N = 37$). Over the succeeding year, he was with other rhinos on only 7.5% of days seen ($N = 92$). Between the ages of 10 and 11 years, he was essentially solitary, being associated with other rhinos in only 3.5% of sightings ($N = 83$).

At Umfolozi, the three youngest females evidently in estrus, as judged by hic-throbbing advances by an accompanying male, were aged 3.8 ± 0.2 years, about 4.0 years, and about 4.5 years, respectively. The youngest age at first parturition was 6.5 ± 0.5 years. Three other females were estimated

Fig. 8.3 Nasonasal meeting between two white rhino subadults.

to be about 7 years of age (± 1 year) when they produced their first offspring. Among zoo-kept animals, the mean age at first parturition is 5.6 years (range 5.1–6.2 years) for known-age animals, and about 8 years for animals for which only the year of birth was known (Lindemann 1982).

Young females remained attached to their companions, including in some cases similar-aged young males, through their first estrus periods. Subadult male companions displayed no sexual interest, and were not driven away by the courting bull. Around the time of parturition females separated themselves from their group associates, and did not rejoin them. Their behavior patterns were thereafter similar to those of adult females.

Subadult males aged up to 8–9 years generally showed no indication of sexual interest in females. However at Hluhluwe North, in the absence of a dominant bull, three young males aged 8–10 years made repeated hic-throbbing advances towards a subadult female, but did not proceed beyond the chin-on-rump posture. The youngest dominant territory holder in the Umfolozi study area was estimated to be about 12 years old. One territorial bull in the Kyle Game Park in Zimbabwe was known to be 12.5 years old (Condy 1973).

In the Pretorius Kop enclosure in the Kruger Park in South Africa, there is a record of a young male siring his first offspring at an age of 7.9 years, incidentally mating with his mother. No mature bull was present (M. C. Mostert personal communication). From zoos it is reported that one male sired a calf when aged 5.5 years (Lindemann 1982).

Summary

In elephants and hippos adolescents remain associated with their natal groups, while in rhinos adolescents of both sexes are driven away by the mother when she next calves. Females attain puberty at 2–3 years in giraffe, 4–7 years among rhinos, and 7–14 years among elephants. Puberty among males generally occurs 1–4 years later than it does in females. Young elephant males leave their maternal groups at puberty. Males of all species do not attain full weight and hence social maturity until well after puberty. Captive animals may attain maturity younger than wild animals. Hippos appear especially variable in the ages at puberty of both males and females.

Reproduction by females

Females remain reproductively active throughout the adult period. Features of interest are the intervals between successive births, changes in fertility with age, and the sex ratio of the offspring produced.

Elephants

For African elephants mean conception intervals vary between 3.3 and 5.5 years in various regions, including Lake Manyara in Tanzania, Luangwa Valley in Zambia, Kruger Park in South Africa, Gonarezhou and Hwange in Zimbabwe, and Tsavo and Amboseli in Kenya. The shortest mean interval is that for the Luangwa Valley population prior to 1968. These are long-term means based on the relationship between the number of placental scars and the age of the female, determined from culled samples.

Short term natality rates, estimated from the ratio of pregnant to nonpregnant females, show a wider variability, due to year to year variations in the proportion of females conceiving. In the Murchison Falls Park in Uganda, the mean birth interval was estimated to be 9.1 years for the population north of the Nile River, and 5.6 years for the population to the south of this river, based on shot samples spanning 1–2 years during the 1960s. However, samples obtained from the same regions over 1973/74, following population reduction, indicate a mean birth interval of 5.1 years for the northern population and 3.6 years for the southern one. An analysis of placental scars in the same samples yielded a mean birth interval of 4.9 years for the Murchison Falls Park South population; while a mean calving interval of 3.8 years was estimated based on animals killed in this region over the period 1947–51. At Tsavo in Kenya, a mean birth interval of 6–7 years was estimated from the ratio of pregnant to nonpregnant females just prior to major starvation mortality during a drought. At Amboseli in Kenya, individually recorded birth intervals averaged 5.6 years during a series of dry years, and 3.5 years over a subsequent sequence of wet years. At Kasungu in Malawi, the mean birth interval between surviving calves was 3.3 years. The shortest birth interval for a female in the wild where the previous calf survived at least one year is 2.8 years (Douglas-Hamilton 1972; Hanks 1972; Jachmann 1986; Laws 1969a; Laws, Parker & Johnstone 1970, 1975; Lee & Moss 1986; Malpas 1978, cited by Croze *et al.* 1981; Moss 1983; Perry 1953; Sherry 1975; Smith & Buss 1973; Smuts 1975c; Williamson 1976).

The proportion of African elephant females pregnant is highest (43%) in the age group 31–40 years. Fertility declines rapidly after 50 years of age. Fetal sex ratios for African elephants vary from 111 males : 100 females in the Kruger Park ($N = 298$) to 95 males : 100 females ($N = 188$) in Gonarezhou in Zimbabwe. In the combined data from all sources, the primary (fetal) sex ratio is 102 males : 100 females ($N = 710$) (Hanks 1972; Sherry 1975; Smuts 1975; Williamson 1976).

Birth intervals of Asian elephant females are about 4 years both in the wild and in captivity. A minimum birth interval of 23 months was recorded following the death of a calf, but the shortest interval after a surviving calf is 36 months (Kurt 1974).

Hippopotamus

Hippo females typically show mean calving intervals of about 2 years. However, in the Kruger Park in South Africa only 5.6% of females were pregnant during a severe drought, compared with 37% in other years. In Uganda about 15% of females examined were reproductively inactive, i.e. neither pregnant nor lactating. Fetal sex ratios show equal proportions of males and females (96 males : 100 females, $N = 269$). Females show little indication of reproductive senesence (Laws & Clough 1966; Smuts & Whyte 1981).

Giraffe

Giraffe exhibit a mean calving interval of 20 months in wild populations, with a range between different areas of 16 to 25 months. For captive animals the mean birth interval is 21.5 months (range 15.6–40 months). However the modal birth interval from all sources is 16–18 months. In the Nairobi Park in Kenya, the birth interval averaged 17 months when the calf died shortly after birth, but 23 months when it survived. The secondary (i.e. birth) sex ratio of giraffe generally shows an excess of males. The sex ratio of calves born in zoos is 160 males :100 females ($N = 115$). Wild populations at Tsavo and Nairobi in Kenya yielded a juvenile sex ratio of 159 males : 100 females ($N = 467$) (Bourliere 1961; Foster & Dagg 1972; Hall-Martin 1975; Hall-Martin & Skinner 1978; Leuthold & Leuthold 1978; Pellew 1983a).

Rhinoceroses

Indian rhinos at Chitwan in Nepal show a median calving interval of 2.8 years. A female that had her newborn infant killed by a tiger calved again after an interval of 18 months. Two aged-looking females did not give birth over a four year period, while the third produced a calf, but lost it within 6 weeks. One female estimated (from cementum lines in her tusks) to be about 30 years old was not pregnant when she died, despite a 3 year interval since her last calf. The sex ratio of calves sighted on more than five occasions was 18 males : 16 females (Laurie 1978).

Black rhinos at Tsavo in Kenya show a mean calving interval of about 2.5 years; while at Ngorongoro and Olduvai in Tanzania, and Amboseli in Kenya, the mean birth interval is about 4 years, including females that did not produce a calf during the observation period. The shortest observed calving interval is 25 months. Black rhinos at Hluhluwe exhibit a mean birth interval of 2.7 years, and those in the adjoining Corridor and Umfolozi a mean interval of 2.3 years, excluding females that did not appear with a calf over the 3.3 year observation period. Including the latter cases, the mean birth intervals become 3.9 years and 2.5 years respectively. The shortest observed calving interval was 20 months. Black rhinos introduced from East Africa into the Addo Park in South Africa show a mean calving interval of 32 months (excluding a few very long intervals), with a minimum of 27 months when the previous calf had survived and 24 months after the calf had died. The birth sex ratio at Hluhluwe is 146 males : 100 females ($N = 86$). At Tsavo the juvenile sex ratio is 129 males : 100 females ($N = 119$). Black rhinos born in zoos include 35 males and 39 females (Goddard 1967, 1970a; Hall-Martin & Penzhorn 1977; Hitchins & Anderson 1983; Klös & Frese 1981, reported in Lindemann 1982).

White rhinoceros

Birth intervals recorded for individually known white rhino cows in the main Umfolozi study area varied between 22 months and 3 years 5 months. The modal birth interval was 2.0–2.5 years (Fig. 8.4), and the mean

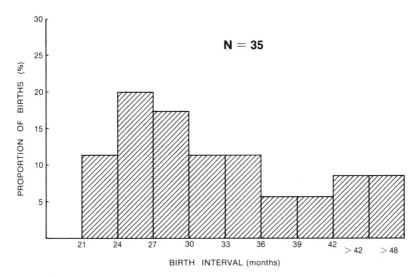

Fig. 8.4 Distribution of individual birth intervals of white rhino cows in the Madlozi study area.

2.6 years ($N = 32$, Table 8.5). Age-related differences in calving intervals were apparent. Three out of the four primaparous females in the sample exhibited calving intervals of less than 2.5 years; while 6 of the 8 cows that appeared elderly either had calving intervals exceeding 3 years, or failed to appear with a calf during the observation period of nearly 3 years.

Estimates of calving rates were also available from the proportion of females with young calves less than a year old in population samples from four different areas in Umfolozi in 1968. This ratio underestimates the calving rate to the extent that infant mortality had occurred. These data suggest that the calving rate was higher elsewhere in Umfolozi than in the main study area at Madlozi (Table 8.5). Based on all study areas combined, the mean natality rate was 0.40, equivalent to a mean calving interval of 2.5 years.

Data for white rhino populations introduced from Umfolozi into other wildlife parks show mean calving intervals varying between 2.7 and 3.5 years (Table 8.6). Zoo records indicate a mean birth interval of 24 months (range 17–44 months), but this includes cases where the previous calf died shortly after birth. Patterns vary among zoos, San Diego exhibiting a mean birth interval of 20 months ($N = 29$), compared with 29 months at Whipsnade ($N = 11$) and 28 months for all other zoos ($N = 21$). One individual cow at San Diego produced six calves in nine years at an average interval of 18.5 months (Lindemann 1982).

The birth sex ratio at Umfolozi was obtained from the sexes of all calves

Table 8.5. *Variation in natality rate among white rhino cows in different years and study areas*

Study area	Year	No. of cows observed	No. of calves born	Specific natality per Ad♀ per year
Madlozi	1968	42	13	0.31
Madlozi	1969	45	15	0.33
Madlozi	1970	43	13	0.30
Madlozi	1971	40	17	0.43
Madlozi	All years combined	(170)	58	0.34
Nqutsheni	1968	55	21	0.38
Gqoyini	1968	35	20	0.57
Dengezi	1968	28	10	0.36
All study areas	1968	160	64	0.40

Table 8.6. *Birth intervals and natal sex ratios recorded for different populations of white rhino*

Area	Period	N	Mean calving interval (y)	Sex ratio ♂:♀
Madlozi area, Umfolozi G.R.	1968–71	53	2.63	35 : 18
Pretorius Kop enclosure, Kruger Park	1965–78	19	2.70	12 : 7
Matopos Game Park, Zimbabwe	1967–77	23	2.85	9 : 14
Kyle Game Park, Zimbabwe	1967–75	23	3.45	7 : 16

Sources: Data for other areas supplied by M. C. Mostert, National Park Board of South Africa, for Pretorius Kop; and by J. H. Grobler and C. J. Lightfoot, Zimbabwe Department of National Parks, for Matopos and Kyle respectively.

born at Madlozi during my study, supplemented by the sexes of juveniles less than a year old seen elsewhere. Since juvenile mortality between birth and one year of age was less than 8%, the latter sample should deviate insignificantly from the sex ratio at birth. For the total sample the birth sex ratio was 173 males : 100 females, which is significantly different from parity (Table 8.7, $N = 139$, Chi square $= 4.7$, $P < 0.05$). A similar preponderance of males among subyearling calves was recorded by Hitchins during the 1972 helicopter census of the Umfolozi–Hluhluwe complex (26 males: 14 females), and by Brooks during a horseback survey of Umfolozi in 1974 (32 males: 21 females).

Table 8.7. *Sex ratio of white rhino calves born at Umfolozi in different years and regions*

| | Year of birth | | | | |
Area	1968 ♂:♀	1969 ♂:♀	1970 ♂:♀	1971 ♂:♀	All years ♂:♀
Madlozi	8:5	11:4	9:4	7:5	35:18
Nqutsheni	14:7	4:1	2:1	0:1	20:10
Gqoyini	7:10	3:2	0:1	—	10:13
Dengezi	5:5	2:0	—	2:0	9:5
Hluhluwe North	1:0	0:1	—	—	1:1
Umfolozi west environs	4:2	2:0	—	—	6:2
Umfolozi south	—	—	7:2	—	7:2
Combined	39:29	22:8	18:8	9:6	88:51
				Overall ratio	173:100

Table 8.8. *Sex ratio of white rhino calves in relation to preceding calving interval*

| | Area | | | | |
Preceding calving interval	Madlozi ♂:♀	Pretorius Kop ♂:♀	Matopos ♂:♀	Kyle ♂:♀	Combined ♂:♀
1.8–2.3 y	6:1	2:2	1:1	0:1	9:5
2.3–2.8 y	5:3	1:0	2:2	0:2	8:7
2.8–4.6 y	4:5	2:3	1:4	4:6	11:18

Sources: Acknowledgements as in Table 8.6.

The sex ratio of white rhinos born in zoos is 147 males : 100 females ($N = 121$, Lindemann 1982). Records from white rhino populations introduced from Umfolozi into other wildlife areas show that a preponderance of male calves is not a general feature. In the Pretorius Kop enclosure in the Kruger Park in South Africa more males than females were produced, but in the Kyle and Matopos Parks in Zimbabwe an excess of female calves was recorded (Table 8.6). If the natal sex ratio is analyzed in relation to the preceding birth interval, for all records combined, there is an apparent tendency for male calves to predominate after short birth intervals, and female calves after long ones (Table 8.8). However this tendency is not statistically significant, due to the small sample size. At Umfolozi younger females tended to give birth at shorter intervals than older ones. If the sexes of the calves born at Madlozi is related to the age category of the mother,

Table 8.9. *Sex ratio of white rhino calves in relation to age of the mother*
Based on calves recorded at Madlozi, 1966–71

| | No. of individuals | | Sex ratio |
	Cows	Calves	♂ : ♀
Young cow, with first or second calf	7	11	9 : 2
Prime cows	28	51	32 : 19
Ageing cows	8	14	5 : 9
All cows combined	43	76	46 : 30

males predominate strongly among first and second-born offspring, while ageing cows gave birth mainly to daughters (Table 8.9).

One cow clearly of advanced age at Umfolozi had no calf with her in 1966 nor in 1968, but produced a calf in 1970. Another elderly cow was accompanied by a calf born in 1965 when first seen. In 1970 she had no calf with her, but was in estrus. Possibly she had given birth in 1968 or 1969, but had lost the calf. Four other elderly cows failed to appear with a calf over the total period of nearly four years spanned by records. Thus it seems that white rhino cows continue producing calves throughout their lifespan, but that with advancing age birth intervals lengthen while infant mortality increases.

Summary

Among megaherbivores, typical birth intervals vary from about 2 years in giraffe and hippo to about 4.5 years in elephants. Shortest birth intervals, where the preceding calf survived, are 16 months for giraffe and 2.8 years for elephants. Rhinoceroses are intermediate. In all species natality rates can vary widely between years and among different populations. White rhino, black rhino and giraffe show variable, but generally male-biased sex ratios at birth; while elephants and hippo produce equal proportions of males and females. There is clear evidence of reproductive senescence among female elephants, and suggestive evidence for female white rhinos and hippos.

Reproduction by males

Males vary in the period over which they are reproductively active during adulthood, as influenced by their ability to maintain dominant status in relation to other males. Males may not attain prime competitive ability until some time after puberty. With advancing age males decline in physical prowess, even though they may remain sexually potent.

Elephants

Among African elephants, musth seems to be restricted to bulls over 30 years in age, although younger bulls may mate if given an opportunity. At Amboseli in Kenya, the success rate of courtship chases by bulls increases with age, from 15% for young adult males aged 21–25 years to 30% for males aged 26–35 years, and 54% for males over 35 years. This was largely due to evasive action by females against younger males. As a result large bulls over 35 years perform 65% of all copulations, although they form only 12% of the population of sexually potent males (> 12 years). Bulls that appear notably aged, and are thus probably over about 50 years, spend little time with females (Hall-Martin 1984; Moss 1983; Poole & Moss 1981; Western & Lindsay 1984).

Most Asian elephant bulls have undergone musth by 30 years of age. The majority of males show musth once annually, but males in poor condition may not show musth for periods of up to 4 years (Jainudeen, McKay & Eisenberg 1972).

Hippopotamus

Hippo males do not achieve asymptotic testis weights until 26–30 years of age, and dominant herd bulls are generally over this age. However, some culled groups contained two males in this age class, suggesting that some prime-aged males hold subordinate status (Skinner, Scorer & Millar 1975; Smuts & Whyte 1981).

Giraffe

Giraffe bulls do not obtain mating opportunities until 8 or more years old. A relatively small number of high-ranking bulls perform most of the copulations. Old bulls show declining sexual activity.

Rhinoceroses

Among Indian rhinos, some males were frequently associated with females, while others were rarely with females (Laurie 1978). No information is available for black rhinos on differing statuses among adult males.

White rhinoceros

Among white rhinos, dominant territory holders consisted mostly of prime-aged males, while subordinate bulls sharing territories included a preponderance of either young adults or ageing animals (Table 8.10). Overall two-thirds of adult males were dominant territory holders, while

Table 8.10. *Comparative age distribution of white rhino territorial bulls and subordinate bulls*

	No. of individ. classif.	Young adults	Prime adults	Old adults
Territorial bulls	36	3	30	3
Subordinate bulls	23	7	7	9
Combined	59	10	37	12

one third were subordinates, but among prime-aged males 80% were territory holders. In the Madlozi study area bulls held dominant status in the same territory for a mean duration of 5.4 years, although elsewhere in Umfolozi territory changes appeared to occur somewhat less frequently. Most prime males claimed another territory after being deposed. Young or old bulls either dispersed to occupy territories in less contested regions, or became subordinate bulls in favorable territory locations.

Summary
In all megaherbivores males do not attain full weight and hence social maturity until several years after puberty. Prime-aged males dominate reproduction, but some prime males may be relegated temporarily to subordinate status. Old males become relegated to peripheral status.

Mortality and lifespan
The risk of mortality varies over the lifespan, being high during infancy and old age and low during the prime period of maturity. Males commonly incur higher mortality rates than females, part of which may derive from reproductive competition. Episodes of higher than average mortality occur during periods of severe drought and reduced food availability. Potential lifespans are difficult to record in the wild. However, ages may be estimated in culled specimens from patterns of tooth wear, and counts of cementum lines evident in sections of teeth. A limitation in the assignment of chronological ages is the availability of representatives of known age. Zoo records reveal upper limits to potential lifespans.

Elephants
For African elephants at Manyara in Tanzania, annual mortality averaged 10% during the first year of life, thereafter declining to 3–4%, but about half of this mortality was related to hunting. At Amboseli in Kenya, where there was no hunting, calf mortality totalled 7.5% between birth and

one year of age, and 15% by 2.5 years, in wet years. During dry years there was a sex difference in calf mortality: first year mortality among males averaged 25% compared with 10% among females, while 55% of males had died by 2.5 years compared with 30% of females. Above 2.5 years, annual mortality rates drop to levels similar to those of adults. In Uganda, mortality among adult elephants was estimated to be 5–6% per annum, due mostly to shooting. At Tsavo in Kenya, annual mortality was 2–2.5% for prime females aged 15–40 years and for males up to 25 years of age; among older males mortality accelerated to 7.5% per annum, but probably incorporated past hunting. At Luangwa Valley in Zambia, the mean mortality rate of animals aged 10–50 years was 4.4% per annum. All of these estimates are based on the age structures found in shot samples, and thus assume a stationary population. They overestimate true mortality rate to the extent that these populations had been increasing rather than stationary (Douglas-Hamilton 1972; Hanks 1972; Laws 1969b, 1974; Laws, Parker & Johnstone 1975; Lee & Moss 1986).

Predation on calves by lions has been documented at Manyara in Tanzania and Kasungu in Malawi, but is probably more widespread. At Tsavo there is a record of a 7–8 year old elephant killed by lions (Douglas-Hamilton 1972; R. H. V. Bell personal communication).

Catastrophic elephant mortality occurred in the Tsavo East National Park in Kenya over 1970–71, when the rains failed over two consecutive years. At least 6000 elephants, or 15% of the total population of 40 000, died over this period. Most of this mortality was concentrated in the northern and central region of Tsavo East National Park, where rainfall remained under 200 mm per annum over two successive years. In this region mortality was estimated to be 70% for mature females, and even higher among dependent calves under 3 years. However, mortality among adult males was only 10% over these two years. Deaths were due to starvation following the elimination of most of the edible vegetation around remaining water sources. At Amboseli in Kenya, 76% of calves born in 1977 died during a severe drought. In the Ruaha Park in southern Tanzania, elephant mortality as indicated by found tusks was concentrated in the second half of the dry season, and increased in a drought year. Tusk sizes indicated that a higher proportion of young animals died during droughts than in other years (Barnes, 1982c; Corfield 1973; Moss, cited in Croze *et al.* 1981; Phillipson 1975).

However the elephant population in the Tsavo West Park, an area with a somewhat higher mean rainfall than that in Tsavo East, failed to show any notable increase in mortality during the 1970–71 drought. During the severe

1982–83 drought experienced in the Kruger Park in South Africa, the elephant population showed no decline, although calf proportions were subsequently low. The desert elephants occupying the Damaraland region of Namibia suffered no drought-related mortality at times when populations of smaller ungulates were decimated, but calf production suffered (G. Owen-Smith 1986 and personal communication; Phillipson 1975; Walker *et al.* 1987).

Life tables for African elephants show a steep rise in annual mortality after an estimated age of 50 years, associated with a decrease in the grinding area of the teeth. However, no captive African elephant has survived longer than about 44 years. For Asian elephants potential longevity in captivity is typically 50–60 years, with a maximum age of 67 years recorded. Hence it is estimated that African elephants in the wild have a potential lifespan of 55–60 years (Laws 1966).

Hippopotamus

For hippos in Uganda, mortality rates are estimated to be 45% between birth and one year, and 4% per annum thereafter. In the Virunga Park in Zaire, young hippos made up about 20% of recorded lion kills. They are also a significant prey of lions in the Kafue Park in Zambia, although elsewhere they generally form a negligible proportion of lion kills. The longevity record for a zoo-kept hippo is 49 years.

Hippos may suffer severe mortality during drought periods. In the Kruger Park in South Africa at least 170 hippos died during the 1970–71 drought. In the 1982–83 drought the population in the Levhuvu River in Kruger Park declined to 30% of its predrought level, with very few calves remaining in the surviving population (Bourliere & Verschuren 1960; Laws 1968b; Schaller 1972; Smuts & Whyte 1981).

Giraffe

For giraffe, calf mortality during the first year varies between 33% and 55%, with most losses occurring during the first month post-partum. Among adults, females experience an annual mortality of about 3–4%, and males about 8–9%, with young males being especially vulnerable. In the Central District of the Kruger Park, giraffe made up 11% of recorded lion kills, but due to their large size formed an estimated 43% of the food intake of lions. Based on the estimated total populations of lion and giraffe, the annual kill rate of giraffe amounts to perhaps 20% of the giraffe population. However, it is likely that giraffe are overestimated in kill records because of their large, conspicuous carcasses. In the Timbavati Reserve adjoining the Kruger Park, most giraffe kills occur during the late dry season, indicating malnutrition as a contributing factor. Giraffe appear to be more important as lion prey in South Africa than they are in East Africa. The highest recorded age for a captive giraffe is 28 years. There are records of both males and females exceeding 25 years in age in the wild (Foster & Dagg 1972; Hall-Martin 1975; Pellew 1983a; Schaller 1972; Smuts 1978, 1979).

Giraffe suffered little mortality either in the Kruger Park or in the adjoining Klaserie Reserve during the severe 1982–83 drought. However, in September 1981 a die-off of giraffe amounting to about half the population occurred in Klaserie, following an exceptionally cold spell. An important consequence of unusually severe cold in the South African lowveld is the frosting of evergreen tree foliage in bottomland areas, resulting in the loss of this critical food source (Walker *et al.* 1987).

Rhinoceroses

For Indian rhinos at Chitwan in Nepal, mortality rates were 3.4% per annum for adults (about one quarter of this due to poaching), 1.2% per annum for subadults, and 8.5% per annum for juveniles. Perinatal losses amounted to 5.6%. Predation by tigers was responsible for about half of the recorded calf mortality. Fighting accounted for nearly 30% of the deaths due to causes other than poaching. The greatest longevity recorded for an Indian rhino in a zoo is 47 years. The oldest animal recorded in the Chitwan population, based on counts of cementum lines in the teeth of animals found dead, was estimated to be about 30 years (Laurie 1978, 1982; Reynolds 1960).

Annual mortality rates of black rhinos at Hluhluwe were 3.5% for adult females, 7.3% for adult males, and 5.7% for immature animals. Neonate losses, incorporating barren females and prenatal mortality as well as postnatal deaths, amounted to 69% at Hluhluwe, but only 9% at Umfolozi. At Tsavo in Kenya, annual mortality rates average about 10% for prime-aged adults and 16% for calves, but these estimates based on found skulls assume a stable population. Black rhinos rarely feature in lion kill records, though there is an instance of a yearling killed and eaten by lions in the Serengeti. Hyena predation seems to be largely responsible for the poor survival of black rhino calves at Hluhluwe, and hyenas may be important predators on calves under 4 months old in East Africa. A black rhino female in the Chicago zoo reached an age of 49 years, before she was destroyed; while other records indicate zoo longevities of up to 38 years (Goddard 1970a; Hitchins & Anderson 1983; Lindemann 1982; Schaller 1972).

Table 8.11. *Mortality estimates based on histories of individually known white rhinos in the Madlozi study area in Umfolozi Game Reserve*

Age class	No. of dif. indiv.	Animal-years observed	No. dying	Mortality % p.a.
AD♂	25	54.8	2	3.6
AD♀	42	98.5	1	1.0
Calves	40	57.8	2	3.5

At Hluhluwe a die-off of about 15% of the black rhino population occurred in 1961, affecting all age classes. Some animals showed partial paralysis of the hindquarters before dying, suggesting the possibility of plant poisoning. Nearly 300 black rhinos died from starvation along a 64 km section of the Athi River in Tsavo National Park in Kenya in 1961, and several hundred more died during the 1970–71 drought in the same region. However, the black rhino population occupying the Namib desert region appeared resistant to droughts that affected other ungulate species (Goddard 1970a: Hitchins & Anderson 1983; G. Owen-Smith 1986).

White rhinoceros

I estimated mortality rates from the disappearance of known individuals from the study population, in relation to the total period over which animals of each age/sex category were monitored. Only five animals disappeared, presumed dead, over the 2.8 year study period, including two adult males, one adult female, and two calves aged 12 months and 18 months respectively. The resulting estimates of annual mortality are adult males 3.6%, adult females 1%, and calves 3.5% (Table 8.11). Only a proportion of the subadults in the study was individually recognizable, and none of these animals died. However, 3 unidentified subadults were found dead over the 3.3 year period. Since the mean number of subadults in the area was 15, this suggests a mortality rate among subadults of 6% per annum.

Perinatal mortality is difficult to detect, due to the secretive habits of cows with newborn calves. One calf was known to have disappeared between 2 and 7 days of age, but only because its mother was fitted with a radio transmitter. Indirect evidence of infant mortality is available from the calving histories of individual females. Documented calving intervals varied between just under 2 years and a little over 3.5 years. Thus if a cow did not produce a calf over a 4 year period, it could be presumed that she was either infertile, or had aborted during pregnancy, or had lost the calf shortly

Table 8.12. *White rhino deaths recorded in Umfolozi Game Reserve in Natal Parks Board files*

Period	Years	Adults + Subadults			Juveniles			Total
		♂	♀	Total	♂	♀	Total	
Aug 1962–Aug 1965	3	19	8	31	5	3	13	44
Aug 1965–June 1967	2	23	13	49	0	1	5	54
June 1967–July 1968	1	6	6	14	1	0	1	15
July 1968–Sept 1969	1	11	4	18	1	0	4	22
Oct 1969–Sept 1970	1	3	2	12	0	1	2	14
Sept 1970–Mar 1971	0.5	—	—	—	—	—	—	11
Total	8.5	62	33	124	7	5	25	160
%		54	29				17	

Sources: From Vincent 1969, and unpublished records.

after birth (the observation period was extended to four years by considering the age of the accompanying calf when each cow was first seen). Only three cows at Madlozi failed to appear with a new calf during this period; while 45 calves were born, 44 of which survived beyond one week of age. The effective calf loss is thus 4/48, or 8.3%. This represents the upper limit for neonate mortality, since it includes cases of infertility and prenatal losses as well as postnatal losses.

I also recorded the age classes of white rhinos found dead outside the study areas, and assigned found skulls to age classes. My records were supplemented by the records of white rhinos found dead reported in Natal Parks Board files. Of the adults plus subadults found dead, two thirds were male (Table 8.12). Juveniles are almost certainly underrepresented because of the rapid disappearance of their small carcasses and skulls. Some adult and subadult carcasses are probably also missed, despite the extensive coverage of the area by rangers, and the large numbers of vultures that congregate on white rhino remains. Arbitrarily it will be assumed that two-thirds of all adult and subadults deaths are recorded, yielding an estimated total of 25 adults plus subadults dying annually. Accepting the age and sex structure of the reconstructed 1969 population (Table 11.6), the following estimates of annual mortality were obtained: adult males 12, or 3.0%; adult females 6, or 1.2%; subadults 7, or 1.1%. These estimates are in close agreement with those calculated for the study population, except in the case of subadults.

Of the 16 cases in which cause of death could be established, 5 were related to injuries received during fighting. All 5 of these animals were males (three adult, two subadult). Six deaths were due to accidents such as falling over a cliff or becoming stuck in mud, two were possibly due to illness, and three could be related to senescence. Neither at Umfolozi nor Hluhluwe did white rhino calves show torn ears to indicate attacks by hyenas. A white rhino male was killed by lions in Umfolozi shortly after my departure (P. M. Hitchins, personal communication). From the Kruger Park in South Africa there are records of a white rhino calf killed by lions, and of a bull attacked and mauled so badly by lions that he had to be destroyed (Pienaar, 1970).

Low rainfall conditions occurred in Umfolozi over 1967–70, 1972–73, and even more severely through 1978–83. Increases in white rhino deaths recorded during these periods were relatively minor, amounting to no more than 50–100% above normal levels. Likewise, in the Kruger Park no increase in white rhino mortality was recorded during the 1982–83 drought. However, over 100 white rhinos reportedly died in Umfolozi during the very severe drought of 1933, when the total population numbered only about 300 animals (Player & Feely 1960; Walker *et al.* 1987).

The highest cementum line count obtained in tooth sections from white rhinos from the wild indicated an age of about 40 years (Hillman-Smith *et al.* 1986). The oldest zoo-kept animal, a female in the Pretoria zoo, was still alive aged 39 years in 1986, although appearing somewhat aged. Thus potential longevity may be estimated as about 45 years.

Summary

Megaherbivores generally show low adult mortality rates from natural causes, of the order of 2–5% per annum. Male mortality rates may be somewhat higher due to fighting injuries, and may be doubled by human hunting. Only giraffe are subject to significant predation as adults. Juveniles of all species are vulnerable to lions, tigers and perhaps hyenas for at least the first month or two. While all species may show episodically high mortality during severe droughts, the main effect is generally on calf survival. However, catastrophic mortality of adults has been recorded among elephants at Tsavo East in Kenya, and among hippos in the Kruger Park in South Africa. Potential longevity is 25 years for giraffe, but 35–60 years for other species.

Comparisons with smaller ungulates

Among many bovids the young lie out for the first month or two following birth, while among megaherbivores only giraffe show this pattern. In medium-sized ungulates nursing generally lasts about six months, compared with a year or longer among megaherbivores. Female bovids generally produce their first offspring between 1 and 3 years of age; and males become sexually potent at a similar age, although social maturity may be delayed until 4–8 years. Most female ungulates breed once annually, although in buffalo and zebra females fail to conceive some years; while among megaherbivores calving intervals span several years. Male bovids generally maintain prime breeding status for no longer than 2–3 years. Medium-sized ungulates show potential lifespans of 12–20 years, half as long as is typical of megaherbivores (Lent 1974; Mentis 1972; Murray 1982a; Owen-Smith 1984; Sinclair 1974).

Predation by carnivores is a significant source of mortality among adults of all ungulate species up to and including the size of African buffalo. Among megaherbivores only giraffe are vulnerable to predation mortality as adults, if human predation is excluded. Among elephants, rhinos and hippos, fighting among males and accidents such as becoming stuck in mud are the major source of mortality, apart from hunting. Population crashes have been documented for a number of medium-sized ungulate species during severe droughts (Schaller 1972; Sinclair 1977; Walker *et al.* 1987).

9

Body size and sociobiology

Introduction

This chapter considers the effects of large body size on social patterns, in particular (i) group size and structure; (ii) male dominance systems; (iii) female mate choice. Except for group size, these features are not readily characterized in numerical terms, and so cannot be related allometrically to body mass. Instead I will employ a cost/benefit analysis, assessing likely gains and losses in the factors determining evolutionary fitness, i.e. survival chances, reproductive contributions, and offspring survival. An inherent shortcoming of such an approach is that it does not adequately allow for possible interactions between these components (Crook & Goss-Custard 1972; Wilson 1975).

Grouping patterns

Jarman (1974) pointed out that among African bovids group size tends to increase with increasing body size. He explained this pattern in terms of the trade-off between the feeding costs of group formation, and the resultant anti-predation benefits. Because of their high specific metabolic rates, small antelopes are selective feeders on high quality plant parts. These are thinly scattered and quickly depleted. Large ungulates in contrast are relatively fiber-tolerant. They experience a much higher density of acceptable food, which is more uniformly distributed and depleted less by other animals foraging in the same area than is the food of small antelope. Intermediate sized species exhibit a gradient between these extremes. Hence the feeding cost of having close companions decreases with increasing body size.

This explains why large ungulates should be more tolerant of nearby conspecifics while foraging than smaller ungulates, but not why they should actively remain in a group. Jarman thus considered predation risks. Animals may benefit from having other animals nearby due to (i) more eyes to detect an approaching predator; (ii) the diluting effect of companions on the likelihood of being seized in an attack; (iii) the confusing effects of companions in disrupting the predator's attack; (iv) cooperative defense in warding off an attack. For small antelope (iv) is of no consequence, while (i) is of reduced benefit because most small ungulates occupy dense habitats. Furthermore, in thick vegetation animals tend to rely on concealment for predator evasion, and crypsis would be less effective if there were other animals nearby to attract the predator's attention. In contrast large ungulates, and in particular species occupying open grassland habitats, have nowhere to hide except amongst other animals. It is thus of greater benefit to be able to detect a stalking predator before it comes within attack range. The very largest species may furthermore be able to cooperatively ward off a predator by closing ranks and presenting powerfully backed horns outwards and vulnerable rears inwards.

Wittenberger (1980) emphasized the trade-off for females between the fitness gains from group formation in terms of individual survival, and the costs resulting from reduced offspring survival due to increased intraspecific competition for high quality food. For social ungulates he proposed that both adult survival and calf survival are raised by small increases in group size, but that beyond a certain group size the effects of food deprivation on offspring survival outweigh the predation benefits of group membership, at least for subordinate animals. He suggested that the particular form of these cost–benefit functions with increasing group size determined the optimal group size for a female of the species. Since male ungulates generally move independently of nursery groups, optimal group sizes should differ for males and females of the same species. Since males do not experience the effects of reduced feeding efficiency on offspring survival, Wittenberger's model implies that the optimal group size for a male should be somewhat larger than that for a female, unless other costs affect males.

Both Jarman's and Wittenberger's conceptual frameworks are based on the tradeoff between feeding costs and anti-predation benefits in relation to group size. Four possible patterns for these functions are sketched in Fig. 9.1. In all cases it is assumed that (i) predation risks decrease asymptotically as a function of group size, (ii) feeding effects on the survival of adults or offspring are linearly related to group size, and (iii) these factors interact multiplicatively, so that axes are appropriately scaled logarithmically. Fig. 9.1(a) is intended to depict the situation for a fairly small selectively feeding ungulate: the fitness component arising from anti-predation benefits rises quite rapidly with increasing group size, while the fitness component related to the effects of foraging efficiency on adult and offspring survival decreases

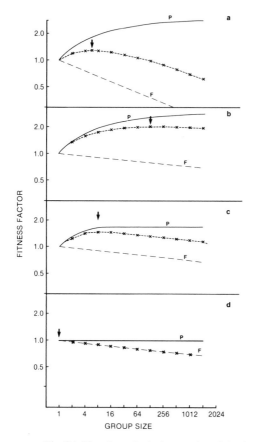

Fig. 9.1 Four hypothetical examples of the functional dependence of fitness components on group size. Solid lines (P) indicate the fitness factor resulting from predation risks; dashed line (F) indicates the fitness factor related to changing foraging efficiency. The product of these fitness components is indicated by the dotted-x curve. Arrows indicate the group size conferring maximum fitness. Example (a) is intended to represent a selectively feeding ungulate, (b) a less selective feeder, (c) a species with reduced anti-predation benefits, and (d) a species invulnerable to predation.

rather steeply with increasing group size. The optimal group size in this example lies in the range 4–8. Fig. 9.1(b) represents a larger ungulate, with similar anti-predation benefits, but with feeding costs less strongly influenced by group size. This leads to a rather larger optimal group size of about 100. In Fig. 9.1(c) feeding costs remain low, but anti-predation benefits are reduced, perhaps because they influence only adult survival, or solely offspring survival, rather than both. The outcome in this case is a small optimal group size of about 10. In Fig. 9.1(d) there are no anti-predation benefits associated with group formation, and the optimal group size is 1.

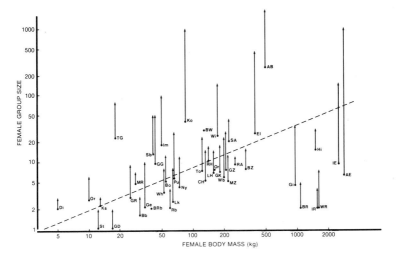

Fig. 9.2 Female group sizes for mainly African large herbivores in relation to body mass. Dot indicates the mean size of female–young groups, arrow the maximum group size recorded. Key to species labels given in Appendix I. Data from Anderson (1980), Attwell (1977, 1982), Backhaus (1959), Bigalke (1972), Conybeare (1980), Cumming (1975), David (1973, 1978), De Vos & Dowsett (1966), Douglas-Hamilton (1972), Dowsett (1966), Duckworth (1972), Dunbar (1979), Dunbar & Dunbar (1974), Duncan (1975), Eltringham (1977), Eltringham & Woodford (1973), Estes (1966, 1967), Estes & Estes (1974), Goddard (1967), Grobler (1973, 1974), Guy (1974), Hall-Martin (1975), Hendrichs (1971, 1975), Hendrichs & Hendrichs (1971), Hillman (1979), Irby (1977), Jacobsen (1973), Jarman & Jarman (1974), Jewell (1972), Joubert (1972, 1974, 1975), Joubert & Bronkhorst (1977), Jungius (1971b), Klingel (1967, 1974), Kok (1975), Kurt (1974), Laurie (1978), Laws *et al.* (1975), Leuthold (1970, 1971, 1974, 1976b, 1977b, 1979), Leuthold & Leuthold (1975), Murray (1980), Novellie (1975), Oliver *et al.* (1978), Olivier & Laurie (1974), Owen-Smith (1984), Penzhorn (1975), Rodgers (1977), Schenkel & Schenkel-Hulliger (1969a), Simpson (1974), Sinclair (1977), Spinage (1974), Van Lavieren & Esser (1980), Viljoen (1980), Von Richter (1972), Walther (1972b, 1978), Waser (1974), Watson (1969).
Regression (based on means):
  (i) for species up to the size of African buffalo (AB), indicated by
     dashed line: $FEMGP = 0.43M^{0.64}$ ($R^2 = 0.38$, $N = 38$, $P < 0.001$).
  (ii) all species: $FEMGP = 2.28M^{0.21}$ ($R^2 = 0.08$, $N = 45$, $P = 0.05$).

All bovids up to African buffalo size are vulnerable to predation even as adults. However, among megaherbivores (excepting giraffe) only immature animals are susceptible to non-human predation. Thus one might anticipate a dramatic drop in group size above a body mass of 1000 kg.

This prediction is indeed supported by the data on female group sizes for a wide range of large herbivores (Fig. 9.2). There is a significant positive correlation between mean female group size and body mass for species up to the size of African buffalo. But while African buffalo form large herds averaging about 250 animals, giraffe occur in small groups numbering typically about 4–5 animals, while rhino females of all species are generally accompanied only by a single offspring. However, hippos and elephants appear anomalous in forming somewhat larger groups.

The hippo situation is a special one in that feeding costs are not a factor during the daytime when groups are formed. Daytime aggregations are simply related to the restricted availability of suitable pools relative to the densities of animals supported by adjacent grasslands. At night hippos forage solitarily.

The formation of small but cohesive family units by elephants of both species could be due to (i) a greater susceptibility of elephant calves to predation, relative to other megaherbivores, (ii) an increased benefit from cooperative defense in warding off a predator attack, or (iii) some other factor influencing group formation not taken into account in the model outlined above.

Elephant calves are indeed likely to benefit from being protected in the middle of family units while the adult females form a protective ring. However, it is not obvious why rhino calves would not gain a similar advantage. White rhino groups including subadults do adopt a defensive rump-to-rump formation, which is clearly advantageous in the event of attacks by lions or hyenas. However, a white rhino cow drives away an older offspring as if the presence of the latter would be more of a detriment than a benefit to the protection of a small calf.

The sizes of all-male groups provide a test as to whether some factor besides offspring protection from predation might be involved in group formation. The mean size of all-male groups tends to rise with increasing body size, but the trend is much weaker than that shown by female–young groups, and is marginally non-significant statistically (Fig. 9.3). The so-called bachelor groups formed by Thomson's gazelle and African buffalo are similar in size, despite a 25-fold difference in body mass; while males of species as large as roan antelope are quite commonly solitary. Since males invariably form much smaller groups than females with young, it is evident that the anti-predation benefits of group formation for offspring survival are a more important factor than feeding costs, contrary to Wittenberger's (1980) emphasis.

Among megaherbivores, adult males are generally solitary in giraffe, rhinos, and hippos on land. African elephant bulls frequently occur in small all-male groups, while Asian elephant males also sometimes join up together. Unlike family units, these bull groups are open in membership.

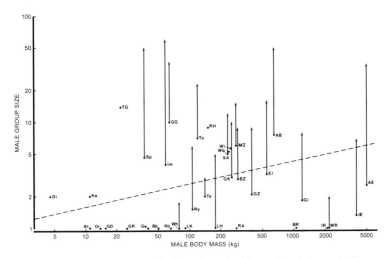

Fig. 9.3 Male group sizes of African large herbivores in relation to body mass. Dots indicate the mean size of all-male groups, arrow the maximum recorded group size. Key to species labels given in Appendix I, references as in Fig. 9.2.

Regression excluding megaherbivores (dashed line, based on means): $MALGP = 0.96M^{0.21}$ ($R^2 = 0.11$, $N = 30$, $0.1 > P > 0.05$). Including megaherbivores no correlation results.

The temporary associations among elephant males seem to be related mainly to the establishment and maintenance of dominance rankings among individuals sharing a common home region. Maintenance of extended social relationships could be a factor in the amalgamation of different family units of elephants, but this does not explain the formation of the cohesive family units.

Other factors besides predation risks to be considered include:

(a) *Habitat structure.* Populations in open habitats tend to form larger groups than populations in closed habitats. This is promoted both proximally, because it is easier for animals to maintain contact where visibility range is high, and ultimately, because predator detection benefits are greater. Notably, browsers occupying wooded habitats, such as tragelaphine antelope species, tend to form smaller groups than grazers of similar body mass (Fig. 9.2). However, this factor would not explain the differences in group size between white rhinos, which are open savanna grazers, and elephants, which are mixed feeders occupying a range of habitats from forests to open savanna woodlands.

(b) *Mobility.* For animals wandering over extensive home ranges, there are potential advantages for younger individuals to remain attached to older ones in order to gain from the experience of the latter regarding the

changing availabilities of habitat resources in different parts of this area
(Laws 1969b; Geist 1971; McKay 1973). African elephants move over much
larger home ranges than other megaherbivores, with their patterns of
movement highly variable in relation to changing vegetation conditions and
water availability (Chapter 4).

The home ranges of certain medium-sized ungulates (e.g. African buffalo, wildebeest and zebra) cover several hundred square kilometers. These
species are distinguished from others not so reliably by the size of their
groups, as by the degree of cohesion displayed in the individual membership
of these groups. This has been most clearly documented for plains zebra
(Klingel 1967) and African buffalo (Sinclair 1974). For other species with
large home ranges, individual cohesion has generally only been inferred
from the consistency in group structure observed over short periods (Estes
1974). Species with medium-sized home ranges like roan and sable antelope
show a similar group stability, as also do greater kudu (Owen–Smith 1984).
In contrast, species with small home ranges covering less than 10 km^2, for
example waterbuck, kob and impala, have groups that change in membership from day to day (Leuthold 1966; Murray 1982b; Spinage 1982).

For elephants, the prime factor promoting the formation of cohesive
family units seems likely to be the continuing benefit that young females
derive from the traditional knowledge of changing resource availability
built up over time by older cows. Such benefits are less for a rhinoceros
moving over a home range of only a few score square kilometers. Nevertheless, the tendency of white rhino subadults to attach themselves to cows
lacking calves could be related to the experience gained thereby. That cow–
subadult associations are not more common in white rhinos is due to the
antagonism shown by cows with small calves to other companions. It seems
that a white rhino cow can defend a single calf effectively against a predator
attack, but that her effectiveness would be diminished if an older offspring
had to be protected at the same time.

Elephant males benefit from the experience of their maternal group up to
the age of puberty, at which time the need to develop dominance relations
with other males causes them to leave. Part of the higher mortality rate that
subadult males experience relative to young females may represent the
survival cost of wandering without such maternal guidance.

Summary

The basic factors influencing group formation by ungulates are
predation benefits and feeding costs. Feeding costs diminish with increasing
body size, due to decreased plant part selectivity. Anti-predation benefits

tend to rise with increasing body size, but fall away in the megaherbivore size range. The cohesive family units formed by elephants are probably due to the benefits to offspring arising from the learning of changing resource distributions within the vast and changing home ranges that they cover.

Male dominance systems

Male dominance systems arise out of the fact that individuals differ in their fighting ability, or 'resource holding potential' (Parker 1974). Thus certain animals have the power to monopolize access to resources such as receptive females.

In many African ungulates, male dominance relations are structured within the spatial framework of territoriality. Alternatives to territoriality include dominance of the environs of a particular female group (as in plains zebra, Klingel 1967), and a rank ordering in dominance among a set of males sharing a common home area (as in African buffalo, Sinclair 1974). Elephants display an unusual system, with males in musth being temporarily dominant over all other males, while avoiding competitive interactions with other males in like condition. The important feature in all of these systems is that a dominant male – within his own territory, with his female group, or among his regular male associates – is challenged infrequently by rival males. This distinguishes these dominance systems from looser patterns in which dominance is transient and frequently contested, at least in the presence of a receptive female. Examples of the latter are to be found among small mammals such as squirrels (Farentinos 1972) and voles (Getz 1972).

Jarman (1974) related territoriality among African bovids to the size and mobility of female groups. In species with females forming small to medium-sized groups that are relatively sedentary (at least during the breeding season), prime males dominate territories in the localities where these groups are most likely to be. Where groups are large and mobile, as in African buffalo, several males are attached to the female herd, with a rank hierarchy governing mate access among them. In plains zebra, a species with small female groups but nomadic habits, males exhibit a female-group-dependent dominance. Comparing bats and ungulates, Bradbury & Vehrencamp (1977) suggested that males defend territories where female groups are least stable, and harems where female groups are most stable.

These arguments predict that territoriality should be the predominant system among sedentary megaherbivores with small group sizes, such as rhinos. A rank dominance system should prevail amongst wide-ranging species like elephants and giraffe. Although elephant females form cohesive groups, these tend to aggregate in large herds during the breeding season.

Geist (1974a and b) suggested that territoriality among ungulates is associated with conditions where (i) habitat productivity is high, and relatively stable over time; (ii) spatial variations in habitat favorability are high; and (ii) females are relatively sedentary during the breeding season. Hence ungulates inhabiting north temperate regions, where habitat conditions vary unpredictably over time, are mostly non-territorial. Emlen & Oring (1977) also emphasized the dependence of resource defense polygyny (their term for territoriality) on circumstances where there was a high spatial variability in food abundance or quality. These arguments imply that territoriality should not be prevalent among megaherbivores, since food quality is more uniform for such relatively unselective feeders than it is for smaller ungulates.

Owen-Smith (1977) suggested that, for ungulates, territoriality is a low benefit–low cost male mating strategy, favored under conditions where the survival costs associated with alternative mating systems are likely to be particularly high. These conditions were listed as being high population density (increasing the frequency of male–male conflicts), and high predator density (increasing the risk of fighting injuries proving fatal). Neither of these two conditons is especially applicable to megaherbivores.

Geist (1974b) proposed further that, because large ungulates have longer potential lifespans, there is selection for a lower frequency of combat, or for less damaging fights, than is typical of smaller ungulates. Owen-Smith (1977) pointed out that, by restricting their aggressive assertions of dominance to discrete spatial areas, territory holders reduce the incidence of escalated contests involving equally powerful rivals. A further factor restricting contesting among evenly matched rivals is prolonged male growth to full size. This results in a smaller proportion of males being of high resource holding potential (Wiley 1974).

Mortality costs

Fights among megaherbivores do not seem to be less damaging than those among male ungulates of smaller species. Laurie (1978) documents the severe and sometimes fatal injuries inflicted by Indian rhino bulls in fights. Hippos are notorious for their damaging conflicts (Verheyen 1954). Among both white rhinos and black rhinos, about half of male deaths could be attributed to injuries sustained in fights (Chapter 8). Serious fights between elephant bulls are less frequently observed, but can prove fatal (Kingdon 1979).

In Javan rhinos, Sumatran rhinos, and black rhinos in most areas, population densities are low. Thus encounters between equally powerful rival males are rare enough for a strict assertion of spatial dominance to be unnecessary. Nevertheless territoriality is clearly expressed in the high density population (nearly 1 per km²) of black rhinos in the Hluhluwe Reserve, and there is suggestive evidence of its existence even in the relatively sparse black rhino population inhabiting the Namib Desert fringe. White rhinos and hippos, which commonly occur in densities exceeding 1 per km², exhibit territorial dominance systems similar in form to those typical of many savanna antelope species.

Both African and Asian elephants tend to aggregate in large herds during the breeding season. Because of the mobility of female groups, a territorial restriction by males would not be advantageous. Thus an individually based rank dominance system similar to that of African buffalo would theoretically be predicted. Elephants do indeed show fairly stable dominance rankings, but superimposed on this is the temporal restriction on dominance imposed by musth periods. It seems that the reduction in contesting brought about by a rank dominance system alone is inadequate.

The survival costs associated with alternative male dominance systems are indicated by the surplus mortality that adult males experience relative to females. This was estimated to be an extra 2% per annum for white rhino bulls, and 3.8% per annum for black rhino bulls (Chapter 8). Notably, for both of these species over half of male deaths could be ascribed to injuries sustained in fights. The surplus mortality incurred by African elephant bulls relative to cows appears to be less than 2% per annum, but this estimate is derived from comparing mortality rates in two different populations. Furthermore, half of the mortality was inflicted by human hunting, which falls more heavily on males than on females.

Among north temperate ungulates, where territorial restrictions are generally lacking, considerably higher mortality costs are shown by males. Rutting injuries accounted for an annual mortality of 10% in Rocky Mountain goats, and 4% in moose. For bighorn sheep the annual mortality rate of males rose from 4% to 16% per annum on attainment of prime breeding age at 8 years (Geist 1971). For greater kudu, a non-territorial African antelope, the annual mortality rate of prime-aged males exceeded 50% per annum, compared with 7% per annum for prime-aged females. However, this surplus male mortality seemed to be a result of the indirect costs associated with increased male body size relative to females, rather than due directly to fighting injuries (Owen-Smith 1984, in preparation). At the other extreme, in African buffalo the annual mortality rate of prime-aged bulls exceeds that of prime-aged females by less than 1% (Sinclair

1977). Nevertheless, it must be born in mind that females experience additional nutritional costs due to the demands of pregnancy and lactation. Thus in the absence of costs associated with dominance contesting, males would be expected to survive better than females.

In terms of fitness costs it is not the annual mortality rate that is of significance, but rather the accumulated loss in survivorship through the reproductive period. In most medium-sized antelope species, males remain in prime condition for only a 2–4 year period. The decreased life expectancy resulting from a 12% mortality cost over a 3 year period is 33%. A white rhino bull can potentially remain an active contestant for territorial dominance from 10 years to about 30–35 years of age. The accumulated loss in survivorship from a 2% annual mortality cost integrated over a 20 year period is almost identical at 32%.

In summary, male megaherbivores reduce the annual mortality costs of competitive interactions by restricting the frequency of escalating confrontations between evenly matched rivals. This may be achieved by spatial restrictions on dominance, as in the territorial systems manifested by white rhinos and hippos; or by temporal restrictions on dominance, as in elephants. Giraffe, with a much shorter potential lifespan than that of elephants, exhibit a rank dominance system alone. Rhinos in low density populations may show a less formally constrained pattern of male dominance relations.

Satellite males

Rubenstein (1980) suggested that sexual selection could lead to a polymorphism in male mating strategies, if the mating gains conferred by the predominant strategy decline sufficiently as its frequency in the population increases. Such conditions exist where the costs associated with the prevailing mating strategy are high, the death rate due to extrinsic causes is low, and the sex ratio close to 1 : 1. These requirements are fulfilled by all megaherbivore species, except perhaps giraffe.

Waltz (1982) suggested that opportunities for satellite males to establish themselves within territories are created by the 'diminishing returns effect': a dominant male's opportunities for matings are a decelerating function of his own or his territory's attractiveness to females. Thus subordinate males occupying the more favorable territories secure some mating opportunities, without the energetic costs and risks associated with maintaining dominant status. The satellite male phenomenon is accordingly most likely under these conditions: (i) a steep gradient in habitat quality; (ii) females coming into estrus synchronously; (iii) little male parental care; (iv) prevailing mating strategy based on site-quality selection; (v) dominant males spending a high proportion of their time consorting with receptive females, or in the maintenance of their dominance. Conditions (i) and (ii) apply less strongly to megaherbivores than to many smaller ungulate species, and none of the other requirements is a special feature of very large animals.

Wirtz (1982) proposed a 'net benefit hypothesis' to account for the presence of satellite males sharing territories in waterbuck. Satellite males gain from future opportunities to take over the territory, while the dominant territory holder gains through having the satellite male deflect some potential rivals. This strategy is favored by a steep gradient in habitat quality, with females concentrating in the best territories. As a result there is a high likelihood of two females being in estrus simultaneously in the same territory.

The subordinate or satellite male phenomenon is a feature of all megaherbivore species. Among elephants, all bulls whatever their age or size act subordinately to a bull in musth. Nevertheless non-musth bulls gather in the vicinity of females in estrus, and are successful in securing some copulations (Moss 1983). Among rhinos and hippos, subordinate bulls share the territories or home ranges of dominant bulls. Although a large proportion of these subordinate bulls are either young or ageing animals, a number of prime males are included. Among white rhinos, instances were recorded of subordinate bulls attempting courtship or copulation in the absence of the dominant bull in the territory. In giraffe, subordinate-ranking bulls gather around potentially receptive females, and attempt copulations in the absence of the dominant bulls in the area.

Among medium-sized antelope species showing male territoriality, some males adopt subordinate or bachelor male status. Except in the case of waterbuck, bachelor males do not restrict their movements to any one territory. Nevertheless, there may be gaps in the territorial mosaic, and bachelor males may secure some mating opportunities when they encounter females in heat in such regions (e.g. Grevy's zebra, Klingel 1974; Coke's hartebeest, Gosling 1974).

It is not the existence of subordinate bulls that distinguishes those megaherbivore species that are territorial, but rather the restriction in movements of these males to a single territory. This is a behavioral response on the part of subdominant males that serves to reduce the incidence of aggressive challenges that they experience from more dominant males. It is more a waiting strategy than an alternative mating tactic, since mating opportunities for such subordinates seem few and far between. For white rhinos there are no vacant spaces in the territorial mosaic. A male unable to claim territorial dominance must either emigrate to peripheral less contested regions, where there are fewer females; or remain on in a more favorable region, temporarily foregoing reproductive opportunities while awaiting an opportunity to claim a territory. Both of these tactics are exhibited by white rhino bulls; but the tactic of wandering from one territory to another, as shown by bachelor males in some antelope species, is too risky and is restricted to short transitional periods. Hippo bulls are restricted by the availability of suitable refuge pools, with subdominant males dispersing into small pools when these are available (Laws & Clough 1966).

Mating gains

Many workers have related territoriality to defense of the food resources contained within territories, serving either to promote the survival of territory holders, or, in the case of male-only territories, to attract females by monopolizing sites of resource enrichment (e.g. Brown & Orians 1970; Emlen & Oring 1977). Owen-Smith (1977) suggested that territoriality among ungulates is promoted by intrasexual selection among males competing for mating opportunities. He suggested that the strength of such sexual selection could be estimated in terms of the 'Potential Mating Enhancement Factor', better termed the 'Potential Seasonal Mating Enhancement' (PSME). This was defined as the annual number of matings achieved by the most successful male in a local population of competing males, relative to the sex ratio of fertile females to potent males. He suggested further that the ultimate effects on fitness are indicated by the 'Potential Lifetime Mating Enhancement' (PLME). This is calculated by integrating seasonal mating performances over the reproductive lifespan, and dividing by the maximum lifetime reproductive output expected for a female. A PLME significantly greater than unity indicates that the variance in male reproductive output is greater than that for females, i.e. there is effective scope for intrasexual selection among males.

It has been assumed by many authors, from Darwin (1871) to Georgiadis (1985), that the degree of body size dimorphism between the sexes reflects the strength of sexual selection operating in a species. Wiley (1974) noted for grouse (Tetraonidae) that delayed sexual maturation by males is a feature of larger species, and that this automatically leads to strong sexual selection because of its consequences for the functional sex ratio. Ralls (1977), comparing birds and mammals, suggested that there was a predictable relation between extreme polygyny and extreme sexual dimorphism, but cautioned that sexual selection could not account for all of the observed

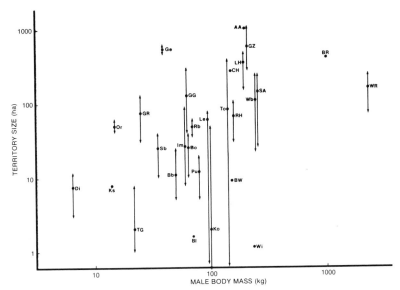

Fig. 9.4 Male territory sizes of ungulates in relation to body mass. The mean and range of territory sizes reported are shown. Key to species labels given in Appendix I. Data from David (1978), Duncan (1975), Murray (1982a), Oliver *et al.* (1978), Owen–Smith (1977).
Regression line (based on means): TER (ha) $= 2.63M^{0.63}$ (SE(b) $= 0.258$, $R^2 = 0.18$, $N = 30$, $P = 0.021$).

variability among species. Clutton-Brock *el al.* (1977) found for primates that, if monogamous species were excluded, there was no correlation between size dimorphism and degree of polygyny, as measured by the socionomic sex ratio. Instead there was a positive correlation between size dimorphism and body mass. Reiss (1986) suggests that the main reason for the association between size and sexual dimorphism is because factors such as food distribution influence both size and opportunities for polygyny.

If the function of a territory was to hold a population of potential mates, territory size should increase with body mass similarly to female home range size, assuming that the resultant degree of polygyny remained independent of body mass. However, the correlation between male territory size and body mass (Fig. 9.4) is much weaker than the correlation between female home range size and body mass (Fig. 6.3) This suggests that either (i) resource gradients are steeper for larger species than for smaller ones, so that it is easier for males of larger species to hold an adequate 'harem' of females within a relatively smaller area; (ii) the effective degree of polygyny decreases with increasing body mass, or (iii) there are other constraints affecting territory size besides bioenergetic demands.

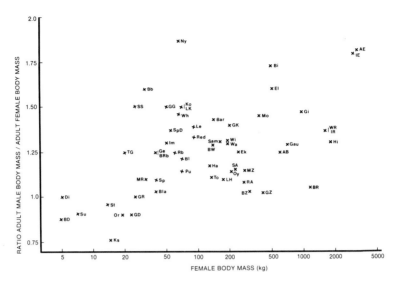

Fig. 9.5 Size dimorphism in adult mass between the two sexes in relation to body mass for various large herbivores. Sources of body mass data are listed in Appendix I.

For ungulates there is an overall positive correlation between the size dimorphism between the sexes and body mass (Fig. 9.5). However, this is due simply to the fact that antelope species under about 25 kg in body mass generally exhibit reverse sexual dimorphism, i.e. females larger than males. If these species are excluded, no significant correlation with body mass remains. Instead tribal or family patterns are apparent. The tragelaphine antelopes are all strongly dimorphic, irrespective of body size. All equids are monomorphic. Among megaherbivores there is a complete range in pattern. Both species of elephant exhibit extreme size dimorphism; white rhino, Indian rhino and hippo are moderately dimorphic; while in black rhino adults of both sexes are of similar size.

There is furthermore no correlation among ungulates between size dimorphism and the degree of polygyny as estimated by PSME (Fig. 9.6). For white rhinos the estimated PSME, based on the mean female population in the most favorable territory in the Madlozi study area, is somewhat on the low side; but it is no lower than that estimated for greater kudu, a strongly dimorphic species. No relation is apparent between estimated PSME and the system of male dominance, whether territorial or other.

The assumed relation between sexual dimorphism and degree of polygyny incorporates the further assumption that male success in combat is directly related to intraspecific variation in body size. However, this need

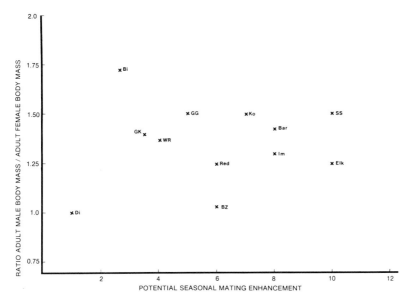

Fig. 9.6 Size dimorphism between the sexes in relation to the Potential Seasonal Mating Enhancement for ungulate males. Data from Clutton-Brock *et al.* (1982); Floody & Arnold (1975); Grubb (1974); Hendrichs & Hendrichs (1971); Klingel (1969); Lott (1979); Murray (1982a); McCullough (1969); Owen-Smith (1984); Schaller (1967); Walther (1972b).

not be the case. Fights amongst zebras and other equids, for example, test agility in both offence and defense, rather than strength (Klingel 1975; Berger 1981). Correspondingly there is a lack of size dimorphism in these species. In species where fights involve bashing or pushing contests, as in sheep (Geist 1971), buffalo (Sinclair 1974), kudu (Owen-Smith 1984), and elephants (Kuhme 1963; Short 1966), mass is clearly a factor. In general, as body size increases, fights tend to be decided more on body mass and strength and less on agility (Geist 1966). This accounts for the common tendency for size dimorphism to increase with increasing body mass.

The fighting technique explanation does not account for the variation in sexual dimorphism among megaherbivore species. In particular both species of African rhino fight in a similar way, using their horns to direct blows into an opponent's body. We thus need to estimate more precisely the effective degree of sexual selection among these species in terms of the potential difference between males and females in their reproductive outputs.

African elephants at Amboseli in Kenya have a mean calving interval of about 5 years, so that on average only 20% of adult females conceive in any one year. Large dominant bulls older than 35 years perform 65% of copulations, although they form only 12% of the adult male population (Moss 1983). Thus each large bull can expect to sire on average 1.1 offspring per year (assuming an even adult sex ratio, and also an equal likelihood of copulations resulting in conception). While growing up, a male can expect to achieve 0.05 copulations per year between the ages of 21 and 25 years, and 0.4 copulations per year between the ages of 26 and 35 years (from Moss' data). A male living to 50 years could thus expect to sire a total of about 20 offspring. However, a female producing a calf every 4 years between the ages of 10 and 55 would contribute only 12 offspring. Thus for African elephants the potential sexual selection factor is about 1.7.

For white rhinos, the most favorable territory in the Madlozi study area held a mean population of 4.2 adult females during the wet season when most conceptions occurred. With a 2.5 year mean birth interval, only 40% of females conceive in any one year. Thus the top white rhino male could expect to sire at most 1.6 offspring per annum. The mean duration of territory occupancy was 5.4 years, though a prime male may subsequently occupy another territory after being displaced from the first one. Since the ratio of territory holders to subordinate bulls is 2 : 1, a male could expect to be a territory holder for a total period of about 20 years between the ages of 10 and 40 years. Let us assume that, of this period, 5 years are spent in the most favorable territory (1.6 offspring per year), 10 years in a territory only half as favorable (0.8 offspring per year), and 5 years in a poor territory (only 0.4 offspring per year). A male experiencing this regime would sire 18 offspring during his lifetime. A female giving birth every 2.5 years between the ages of 6 and 40 would produce 14 offspring. Thus for white rhino the potential sexual selection factor is 1.3, somewhat less than that calculated for African elephants.

For black rhinos at Hluhluwe, highest local densities were about 1.5 per km^2. The territories occupied by black rhino males in this region covered about 4 km^2 (Hitchins & Anderson 1983). With one third of the population consisting of adult females, a territory this size would include a mean population of 2.0 potentially fertile females. At Mara where the maximum home range size of a black rhino bull was 18.6 km^2, the mean population density in occupied habit was 0.14 per km^2 (Mukinya 1973). Assuming local population densities up to twice the mean, such a territory would include on average about 2 adult females, as at Hluhluwe. A black rhino bull remaining in such a territory for two-thirds of his reproductive lifespan between the ages of 10 and 35 years would sire only 11 offspring, while a female surviving to 35 years and calving every 2.5 years would produce 12 offspring. The resultant sexual selection factor is close to unity. For sexual selection to be effective in black rhinos, territory sizes would need to be somewhat larger for a given population density, or female distribution would need to be somewhat more patchy. Such requirements seem to be atypical of most black rhino populations.

Thus variations in the degree of size dimorphism among megaherbivores do seem to be related to the degree of sexual selection operating. For size dimorphism to evolve, it is necessary both that male fighting success be dependent on body size, and that males gain sufficient benefit in terms of mating enhancement as a result of the dominance they acquire. This explains the varying patterns found among different groups of animals (Reiss 1986).

Female mate choice

In selecting a mate, females need firstly to recognize conspecific features. Signals used in courtship, whether visual, auditory or olfactory, may serve this basic specific-mate recognition function (Patterson 1985). Nevertheless females may also vary in their responses to different courting males, even though all are of the same species. A female may choose to mate with a particular male (i) because of the superior parental care that he provides, (ii) because of the quality of the resources that he holds, (iii) because of his good genetic constitution, or (iv) because he protects her from the energy-draining attentions of other males (Rubenstein 1980). Females could even favor a particular phenotypic characteristic in a male simply because it has come to serve as a conventional attractant to females, without any other functional value, since sons exhibiting this feature will also be attractive to females (the 'sexy son' hypothesis, Weatherhead 1984).

Among ungulates parental care is not a factor. Superior territory resources are of significance only where the female remains in the male's territory during the critical periods of late pregnancy and lactation. This occurs in some territorial antelope species, for example tsessebe in some regions (Joubert 1972), but is unusual among ungulates. This leaves only hypotheses (iii) and (iv) to be considered.

Males occupying the most favorable territories, or top dominance ranking, may be of superior genetic quality simply in terms of their proven ability to succeed in male–male competition. This may be due to better fighting ability promoted by large size, greater agility, or an edge in physical strength and endurance. Agility and endurance are qualitites that would generally be beneficial to both male and female offspring, for example in terms of ability to evade predators. However, this latter factor is applicable less to megaherbivores than to smaller species. If a male can maintain physical strength despite restricting his foraging to the limited confines of a territory, his food-processing ability is also well tested (cf. the handicap hypothesis, as suggested by Zahavi 1975). In these circumstances, a female simply needs to identify a high-ranking male.

African elephant females actively select older males over 35 years in age as suitors. They attempt to evade courtship advances from younger males, vocalize loudly to attract attention if caught by a younger male, and keep close by an older male if one is present. Moss (1983) suggests that, by favoring older males, females select for factors promoting longevity, which are of benefit to female progeny as well as to male offspring.

Noisy chases are also a feature of courtship in Indian rhinos. Laurie (1982) suggests that these not only test the suitor's strength, but also attract other males, so that the eventual sire is likely to be the strongest male in the area. Hippo copulations are noisy; but where the suitor is the dominant bull in the pool, other bulls do not interfere. In white rhinos, territory boundary blocking interactions between a territory holder and the female he is accompanying are also noisy, but do not attract rival males. A female white rhino gets another suitor only if the male lags so that she is able to elude him and cross into another territory. The premating consort period shown by white rhinos is unusually prolonged, lasting up to 3 weeks. During this time females proceed as if unaware of territory boundaries, so that it is a repeated challenge to the accompanying male to ensure that the female remains within his territory until she becomes receptive to mating. In black rhinos, noisy attacks by the female on the male are sometimes a feature of courtship; but because of the low population densities typical of black rhinos, there is rarely another male nearby.

In all of these cases females gain from the protective influence of a dominant consorting male, so that this proximate benefit cannot be excluded. However, a peculiar feature of the captive breeding of white rhinos helps resolve the issue of female mate choice.

White rhinos initially proved extraordinarily difficult to get to breed in captivity. Some zoos kept male–female pairs of white rhinos for periods exceeding 10 years, without any sexual activity. Females showed no signs of estrus, and males made no courtship advances. From an analysis of the circumstances, Lindemann (1982) concluded that the main requirement for successful breeding was the presence of more than one male. White rhinos produced offspring in 12 out of 28 zoos where more than one male was present, compared with only 3 out of 14 zoos in which two females were present, but only one male. Where just a single male–female pair was kept, there was only one record of breeding. In this exceptional case, the original male died and was replaced by a new one, and successful mating took place two months later. At Whipsnade near London, the original pair of white rhinos showed no reproductive activity over three years. When further white rhinos were introduced into the enclosure, the original male became the dominant male and sired most of the calves born; but neither he nor any other male mated with the original female. The same pattern was repeated at San Diego zoo in California, where the original pair remained together in a large enclosure for nine years without breeding (Rawlins 1979).

Lindemann (1982) suggested that male–male competition was necessary for male sexual behavior to be stimulated. However, the most notable feature is that females offered no choice of male showed no signs of estrus. This evidence suggests that female white rhinos need to exert some selection among potential mates for estrous cycling to be initiated.

The stimulatory effect of mate choice on estrus seems to be a peculiar feature of white rhinos. Both Indian rhinos and black rhinos breed in zoos without this necessity. Nevertheless, some form of manifestation of female choice is a feature of courtship in all megaherbivore species. The kinds of mate discrimination shown serve to exclude (i) young males, (ii) subdominant males, and (iii) males lacking in physical vigor. All these mechanisms would serve to eliminate as sires those males with any physical manifestations of genetic impediments to survival, likely to be detrimental to both male and female progeny. Such genetic selection might be especially important among long-lived species, for which the gene pool could lag somewhat behind in its adaptiveness to prevailing environmental conditions.

Summary

Megaherbivores differ from smaller ungulates in being less gregarious, a consequence of their invulnerability to predation as adults. Nevertheless, social groupings are probably promoted in elephants by the advantages of the group as a repository of traditional knowledge of resource locations where home range sizes are extensive.

Megaherbivores show a variety of male dominance systems comparable to those of medium-sized ungulates. The common feature of all megaherbivores is some mechanism for restricting the frequency of damaging combat among rival males. Territorial systems with relegation of some males to subordinate status achieve this where females are relatively sedentary, and local population densities sufficiently high. In elephants, where these conditions are not met, a unique system of temporal rather than spatial partitioning of male aggressive behavior has evolved.

The potential for strong sexual dimorphism is promoted by fighting techniques dependent upon strength and weight rather than agility, a feature that increases with increasing body mass. However, this potential is expressed only if intrasexual selection among males is effective.

Mechanisms of female choice are a feature of all megaherbivores. These involve either displays to attract other males, tests of the physical prowess of the suitor, or a favoring of the largest and hence oldest class of male. In white rhinos estrous cycling occurs only after a female has experienced a choice of males. Female choice could be due to the potential lag in the gene pool relative to prevailing ecological conditions.

10

Body size and reproductive patterns

Introduction

Much attention has been devoted by ecological theorists to patterns of variability among life history parameters. These not only influence rates of population increase, but are the features most directly associated with genetic fitness. Hence they are likely to be subject to strong natural selection in relation to the ecological conditions prevailing for the species.

A key concept is that of r- and K-selection (MacArthur & Wilson, 1967; Pianka 1970; Horn 1978). These labels are derived from the logistic equation of population growth: r represents the potential rate of population increase, K the equilibrium density or carrying capacity attained by the population. In unstable environments populations are likely to spend much of the time in phases of population increase following catastrophic reductions in density. Under these conditions density dependent factors exert relatively little influence, and the most successful phenotypes should be those which can multiply their descendants at the greatest rate during the increase phase. Such circumstances favor rapid reproduction, i.e. short breeding intervals, large litters and quick growth to maturity. In contrast, in more stable environments populations remain close to carrying capacity for most of the time. In such circumstances intraspecific competition for limiting resources is strong, and the most successful individuals are likely to be those capable of using these limiting resources efficiently towards maintaining a high population density. Selection should favor greater longevity, slower breeding rates and more parental care. It is not habitat instability *per se* which is of significance, but rather the period of habitat fluctuations in relation to the response time of the population. Thus insect populations may respond to changes within a season, large mammal populations to variations over periods of several years.

Megaherbivores should potentially be among the most extreme 'K-strategists', since larger size automatically confers greater longevity and slower breeding rates. However, first order adaptations (or exaptations, in the sense of Gould & Vrba 1982), imposed directly by body size, need to be distinguished from second order adaptations in the form of deviations in life history parameters from those typical of species of a particular size (Western 1979; Millar & Zammuto 1983). Boyce (1979) suggested that, in unpredictable environments, natural selection might favor adaptations enhancing survival through bad times, rather than high rates of population increase.

Several recent reviews have considered relationships between life history variables and body mass. Generally a proportional increase in adult mass is associated with a corresponding proportional change in a particular life history attribute, so that relations are expressed as power functions of the form $Y = a M^b$, where M represents body mass. For example, Fenchel (1974) showed that for all animals, including unicellular organisms as well as both heterotherms and homeotherms, the intrinsic rate of population increase (r_{max}) varies in relation to $M^{-0.28}$. For mammals alone, Hennemann (1983) obtained an almost identical estimate for the exponent ($b = -0.26$).

However, rate of population increase is the resultant of several components: age-specific birth rates, age schedule of deaths, and the age structure of the population. In terms of the reproductive attributes of individual females, it is the outcome of age at sexual maturity, intervals between successive offspring, and the number of offspring produced per litter. It is these attributes that are directly subject to natural selection.

Blueweiss *et al.* (1978) found for a wide range of mammals that litter weight varied in relation to $M^{0.82}$, gestation time in relation to $M^{0.26}$, average life span in relation to $M^{0.17}$, and time to sexual maturity in relation to $M^{0.27}$. For African artiodactyls alone, Western (1979) reported that gestation periods varied as a function of $M^{0.16}$, age at first reproduction as $M^{0.27}$, longevity as $M^{0.22}$, and birth weight as $M^{0.72}$. Values for exponents tend to cluster around either 0.75 or 0.25. Lindstedt and Calder (1981) pointed out that no adaptive explanation is required for such power relationships; they follow as inevitable consequences of the allometry of the underlying physiological processes. Time periods for unit metabolic action, such as maturation period, birth intervals and lifespans, tend to be scaled in relation to body mass to the power about one quarter; and volume or mass rates, such as litter mass, in relation to body mass to the power three-quarters.

In this Chapter I focus attention on the life history components determining the reproductive investments of female large herbivores. The underlying question to be asked is this: are the low reproductive rates typical of megaherbivores simply an allometric consequence of their large size, or do megaherbivores show second order adaptations towards lowered reproductive investments?

Seasonality of reproduction

Most medium-sized ungulates show narrowly restricted breeding seasons outside the equatorial zone of Africa, and some do so even in equatorial regions. Generally the mating peak occurs in autumn at the start of the dry season. Following a gestation period of 7–9 months, the peak in births falls during the wet season months (Mentis 1972; Spinage 1973).

For most of these species the birth peak seems to be associated with the time of maximum food availability. The nutritional demands of the mother rise rapidly during late pregnancy and peak during early lactation. Thus reproduction seems to be timed so that births occur at the optimal time of the year in terms of the mother's ability to nurture the offspring. Among ungulates inhabiting north temperate regions with strong seasonality in food availability, the mechanism controlling seasonal breeding is related to day-length. In so-called short-day species, ovarian cycling is stimulated by declining daylengths following the fall equinox. In long-day species, estrus follows the spring equinox (Sadleir 1969). There is suggestive evidence for photoperiodic control of reproductive seasonality among African ungulates inhabiting subtropical regions, with impala and wildebeest being examples of short-day species, and plains zebra an example of a long-day species (Spinage 1973).

Estes (1976) suggested that the narrow birth peak of wildebeest in equatorial East Africa is selected for ultimately by predation pressure on neonates. By having most calves born during a brief 2–3 week period, predators (in particular spotted hyenas) become saturated with prey. Out of season calves have a much greater chance of being killed than calves born during the peak.

Laws offered two hypotheses to account for the pattern of seasonal breeding shown by African elephants (Laws, Parker & Johnstone 1975). In this species conceptions peak during the mid-wet season, with the birth peak occurring at the start of the wet season (following a gestation of 22 months). Laws suggested that rainfall acts proximally by inducing new plant growth with a high protein content, which stimulates ovarian activity and promotes fertility. Ultimately this results in calves being born at the optimal time for their survival, i.e. when food abundance is greatest. The lush plant growth of the mid-wet season supports the early lactation period, and weaning two years later also tends to occur during the wet season. Calves born during the dry season also face additional rigors from the daily travelling to and from water.

Hall-Martin, Skinner & Van Dyk (1975) related the wet season conception peak of giraffe to optimal nutritional conditions for the female at the time of estrus, rather than to any influences on the survival of the offspring. They suggested that the extended lactation period reduced the effects of season of birth on calf survival.

In white rhinos the gestation period is 6 months shorter than that of elephants. Thus the birth peak and the conception peak cannot both occur during the same season, as is the case in elephants. For white rhinos the two mating peaks both fall during the wet season, with the resultant birth peaks occurring during the dry season. A similar pattern is shown by both black rhinos and giraffes in southern Africa (Chapter 7). Thus it appears that it is the proximate effect of nutrition on estrous cycling that governs seasonal breeding in megaherbivores, rather than any ultimate consequences of when the calves are born. Hippo follow the early dry season mating – wet season calving pattern typical of smaller ungulates, but their gestation period is only 8 months.

For megaherbivores the critical periods influencing offspring survival, including late pregnancy, early lactation and weaning, are drawn out through different periods of the year, so that there are no strong selective pressures favoring any particular time for reproduction. With no photoperiodic control of estrous cycling, ovarian activity is responsive to the prevailing nutritional regime.

Observations on white rhinos suggest it is the suppressant effect of adverse protein or energy balance on ovarian activity that controls seasonality in reproduction, rather than a stimulatory effect of high nutrition. Failing conception, estrus recurred at intervals of about 30 days. However, females failed to show signs of estrus when the next cycle coincided with a midsummer drought. Some females came into estrus as late as June (mid-dry season) in a year when fresh grass growth occurred in this month. No conceptions occurred during the peak dry season months of August–September. Under equatorial conditions where seasonality in plant growth is less extreme, and may vary between years in relation to the timing of the long and short rains, elephants and black rhinos show no clear seasonal patterns in reproduction. Even in subtropical regions, birth seasons of these species can vary between years in response to rainfall variability. This mechanism explains some of the variability in calving intervals shown by megaherbivore females.

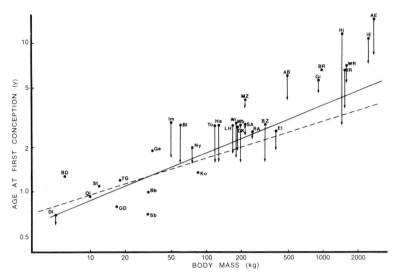

Fig. 10.1 Age at first conception for mainly African large herbivores in relation to body mass. Points indicate the mean age at first conception in natural populations, and arrows the earliest reported age at conception from either zoo or field records. Key to species labels given in Appendix I. References as listed in Figure 10.2, plus Dittrich (1976) and Lang (1967).

Regressions, based on the earliest recorded ages at conception, are as follows:

(i) For all species (solid line): AFC(yrs) $= 0.41M^{0.32}$ (SE(b) $= 0.032$, $R^2 = 0.77$, $N = 32$, $P < 0.0001$).

(ii) With megaherbivores excluded (dashed line): AFC(yrs) $= 0.54M^{0.25}$ (SE(b) $= 0.045$, $R^2 = 0.57$, $N = 25$, $P = < 0.0001$).

Age at first conception

The age at which females first reproduce varies widely among populations and individuals. I have extracted from published data both the earliest reported conception age (whether from the field or under zoo conditions), as well as the median age at first conception recorded in populations under natural conditions. It is the minimum age at first conception that indicates the potential of the species for population increase.

The minimum age at first conception increases in relation to $M^{0.32}$ for all large herbivores, but in relation ot $M^{0.25}$ if megaherbivores are excluded (Fig. 10.1). For artiodactyls alone, Western (1979) found that the age at first reproduction (including gestation time) increased in relation to $M^{0.27}$. Georgiadis (1985) suggests that females begin reproducing at an age that is a constant fraction (18%) of potential lifespan, irrespective of body size, and that this corresponds to the time when 80% of mature weight is attained.

It is evident in Fig. 10.1 that points are not evenly scattered. Rather there are three clusters: (i) small antelopes, which tend to conceive during or at the end of their first or second years of life; (ii) medium- to large-sized antelopes, which under favorable conditions can conceive as yearlings, but more usually experience their first fertile estrus as two-year-olds; (iii) megaherbivores plus buffalo, which show much individual and population variability in age at first conception. Under zoo conditions, conception may occur as early as 2.4 years among hippos (Dittrich 1976), 3.0 years in Indian rhinos (Lang 1967), and 3.7 years in white rhinos (Lindemann 1982). In the wild, first conception may be delayed as late as 8.2 years in black rhinos (Hitchins & Anderson 1983), 12 years in hippos (Laws 1968a) and 18 years in African elephants (Laws *et al.* 1975). Notably, among megaherbivores the youngest observed ages at first conception are close to those expected on the basis of the overall trend with body size.

For medium-sized antelope, potential longevity is typically about 15–20 years (Western 1979). Eighteen percent of this is about 3 years, which is the usual age at first parturition among these species. Most megaherbivores also produce their first offspring at about 18% of their potential lifespan (Table 10.1). The exceptions are African elephant and hippo, for which reported ages at puberty are commonly longer than predicted by this rule. This discrepancy could be due to the suggested overestimation of ages in this range for these two species (see Chapter 7).

Thus megaherbivores show tendencies towards delayed sexual maturity as expected of *K*-strategists. However, rather than being a species-typical constraint, this retardation of reproduction is apparently a flexible response adjusted to the prevailing ecological circumstances.

Birth intervals

Intervals between births incorporate the gestation period, but need not be directly related to it, since there can also be a variable interval between parturition and the next fertile estrous cycle. The data I will use represent the mean calving intervals of prime-aged females in natural populations. Where a range of values for different populations has been reported, I have chosen the shortest reliably estimated mean value. Where data are available for tropical conditions with breeding largely independent of the seasonal cycle, I have selected these in preference to southern African figures. Beyond this, it is not possible to make allowance for variations in demographic vigor between populations.

For African large herbivores, the birth interval (*BI*) in months is related to adult female body mass according to the relation $BI = 3.4 \, M^{0.27}$ (Fig.

Table 10.1. *Potential lifetime reproductive outputs by megaherbivore females*

Species	Age of puberty (y)			Birth interval (y)		Potential longevity (y)	Potential progeny	
	Mean	Min	Predicted[a]	Mean	Min		Mean	Max
African elephant	15	7	9	4.5	3	(60)	10	17
White rhinoceros	5.5	4	6.8	2.5	1.9	(45)	16	20
Indian rhinoceros	6	3	6.8	2.8	(2)	(45)	14	21
Black rhinoceros	6	4.5	6	2.5	2.1	(40)	14	18
Hippopotamus	10	3	6.5	2	(1.5)	40	15	25
Giraffe	4.5	4	3.8	1.7	1.3	28	14	18

Notes: (bracketed) – tentative estimates.
[a] Predicted age at puberty = 18% of longevity minus gestation time.

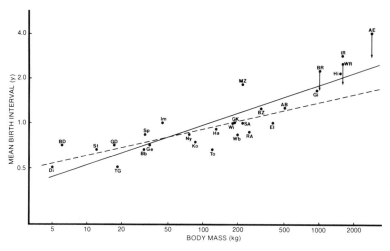

Fig. 10.2 Birth intervals of African large herbivores in relation to body mass. Points represent mean intervals between successive births reported for prime females in natural populations (see text); for white rhino, black rhino and African elephant the shortest reported birth interval is indicated by an arrow. Key to species labels given in Appendix I. Data from Anderson (1978), Bigalke (1970), Buechner *et al.* (1966), Chalmers (1963), Cumming (1975), Duncan (1975), du Plessis (1972), Goddard (1970a). Gosling (1974), Grobler (1980 and personal communication), Hendrichs & Hendrichs (1971), Hillman (1979), Hitchins & Anderson (1983), Huntley (1971), Hvidberg-Hanson (1970), E. Joubert (1974), Laurie (1978), Leuthold & Leuthold (1978), Morris & Hanks (1974), Murray (1980), Roth *et al.* (1972), Simpson (1968, 1974), Sinclair (1977), Skinner & Van Zyl (1969), Smuts (1975c, 1976), Spinage (1970), Von Ketelholdt (1977), Von Richter (1971), Watson (1969), Western (1979), Wilson (1966) and Wilson & Clarke (1962).
Regression lines:
  (i) all species (solid): BI(months) $= 3.4M^{0.27}$ (SE(b) $= 0.029$, $R^2 = 0.77$, $N = 28$, $P = < 0.0001$).
  (ii) excluding megaherbivores (dashed): BI(months) $= 5.0M^{0.17}$ (SE(b) $= 0.033$, $R^2 = 0.56$, $N = 22$, $P = < 0.0001$).

10.2). For megaherbivores, mean birth intervals all fall above the regression line. However, the shortest individual calving intervals for white rhino, black rhino and African elephant lie close to this regression line. If megaherbivores are excluded, the slope of the regression is somewhat less steep ($b = 0.17$), and even the minimum values for megaherbivores fall well above the regression line.

Theory suggests that birth intervals, as a time factor, should vary in relation to an exponent of 0.25. It seems that variability in birth intervals among bovids and equids from impala to buffalo size is constrained by the annual cycle. With this consideration, the overall trend including

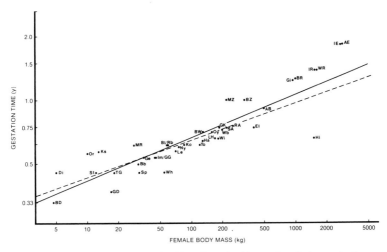

Fig. 10.3 Gestation times of African large herbivores in relation to body mass. Key to species labels given in Appendix I. Data from Dittrich (1972, 1976), Mentis (1972), Smithers (1983), Von Ketelholdt (1977), Western (1979). Regression lines:
    (i) all species (solid); $BM(\text{kg}) = 3.15M^{0.20}$ (SE(b) = 0.018, $R^2 = 0.78$, $N = 39$, $P = < 0.0001$).
    (ii) excluding megaherbivores (dashed): $BM(\text{kg}) = 3.49M^{0.17}$ (SE(b) = 0.019, $R^2 = 0.73$, $N = 33$, $P = < 0.0001$).

megaherbivores may be taken as more generally valid. Thus the shortest calving intervals recorded for megaherbivores are close to the values expected on the basis of their body size, while the mean birth intervals found in populations tend to be somewhat longer than this. Lengthened birth intervals can be interpreted as an ecological adjustment to food restrictions, which come into effect when populations reach densities close to the environmental carrying capacity K.

Gestation times show a closely similar pattern to that exhibited by birth intervals. For all African large herbivores, gestation time scales in relation to an exponent of 0.20; while if megaherbivores are excluded, a slightly lower exponent of 0.17 is obtained (Fig. 10.3). The latter figure is almost identical to the value of 0.16 reported by Western (1979) for artiodactyls alone. However, Martin & MacLarnon (1985) found, using reduced major axis regression, that gestation time for artiodactyls varied in relation to $M^{0.13}$. By considering trends shown by the different grades apparent among mammals, they concluded that the best fit value for the exponent relating gestation time to maternal body mass is 0.10. Megaherbivores apparently show somewhat longer gestation times than expected for their body mass on the basis of the general trend among mammals, with hippo a notable exception. However, the trend in gestation times is somewhat flatter than that expected for a time period parameter. It appears likely that variability in gestation times among medium-sized mammals has been constrained by the need to synchronize reproduction with the seasonal cycle.

In African elephants, fetal growth does not get under way until 20% of gestation time has passed; while among other mammals, this initial lag is generally assumed to occupy no more than 10% of gestation (Craig 1984). If this difference is general among megaherbivores, it could account for most of the extra gestation time shown by these very large species.

Among African ruminants, the relation between mean birth interval and potential longevity is such that the potential lifetime reproductive output of a female is about 14 offspring (Georgiadis 1985). Megaherbivores, except for African elephant, conform to this pattern (Table 10.1). This is a further indication that ages have commonly been underestimated in this species (Chapter 7). An elephant female attaining puberty at 10 years and calving every 3.8 years would produce 13 offspring during a 60 year lifespan, which conforms more closely with the data for other species.

Maternal investment in reproduction

Reiss (1985) suggested that the maternal energy available for reproduction is equal to the difference between the energy assimilation rate and the energy required for maintenance. Both of these factors are scaled allometrically, with estimates for the power coefficient b varying between 0.52 and 0.95. Hence, maternal reproductive investments must also be scaled in relation to body size with the exponent b falling within this range.

However, as was shown in Chapter 3, food intake by females may vary in relation to stage of pregnancy or lactation. Generally the maternal contribution to the young peaks shortly after birth, but before the young secure significant amounts of food from the environment. Millar (1977) derived a formula expressing the peak reproductive effort in terms of the relation between the metabolic demands of the young at commencement of weaning, and the maintenance metabolic demands of the mother, i.e. $n \times m^{0.75}/M^{0.75}$ (where n = litter size, m = mass of individual young at weaning, and M = maternal body mass). For a wide range of mainly small mammals, he found that birth mass scaled in relation to $M^{0.71}$, weaning mass in relation to $M^{0.73}$, and reproductive effort (according to the above formula) in relation to $M^{-0.17}$. Thus larger mammals invest less in reproduction, relative to their maintenance metabolic requirements than smaller mammals.

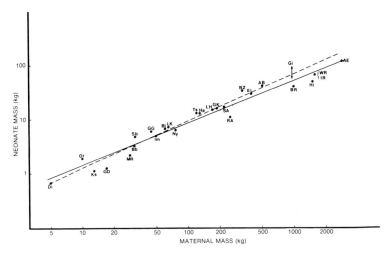

Fig. 10.4 Neonate mass in relation to maternal mass for mainly African large herbivores. Points represent mass at birth for each species, including both zoo and field data (except for giraffe where upper arrow indicates data from wild, lower arrow data from zoos). Key to species in Appendix I. Data from Ansell (1963), Dittrich (1972), Hall-Martin & Skinner (1978), Lang (1961), and references given for Fig. 10.2.

Regression lines:
  (i) all species (solid): BM(kg) $= 0.23M^{0.79}$ (SE(b) $= 0.031$, $R^2 = 0.96$, $N = 27$, $P = < 0.0001$).
  (ii) excluding megaherbivores (dashed): BM(kg) $= 0.17M^{0.87}$ (SE(b) $= 0.044$, $R^2 = 0.95$, $N = 21$, $P = < 0.0001$).

Data on neonate mass are most readily available for zoo-kept animals. I have simply averaged the published records for a species without regard to their source (whether zoo or wild animals) or the sex of the offspring (males generally weigh more than females at birth).

For large herbivores, neonate mass increases as a function of maternal mass to the power 0.79 (Fig. 10.4). This is identical to the value for b reported for various ungulates by Robbins & Robbins (1979). For both white rhinos and African elephants, the weight of the newborn young represents about 4% of the mother's weight, compared with 9% for medium-sized antelope such as kudu, and 13.5% for the smallest antelope, dikdik. In this case the points for rhinos and hippos, but not for African elephants, fall below the regression line. If megaherbivores are excluded, a higher estimate for b of 0.87 is obtained. This is greater than the power coefficient for birth weight of 0.72 found by Western (1979); but almost identical to that reported by Martin & MacLarnon (1985) for artiodactyls. Thus megaherbivores produce somewhat smaller offspring than expected from the trend among other African ungulates. This could simply be a consequence of the tendency for gestation periods to increase less steeply with increasing body mass than expected for a time period parameter.

Martin & MacLarnon (1985) refine the analysis further by taking into account grade differences between precocial and altricial species. They found neonatal mass to be related to $M^{0.91}$ for mammals with precocial young in general; while for artiodactyls alone neonatal mass is related to $M^{0.89}$. However, the overall best fit value for the exponent relating neonate mass to maternal mass is 0.80. This agrees closely with the value of 0.79 that I obtained for all large herbivores in Fig. 10.4. Thus in terms of offspring mass, females of larger species appear to invest slightly more in reproduction, relative to their maintenance metabolic requirements, than smaller mammals; although somewhat less as a fraction of maternal body mass.

Maternal investments can be estimated alternatively in terms of the milk energy supplied to the young at peak lactation. This has been found to scale in direct relation to the maintenance metabolic rate of the mother, i.e. $M^{0.75}$. Thus the peak lactational energy output of female mammals tends to be a constant fraction of their maintenance metabolism. Furthermore, the metabolic demands of sucking young are proportional to their mass to the power of 0.83, rather than 0.75. The total metabolic mass of the litter proves to be a more reliable predictor of milk energy yield than maternal metabolic mass (Oftedal 1984).

Oftedal (1984) suggested that the ratio between the metabolic demands of the young (proportional to $M^{0.83}$), and the maintenance metabolism of the mother (proportional to $M^{0.75}$), was the most widely applicable index of maternal reproductive effort. Among the eight ungulate species producing single young for which data were presented, values for this ratio vary from 0.55 for goats to 0.34 for beef cattle and red deer and 0.28 for reindeer. Megaherbivores show considerably lower values for this ratio (Table 10.2). This implies that females of these very large species make lower peak investments in reproduction, relative to their maintenance metabolic requirements, than do smaller ungulates.

However, the overall lifetime investment by females in reproduction extends over the whole period of each reproduction cycle, including any interim when females are neither pregnant nor lactating. Additional metabolic costs are incurred in transforming maternal energy into offspring biomass or milk. Insufficient data are available to attempt a comparative analysis in terms of lifetime reproductive investments for a range of large herbivore species. Thus I will merely illustrate such calculations for the two ungulate species with which I am personally familiar from my own field research: the white rhino, as a representative megaherbivore; and the greater kudu, as a representative ungulate of medium–large size.

Table 10.2. *Ratio between the metabolic mass of the offspring and maternal metabolic mass for megaherbivores*

Species	Maternal mass (kg)	Neonate mass (kg)	Metabolic mass ratio[a] Birth	Weaning
African elephant	2800	120	0.138	—
White rhinoceros	1600	65	0.126	0.196
Indian rhinoceros	1600	65	0.126	—
Black rhinoceros	1000	40	0.120	—
Hippopotamus	1400	50	0.112	—
Giraffe	850	55[b]–93[c]	0.177–0.273	—

Notes: [a] $m^{0.83}/M^{0.75}$, where m = offspring mass, M = maternal mass.
[b] zoo data.   [c] wild data.

The maternal investment in reproduction can be partitioned between the periods of pregnancy and lactation. Within each of these it can be separated into two components: (i) the maintenance metabolic costs of the offspring, proportional to offspring mass to the power three-quarters; (ii) the growth increment of the young, related to the energy content of this mass. The latter is assumed to be about 25kJ per gram of biomass, while the former is equal to 293 times $M^{0.75}$ kJ on a daily basis. To simplify calculations I will ignore variations in the utilization of metabolizable energy, as well as activity increments. However, the efficiency of utilization of energy for milk production is only about two-thirds of that for maternal growth, and this must be taken into account (Moen 1973).

Since inadequate data are available to integrate the maternal contribution on a daily basis, calculations will be simplified by summing the mean maternal investments over set periods. During the first two-thirds of pregnancy, the energy requirements of the fetus will be assumed to be negligible. During the penultimate sixth of pregnancy, its mean mass will be taken as one fifth of the birth mass; and during the final sixth, as one half of birth mass. These masses will be used to calculate the daily maintenance requirements of the fetus, which are then multiplied by the duration in days of each period. The investment in neonatal growth will be based on the birth mass. The period of infancy leading up to weaning will be divided into four periods, corresponding to those for which estimates of grazing time relative to that of the mother were made (Table 8.2). It will be assumed that the relative grazing time reflects the proportion of its energy requirements that the calf gets from the vegetation, with the mother's milk contribution making up the remainder. Finally, the energy contribution of the mother to the offspring needs to be related to her own basal metabolic requirement.

The latter is equal to 293 $M^{0.75}$ kJ per day, summed over the whole period between successvie births. Based on a calving interval of 30 months, the relative proportion of her own maintenance requirements that a white rhino mother devotes to reproduction is estimated to be 22.7%. For the minimum observed calving interval of 22 months, this proportion increases to 31%.

Adult female kudus attain a body mass of 180 kg. The birth mass of a kudu offspring is 16 kg, the period to the end of weaning is 6 months, and the weight of the offspring at the end of this period is about 60 kg. No data are available on the proportion of its energy requirements that the young secures from the vegetation. These will be assumed to be the same as those recorded for white rhino calves for corresponding segments of the period to weaning. Calculations suggest that an annually breeding kudu female invests 33% of her maintenance energy requirements in reproduction.

Hence it seems that a white rhino female breeding at the shortest observed interval makes the same reproductive effort as a typical large antelope female. However, white rhino females generally do make a lower reproductive investment than is typical of medium-sized ungulates, due to the fact that their calving intervals tend to be about one-third longer than the minimum. On the other hand, the reproductive investment of the typical antelope female is also reduced by perhaps one-third by the fact that on average about 50% of calves die shortly after birth (personal observations); while for white rhinos infant mortality is negligible.

Hence it is evident that, in order to calculate the effective reproductive investments by ungulate females, a number of factors have to be taken into account. These include not only the peak reproductive effort made after offspring are produced, but also the intervals between reproductive episodes, and the likelihood of the young surviving. Typical lifetime investments in reproduction made by female megaherbivores appear to be no different from those shown by medium-sized ungulates. Megaherbivores show great flexibility in birth intervals, in relation to ecological circumstances; if conditions are unfavorable, either conception is delayed, or the fetus is aborted early in pregnancy. The lag in commencement of fetal growth found in African elephants could allow a period for abortion with minimal costs to the mother. Small and medium-sized ungulates generally produce offspring according to schedule, unless conditions are exceptionally unfavorable. Neonatal mortality adjusts birth rates to the prevailing circumstances. The reproductive costs of unsuccessful attempts are minimal where breeding is coupled to the seasonal cycle; a delay in conception would simply result in the young being born at an unfavorable time of the year, and so unlikely to survive anyway.

Hence in terms of reproductive investments, megaherbivores appear no more *K*-adapted than African savanna antelopes. However, African large herbivores, except suids, produce single young; whereas many of the goats, sheep, deer and gazelles inhabiting north temperate regions commonly produce twins or even triplets (Geist 1981). In terms of litter size, all of the large herbivores occupying tropical and subtropical savannas are more *K*-selected than temperate zone species.

Offspring sex ratio

In a classical paper, Fisher (1930) argued that natural selection would counteract any marked deviations from parity in the birth sex ratio. However, if one sex required a greater maternal investment than the other, the sex ratio should be biased against the more expensive sex by the end of the period of maternal care. Among ungulates, male calves are commonly larger than female offspring at birth, and in some species (e.g. red deer) male offspring grow faster and hence nurse more frequently than females (Clutton-Brock, Guinness & Albon 1982). However, male ungulates also tend to show higher mortality rates than females, from conception onwards (Robinette *et al.* 1957). To counteract the high prenatal mortality of males, the sex ratio should be biased in favor of males at conception, but become even by the time of weaning. Clutton-Brock & Albon (1982) point out that where daughters remain in the maternal home range, mothers may incur feeding costs affecting the survival of subsequent offspring. Under such conditions the post-weaning costs of daughters may exceed the pre-weaning costs of sons, tending to promote a male-biased sex ratio even at weaning.

Trivers & Willard (1973) proposed that sex ratio at birth should vary in relation to ecological circumstances. In many vertebrate species the reproductive success of males is strongly influenced by body size or physical prowess, while the reproductive output of females is more constant. Thus mothers in prime condition, and thus able to nurture large offspring, should produce relatively more sons; while mothers in poor condition should produce more daughters. More generally, females in prime condition should invest more heavily in offspring of the sex showing the greatest variance in reproductive success, provided that the subsequent reproductive performance of their offspring is influenced by early maternal investment (Maynard Smith 1980).

While the segregation of X- and Y-chromosomes tends to promote a balanced sex ratio, two mechanisms of adjustment of the birth sex ratio have been documented for white-tailed deer: (i) females conceiving late in the estrous cycle produce more male offspring than females conceiving early in the cycle (Verme & Ozaga 1981); (ii) male fetuses show a higher *in utero* mortality than do female fetuses (Robinette *et al.* 1957). The latter pattern could simply be an effect of the more rapid growth rate, and hence higher nutritional demands, of males relative to females, so that male offspring are more vulnerable under conditions of food deprivation (Clutton-Brock, Albon & Guinness 1985).

In all of the larger ungulates, male reproductive output is influenced by dominance status and thus tends to be more variable than that of females. Furthermore, in most species male reproductive performance is dependent on body size. Thus the potential exists for facultative variations in natal sex ratio, although the mean sex ratio should not deviate far from 1:1.

There is no reason to expect the pattern among megaherbivores to differ from that shown by medium-sized ungulates. Nevertheless, available data on the perinatal (fetal or birth) sex ratio reveal a tendency for the proportion of male offspring to increase with increasing body size (Fig. 10.5). This tendency is statistically significant if the data point for the single small antelope species in the sample, steenbok, in excluded. Steenbok show a more extreme sex ratio bias in favour of males (2:1) than has been recorded for any ungulate, based mostly on a single sample. Notably, steenbok also show reverse sexual dimorphism, i.e. females larger than males. Apart from steenbok, the next highest sex ratio is that shown by white rhino (combining all samples). Other species showing a perinatal sex ratio exceeding 120 males:100 females include two other megaherbivores (black rhino and giraffe) and one large cervid (moose). African elephant and hippo show fetal sex ratios close to parity.

In both rhino species, females disperse from the maternal home range; and neither in moose nor in giraffe are there persistent associations between mothers and daughters. In contrast, young elephant and hippo females remain associated with their mothers. Hence these extremely male-biased sex ratios cannot be explained in terms of the post-weaning costs of female progeny.

In the case of white rhino, young females produced a higher proportion of male calves, and also exhibited shorter birth intervals, than old females (Table 8.9). The predominance of male offspring born at Umfolozi during the study period can be ascribed in part to the high proportion of young females in the expanding population. Over this period the population was increasing at a higher rate than had been shown earlier. Thus it appears that habitat conditions were favorable for white rhinos, despite below-average rainfall. Since no sex ratio bias was evident among the adult and subadult segments at the time of my study, the sex ratio bias among calves born

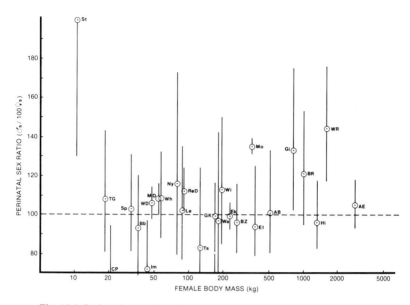

Fig. 10.5 Perinatal sex ratios of various large herbivores in relation to body mass, based on sex ratios recorded for either fetuses or new born calves, combining data from all sources, zoo records as well as wild populations. Lines indicate 95% binomial confidence limits; dashed line indicates even sex ratio. Key to species labels given in Appendix I. Data from Attwell & Hanks (1980), Allen-Rowlandson (1980), Child *et al.* (1968), Clough (1969), Clutton-Brock *et al.* (1982), Frame (1980), Goddard (1970a), Hall-Martin (1975), Hanks (1972b), Hitchins & Anderson (1983), Houston (1982), Huntley (1971), Hvidberg–Hansen (1970), Klingel (1965), Klös & Frese (1981), Laurie (1978), Laws (1969a), Laws & Clough (1966), Leuthold & Leuthold (1978), Lindemann (1982), Mason (1982), Mentis (1972), Mitchell *et al.* (1977), Morris & Hanks (1974), Mossman & Mossman (1962), Pienaar *et al.* (1966a), Reuterwall (1981), Robinette & Archer (1971), Robinette & Child (1964), Roth *et al.* (1972), Sherry (1975), Simpson (1968), Sinclair (1977), Smuts (1974, 1975c), Smuts & Whyte (1981), Sowls (1974), Spinage (1982), Talbot & Talbot (1963a), van Zyl & Skinner (1970), Verme (1983), Watson (1969), Williamson (1976), Wilson & Child (1964, 1965), Wilson & Clarke (1962), Wilson & Kerr (1969).

during the study period must have been a new phenomenon (see Chapter 12). The sex ratio bias towards males among zoo-kept white rhinos can likewise be ascribed to a predominance of young females, since young animals are preferentially acquired and no zoo has held white rhinos for very long.

Among mule deer, primaparous females showed a fetal sex ratio of 122 males : 100 females, compared with 106 males : 100 females for older mothers (Robinette *et al.* 1957). However, red deer showed no difference in natal sex ratio between first-time mothers and older females (Clutton-Brock,

Guinness & Albon 1982). Instead dominant hinds produced a higher proportion of sons than did subordinate hinds (Clutton-Brock, Albon & Guinness 1984). The latter pattern suggests that good female body condition may have a positive influence on the proportion of sons. In contrast, among white tailed deer and mule deer females on a poor plane of nutrition produce a higher proportion of male progeny (Verme 1983). A feature of these small deer is that they commonly produce multiple offspring, with the sex ratio of litters of one significantly more male-biased than litters of two or three. The fact that these deer produce more daughters than sons when in good condition could be explained on the grounds that it is more advantageous for mothers to produce a large litter than a large son in such circumstances. Scandinavian moose, for which the birth sex ratio averages 135 males : 100 females ($N = 160\ 911$), show a high frequency of twinning (Reuterwall 1981).

None of the above hypotheses explains the bias in birth sex ratio towards males shown by black rhinos or giraffes. The fact that certain megaherbivores produce more extremely male-biased sex ratios than is typical of smaller ungulates could simply be a consequence of their variable birth intervals. The flexibility in time to conception provides scope for selective abortion of embryos of the less advantageous sex, with less effect on reproductive output than is the case for smaller species with restricted breeding seasons.

Summary

The fundamental features distinguishing megaherbivores from smaller ungulates are these:

1. Prolonged and highly variable times to puberty are typical of both males and females, with these times sensitive to prevailing nutritional conditions. Among males social maturity is delayed several years after puberty.

2. Gestation periods exceed one year, so that birth intervals usually span two years or longer. With sensitive periods of reproduction spread out over the seasonal cycle, there is no strong selection for a restricted breeding season. Female sexual cycling thus responds directly to the prevailing nutritional conditions.

3. Peak maternal investments in lactation are lower than those of smaller ungulates. However overall lifetime investments in reproduction appear no different from those of medium-sized bovids. Megaherbivores show a flexibility in time to conception following

parturition; while smaller ungulates show variability in perinatal mortality, in relation to the prevailing ecological conditions.

4. Due to the flexible birth interval and weak seasonality, the scope for variation in offspring sex ratio, in relation to factors such as age of mother and prevailing habitat conditions, is greater than it is among smaller species.

11

Demography

Introduction

In this chapter I cover those ecological attributes that are features of populations rather than of individual animals. Population ecology is generally framed in terms of the logistic model of population growth. The parameters of this model include (i) the maximum or 'intrinsic' rate of population growth shown when population density is very low, labelled r_{max}; (ii) the equilibrium density or 'carrying capacity' eventually attained, labelled K.

However, rate of population growth is the difference between recruitment, determined by processes of birth and immigration, and losses, the outcome of deaths and emigration. Since natality and mortality rates vary with age and sex, the realized rate of population increase is influenced by population structure. Strictly, r_{max} and K are defined only for populations that have attained an equilibrium age and sex composition. However, real populations seldom remain at any equilibrium for long, due to environmental fluctuations. Furthermore, density varies spatially over the population range in relation to habitat suitability.

The ecological features to be considered in this chapter include (i) population composition, in terms of age structure and sex ratio; (ii) rates of population change with time; (iii) population densities attained.

Population structure

In considering age structure, the functional age classes include (i) adults, i.e. animals that have passed the age of socio-sexual maturity and are reproductively active, or at least potentially so; (ii) juveniles, i.e. animals that have not attained the age of independence from their mothers; (iii) subadults, i.e. animals that are intermediate between the above two categories. Puberty or physiological sexual potency is attained during the subadult period. I will use the term 'immature' to encompass both juveniles and subadults. Population composition is not a species-specific feature, but varies depending on whether the population is increasing or stable. Expanding populations contain a higher proportion of immature animals than stationary ones.

Elephants

African elephants attain puberty between 10 and 15 years of age. Females complete their growth between 15 and 20 years, but males continue growing until 30 years old. Since the typical calving interval is 4–5 years, animals under 5 years are classed as juveniles. Animals over 15 years of age will be regarded as adults, although males do not attain full socio-sexual maturity until the age of 30 years. This provides a basis for a 5–year grouping of year classes. In practice the proportion of animals assigned to particular age classes is influenced by the ageing technique used. Populations for which age was estimated from dentition using Laws' (1966) method tend to show a higher proportion of animals in the age range 10–15 years than in the range 5–10 years, and relatively few animals aged 15–20 years.

The proportion of adults (> 15 years) varies from 34–38% in increasing populations, such as those at Kruger in South Africa and Manyara in Tanzania, to over 50% for populations that were putatively stable (e.g. Luangwa in Zambia), or recovering from large-scale drought mortality (e.g. Tsavo in 1974) (Table 11.1). However, in expanding populations females are likely to attain maturity at a younger age than they do in stable populations. Thus if functional age classes were defined in terms of reproductive status rather than age, the proportion of animals that were adult would vary relatively little with population status. For example, if females, but not males, attained sexual maturity at 10 years in expanding populations, the proportion of functional adults in such populations would be about 45%. The low proportion of juveniles in the Tsavo East population indicates that recruitment had remained suppressed for some time after the 1971 drought.

Bulls that were solitary or in all-male groups formed between 6% and 16% of the total population in samples from Uganda, Kenya, Tanzania and South Africa. In Uganda, 9% of the animals in family units were adult males, but this proportion may vary seasonally. Adult males associated with family units tend to be younger than those in bull groups. At Amboseli in Kenya, fully-grown males older than 35 years formed only 12% of the

Table 11.1. *Age structure of African elephant populations*

Area	Time	Segment	Status	Ageing technique	Age-class proportions (%)					Reference
					Juvenile (0–5y)	Immature (5–10y)	Subadult (10–15y)	Young adult (15–20y)	Adult (20+y)	
Manyara, Tanzania	1970	All	Increasing	Height	29	19	14	5	33	Douglas-Hamilton 1972
Hwange, Zimbabwe	1971–72	Females	Increasing	Dentition	26.5	15	16.5	6	36	Williamson 1976
Kruger, South Africa	1968–74	Females	Increasing	Dentition	29.5	16.5	19.5	5.5	29	Smuts 1975c
Kasungu, Malawi	1978	All	Increasing	Height	35	24.5	8.5	7.5	24.5	Jachmann 1980
Luangwa, Zambia	1968–69	Females	Stable?	Dentition	18.5	13.5	12	8	48	Hanks 1972
Tsavo East, Kenya	1962–66	All	Stable?	Length	31	20	11	8	29.5	Leuthold 1976a
Tsavo East, Kenya	1974	All	Recovering	Length	11.5	19.5	17	10.5	41.5	Leuthold 1976a
Tsavo West, Kenya	1974	All	Recovering	Length	23.5	21	11	8.5	36.5	Leuthold 1976a
Murchison Falls south, Uganda	1966	All	Declining	Dentition	13	12	12.5	11	51.5	Laws et al. 1975
Mean, all populations					24.2	18.0	13.6	7.8	36.4	

adult male segment of the population (Croze 1972; Hall-Martin 1984; Hanks 1972; Laws, Parker & Johnstone 1975; Leuthold 1976b; Moss 1983; Smuts 1975c).

The life tables presented for the Murchison Falls, Uganda, population indicate an adult sex ratio of 73 males : 100 females. At Tsavo in Kenya, the population sex ratio was 80 males : 100 females. In the Kruger Park in South Africa, the sex ratio of animals up to 14 years (i.e. prior to the age at which males leave family groups) was 98 males : 100 females, while the sex ratio for the total population was estimated to be 95 males : 100 females. Since adults (> 15 years by Laws' method) made up 35% of the Kruger Park population, an adult sex ratio of 92 males : 100 females is indicated. This is based on the period 1968–74. In 1964 an adult sex ratio of 107 males : 100 females was recorded, suggesting a greater preponderance of bulls among the elephants initially colonizing this park. In the Kasungu Park in Malawi, the adult sex ratio was 55 males : 100 females, due largely to selective shooting of males in crop protection and by poachers (Laws 1969a, 1969b; Jachmann 1986; Pienaar, van Wyk & Fairall 1966b; Smuts 1975c).

Only limited data are available for Asian elephants. The proportion of adults in the Ruhunu Park in Sri Lanka was 65%. The adult sex ratio of the combined population sample from various parks in Sri Lanka is 35 males : 100 females, a difference related to greater dispersal and mortality among subadult males (McKay 1973; Santiapillai, Chambers & Ishwaran 1984).

Hippopotamus

Among 932 female hippos culled in Uganda and the adjoining lakeshore region of Zaire, 72.5% were sexually mature, 9% pubertal and 18.5% immature. In a sample of 585 hippos of both sexes shot in the Luangwa River in Zambia, 78% were aged over 10 years. Of 225 hippos culled in the Kruger Park in South Africa in 1974–75, 68% were adult (> 10 years). The Kruger Park population was believed to be increasing following earlier drought losses, while other populations were probably stable. However, young hippos are difficult to shoot and retrieve from the water, so that all samples overestimate the proportion of adults.

In the Uganda sample, the adult sex ratio was 72 males : 100 females ($N = 1421$) for animals culled along lakeshores, but 196 males : 100 females ($N = 740$) among animals culled in ponds away from the lake. At Luangwa, the adult sex ratio was 89 males : 100 females ($N = 312$) for animals shot during the first year of culling; but 63 males : 100 females ($N = 145$) among animals culled from the same section of river the following year, when less effort was expended on small groups in isolated pools. In the Kruger Park, animals culled from various rivers over 1974–77 showed a sex ratio of 52 males : 100 females ($N = 463$) among all age classes. The sex ratio of adults > 10 years was 41 males : 100 females ($N = 154$). Since males do not achieve full growth, and hence dominant breeding status, until at least 26 years old, the functional sex ratio was 25 males : 100 females ($N = 109$). However, the strongly female-based sex ratio in the Kruger Park sample was probably due in part to dispersal by a proportion of males into isolated pools and dams away from the rivers where shooting was carried out. Hippos shot in Kruger Park in 1964 under drought conditions, when animals were crowded in remaining river pools, showed an even sex ratio (Laws & Clough 1966; Pienaar, van Wyk & Fairall 1966; Marshall & Sayer 1976; Smuts & Whyte 1981).

Giraffe

The increasing giraffe population in the Serengeti Park in Tanzania contained 52% adults (> 5 years), while the Nairobi Park population in Kenya contained 56% adults. The adult sex ratio at Serengeti was 70 males : 100 females; at Nairobi Park it was 80 males : 100 females; while at Timbavati in South Africa it was 75 males : 100 females. Since the birth sex ratio is biased in favor of males, this indicates considerably higher mortality or dispersal by males relative to females (Foster & Dagg 1972; Hall-Martin 1975; Pellew 1983a).

Rhinoceroses

The Indian rhino population at Chitwan in Nepal contained 52% adults, 21% subadults and 26.5% juveniles. The sex ratio of adults was 62 males : 100 females. These proportions are based on functional age classes, i.e. females were classed as adult once they had given birth, while males were not classed as adult until fully grown at about 10 years (Laurie 1978).

Healthy black rhino populations in various parts of East Africa contain about 60% adults (fully grown, i.e. older than about 8 years), compared with 82% adults in the declining Hluhluwe Reserve population (Table 11.2). Adult sex ratios generally show an excess of males, except for the Tsavo and Serengeti populations.

White rhinoceros

The composition of the white rhino population in the Umfolozi–Hluhluwe complex, and its regional distribution during both wet and dry seasons, was determined from aerial surveys carried out in August 1970 and

Table 11.2. *Age and sex composition of black rhino populations*

Area	N	Adults %	Subadults %	Juveniles %	Adult sex ratio (♂s : 100 ♀s)	N	Reference
Ngorongoro, Tanzania	108	61	19	20	128	66	Goddard 1967
Olduvai, Tanzania	74	56	24	20	116	41	Goddard 1967
Serengeti, Tanzania	67	66	6	28	83	44	Frame 1980
Tsavo East, Kenya	531	64	13	23	98	338	Goddard 1970a
Amboseli, Kenya	72	61	—	—	126	44	Western & Sindiyo 1972
Hluhluwe, South Africa	199	82	15	3	110	163	Hitchins & Anderson 1983
Umfolozi–Corridor, South Africa	129	72.5	15.5	12	125	94	Hitchins & Anderson 1983

February 1971. All solitary animals were recorded as adult males, while additional adult males associated with cows and calves were distinguished by their larger size. All adults accompanied by calves were recorded as adult females. Groups including two or more similar sized animals were recorded as subadults. Some subadults that had joined a cow lacking a calf could have been recorded erroneously as juveniles; but in compensation some small juveniles could have been overlooked. Adult females associated with large subadults could not be distinguished from the latter.

To calculate regional densities, the total number of rhinos recorded in each of these surveys was compared with the total recorded in the 1970 helicopter census, and a correction factor derived from the ratio of the two totals (assuming the helicopter results to be unbiased). Regional population totals were then corrected by this factor. In addition, a correction was applied to the February 1971 data for the 82 white rhinos known to have been caught and removed between August 1970 and February 1971.

Further data on population structure were available from the mean composition of the white rhino population inhabiting the study areas, and from the animals recorded during foot surveys of surrounding areas. These records are unbiased, but cannot be compared directly with the aerial count data because of the rhino removals that took place in the interim between the ground and aerial surveys.

Consistent regional differences in population structure were evident from the two aerial surveys, and were supported by ground samples (Tables 11.3, 11.4 and 11.5). The proportion of calves tended to be highest in the western section of Umfolozi between the two Umfolozi rivers, where the population density was greatest. The proportion of adult males was highest to the south of the White Umfolozi River, and in the Corridor to the north of the Black Umfolozi River, where prevailing population densities were somewhat lower. These represent regions into which the population had expanded from its original concentration between the two Umfolozi rivers.

The structure of the total white rhino population occupying the Umfolozi–Hluhluwe complex in 1969 can be reconstructed by (i) adding the 241 white rhinos removed between mid-1969 and late 1970 to the aerial count total in their respective age/sex classes; (ii) assuming that the aerial count figures overestimate the proportion of adult males by an amount of 1%, equal to the proportion of lone cows plus solitary subadults in the study area populations; (iii) assuming that the proportion of calves recorded in the aerial surveys underestimates the true proportion by an amount of 1%, perhaps due to some small infants being overlooked; (iv) assuming that the cow : calf ratio recorded in study area populations (86 calves : 100 cows,

Table 11.3. *Population composition and distribution of white rhinos in Umfolozi Game Reserve and the adjoining Corridor*

Region	Area (km²)	August 1970					February 1971				
		Count	Corrected total	Density (km⁻²)	Ad♂ %	calf %	Count	Corrected total	Density (km⁻²)	Ad♂ %	calf %
1. *Umfolozi Game Reserve:*											
(a) Western section between Umfolozi Rivers	248	467	865	3.5	21.8	21.0	380	945	3.8	19.2	21.1
(b) Eastern section between Umfolozi Rivers	113	117	217	1.9	17.2	18.0	76	173	1.5	18.6	19.9
(c) South of White Umfolozi River	94	120	222	2.4	29.2	20.8	114	259	2.8	28.0	19.3
2. *Southern Corridor*	216	157	291	1.4	23.1	22.4	95	216	1.0	29.5	20.0

Table 11.4. *White rhino population density and structure in the study areas*

Area	Period	Extent (km²)	Aver. popul.	Popul. density (km⁻²)	Population composition				Biomass (kg km⁻²)
					Ad♂ (%)	Ad♀ (%)	Subadult (%)	Calf (%)	
1. Western section of Umfolozi Game Reserve:									
(a) Madlozi	Wet season 1968–71	8.9	50.7	5.7	16.5	31.7	26.2	25.6	7478
	Dry season 1969–70	8.9	40.9	4.6	17.8	30.6	30.6	20.8	6175
(b) Nqutsheni	Jan.–Feb. 1969	5.1	35.4	7.0	23.3	27.5	21.9	27.2	9496
(c) Gqoyini	Mar–Apr 1969	7.5	38.7	5.1	20.9	22.8	34.4	22.0	6808
Three western study areas combined		21.5	123.0	5.7	19.8	27.6	27.5	25.0	7579
2. Eastern section of Umfolozi Game Reserve:									
Dengezi	May 1969	6.1	(15.1)	(2.5)	15.1	27.6	31.0	26.4	3183
3. Hluhluwe Game Reserve:									
North-eastern corner	April 1969	66	30	0.5	16.7	6.7	70.0	6.7	643

Notes: Aver. popul. = average population size.
Popul. density = population density.

Table 11.5 *White rhino population structure from ground samples outside the study areas*

Area	Period	N	Ad♂ (%)	Ad♀ (%)	Subadult (%)	Calf (%)
1. Western section of Umfolozi Game Reserve:						
(a) Madlozi environs	Aug–Sept 1969	138	16.0	22.5	43.5	18.1
(b) Nqolothi–Mfulumkhulu	Oct 1969	24	—	—	—	—
(c) Nqutsheni environs	Jan 1969	45	13.3	31.1	26.7	28.9
(d) Gqoyini environs	Mar 1969	18	—	—	—	—
Combined, Umfolozi West		225	16.0	25.8	37.3	20.9
2. South of W. Umfolozi River:						
	Feb 1971	61	27.8	28.5	21.3	21.3

Table 11.6. *Estimated overall composition of the white rhino population in the Umfolozi–Corridor–Hluhluwe complex, as of early 1969*

Upper figure is estimated percent from population samples.
Lower figure is calculated total number of rhinos.

Region	Population size	Ad♂	Ad♀	Subad.	Calf
1. Umfolozi Game Reserve:					
(a) West		18	27	32	23
	1100	198	297	352	253
(b) East		16.5	23	40	20
	177	30	41	72	35
(c) South		27.5	24	27.5	21
	259	71	63	71	54
2. Corridor		25.5	25.5	27	22
	303	77	77	82	67
3. Hluhluwe Game Reserve:					
(a) South		25.5	25.5	27	22
	42	11	11	12	9
(b) North		16	7	70	7
	37	6	3	26	3
Grand total	1918	393	492	615	421
percent		20.5	25.6	32.1	21.9

$N = 461$) applies to the whole population. The reconstructed 1969 population in the whole complex contains 20.5% adult males, 25.6% adult females, 32.1% subadults and 21.9% calves (Table 11.6). However the subpopulation in the western section of Umfolozi, including nearly 60% of the total, included only 18% adult males, with 27% adult females, 32% subadults and 23% calves. In other regions, adult males equalled or outnumbered adult females.

Prior to early 1969 most rhino removals had taken place from outside the boundaries of the Umfolozi Game Reserve. Thus the composition of the reconstructed 1969 population can be regarded as unaffected directly by the capture operations, though influenced by the age structure of those animals dispersing out of the game reserve.

The adult sex ratio in the reconstructed 1969 population is 80 males : 100 females ($N = 684$, based on the total population sample recorded in the two aerial surveys). The adult sex ratio observed in the western study areas was 72 males : 100 females (Table 11.4); while in the population sample from surrounding areas it was 62 males : 100 females ($N = 94$, Table 11.5). This is the functional sex ratio of socially mature males (aged 10–12 years and over) to reproductively active females (aged about 7 years or more).

Table 11.7. *Sex ratio of immature age classes of white rhinos*
Based on number of different individuals recorded in various population samples.

Area	Calves (1–3 yrs)			Subadults Class I (3–6 yrs)			Subadults Class II (6–9 yrs)		
	N	♂	♀	*N*	♂	♀	*N*	♂	♀
Madlozi, Jan 1969	31	15	16	32	16	16	18	11	7
Nqutsheni, 1969	30	11	19	21	9	12	23	20	3
Gqoyini, 1969	18	10	8	28	14	14	5	4	1
Dengezi, 1969	15	10	5	14	7	7	6	2	4
UGR west environs, 1969	19	9	10	13	8	5	9	6	3
UGR south, 1971	3	2	1	8	3	5	4	4	0
Combined	116	57	59	116	57	59	65	47	18
Ratio ♂:♀		97:100			97:100			261:100	

The observed sex ratio among calves and subadults under 6 years was 97 males:100 females (Table 11.7). Among older subadults there was a preponderance of males, due to the delayed sexual maturity of males relative to females. Taking into account the observed proportions of subadults in the population, the sex ratio of all animals over 6 years of age is calculated to be 98.5 males:100 females.

Population samples recorded subsequent to my study showed that a male-biased sex ratio had spread into older age classes. A horseback survey during 1974 yielded a sex ratio for adults plus subadults (> 4 years) of 154 males:100 females (Brooks 1974).

Summary

The proportion of reproductively mature adults is typically about 60–70% in stable populations of megaherbivores, compared with 45–55% in increasing populations. African elephant populations seem to contain a somewhat higher proportion of immatures, and hippo populations a higher proportion of adults, than other species. However, this variability could be due partly to varying interpretations of the ages at which males and females attain reproductive maturity.

The adult sex ratio tends to be even in elephants and white rhinos, except where modified by differential dispersal or shooting. Black rhino populations commonly show an excess of adult males; while in giraffe and hippo the adult sex ratio tend to be female-biased. However, in both hippo and white rhino some males disperse into less suitable habitats, leaving a female-biased sex ratio among the adults remaining in favored localities.

Population growth

Population growth rate can be estimated in two ways: (i) by repeated censuses over a period of time; (ii) by extrapolation from known fecundity and mortality rates and population structure, either by empirical calculation or from a computer model. In the absence of more precise evidence, a rough indication of the rate of population increase can be obtained from the fraction formed by calves in the age range 0–1 years, relative to the population segment older than one year (assuming negligible adult mortality). However, where the offspring are born during a short season, this proportion declines over time following the birth pulse, because infant mortality tends to be highest shortly after birth. The best estimate of recruitment is that obtained shortly before the next birth season, after the mortality of sub-yearlings has dropped to levels similar to that of the post-yearling segment. As populations grow, animals may redistribute by dispersing into previously unoccupied areas. Thus a distinction needs to be made between the overall rate of expansion of the whole population, and the rate of density increase in different sections of its range.

Elephants

Few African elephant populations exist at stable densities. At one extreme is the small population of about 100 elephants in the Addo Park in South Africa, which maintained a growth rate of almost 7% per annum between 1954 and 1979. Since 25 years would be insufficient time for a stable age distribution to be attained, this rate of increase may be a little higher than the sustained r_{max}. At the other extreme is the population in the Murchison Falls Park in Uganda, which had evidently declined from about 22 000 animals in 1946 to 8000 in 1971. This decline was not reflected in density because of the compression effect of surrounding human disturbance. At Manyara in Tanzania, the increase rate shown by the family units was 3.7% per annum, but mortality due to shooting removed about 2% of the population annually. The population in the Ruaha Park in Tanzania showed an increase rate of 9% per annum between 1965 and 1977, but intrinsic increase was augmented by immigration from surrounding regions.

The Kruger Park population in South Africa has been maintained at about 7500 animals since 1968 by annual culling. The natural rate of increase can be calculated by taking account of animals removed from the population. A total of 9044 elephants was culled between 1967 and 1980, yielding a rate of population growth of 10% per annum. However, the population was augmented by immigration from Mozambique prior to the completion of a border fence in 1974. The number of elephants removed over the period 1974–84 indicates a population growth rate of 7.5% per annum, but some immigration from Mozambique may have continued despite the fence.

The proportion of calves less than one year old in the Kruger Park population averaged 7.0% over a ten year period, indicating the approximate upper bound to the rate of increase in the absence of immigration. The proportion of sub-yearling calves recorded in other elephant populations varies between 6% and 8.5% based on long term averages. However, population samples from single years may show calf proportions varying between 2.4% and 10.4%, due to the effects of varying rainfall on short term fertility (Barnes 1983; Croze 1972; Douglas-Hamilton 1972; Eltringham 1977; Hall-Martin 1980, 1984, unpublished reports; Hanks 1972; Laws, Parker & Johnstone 1975; Lewis 1984; Sherry 1975).

A computer simulation by Hanks & McIntosh (1973) suggested a maximum sustained rate of increase by African elephants of about 4% per annum. This is based on a 3.5 year birth interval, 5% per annum juvenile mortality, 1% per annum mortality of prime adults, and puberty at 12 years. However, if females first conceived at 8 years (as recent evidence suggests), and the mortality of prime adults was only 0.5% per annum, a maximum population growth rate of slightly over 6% per annum could be sustained, based on my own simulation modelling.

The colonization history of the Kruger Park reveals the importance of dispersal as a population process. In 1905 the total population was estimated to be only 10 elephants, restricted to the extreme north. By 1959 the estimated elephant total was still under 1000. At about this time the first breeding herds appeared in the southern section of the park. The first helicopter census carried out in 1964 revealed 2374 elephants, with only 61 elephants, mostly males, in the southern district. In 1967, 6586 elephants were counted and culling commenced. By 1983 the population in the southern district numbered over 1200, including breeding herds as well as males. Most of the population increase was due to immigration from neighboring Mozambique (Hall-Martin 1984).

Park populations of elephants in other parts of Africa, e.g. Murchison Falls in Uganda and Tsavo in Kenya, have also been augmented by immigration, due largely to human settlement and hunting in adjoining regions. Elephants first appeared in the Serengeti Park in Tanzania in 1957, and increased to over 2000 by 1965 (Lamprey *et al.* 1967; Laws 1969b; Laws, Parker & Johnstone 1975).

Among Asian elephant populations, juveniles less than a year old formed
6.5% of the combined sample recorded in the Gal Oya, Lahagula and Yala
Parks in Sri Lanka. Thus their potential rate of population increase appears
closely similar to that shown by African elephant (McKay 1973).

Hippopotamus

The hippo density per kilometer of river in the North Luangwa
Park in Zambia increased from 3.8 in 1950 to 22 in 1972, while that in the
South Luangwa Park increased from 5 to 11 over the same period. These
population changes suggest annual growth rates of 11% and 6% per annum
respectively. In the Queen Elizabeth Park in Uganda, the proportion of
calves less than a year old was 6% in 1958 when the population density was
very high, but increased to 14% in 1966 following large-scale population
reduction. Thus it appears that the potential rate of population increase of
hippos exceeds 10% per annum.

At Luangwa an attempt was made to eliminate all hippos from a 24 km
section of river. A total of 652 hippos was shot, but 200 hippos moved in to
recolonize this section during the following rainy season. The majority of
the dispersing animals were adult females (Laws 1968b; Marshall & Sayer
1976).

Giraffe

The giraffe population in the Serengeti Park in Tanzania maintained a growth rate of 6% per annum over the five year period 1971–76.
Juveniles less than one year of age formed 15.5% of this population,
compared with 11% in the Nairobi Park population. In the Kruger Park
population the proportion of juveniles under a year old recorded in dry
season surveys was 5.2% when the population was declining, but 7.1% the
following year when a population increase was recorded (Mason 1984,
1985b; Pellew 1983a).

Rhinoceroses

The Indian rhino population at Chitwan was estimated to be
increasing at a rate of perhaps 6% per annum, based on the population
structure, calving rate and estimated mortality (Laurie 1978).

For the black rhino populations at Ngorongoro and Olduvai in Tanzania, recruitment based on calf proportions was estimated to be 7% per
annum over the period 1962–66. At Tsavo in Kenya, subyearling calves
formed 9.0% of the population; while at Masai Mara this proportion was
7%. Black rhinos introduced into the Kruger Park have maintained a
steady rate of increase of 9% per annum.

East African black rhinos are reportedly slow to disperse into new areas. However, the Corridor and Umfolozi Game Reserve have been steadily colonized by black rhinos dispersing from the adjoining Hluhluwe Reserve (Goddard 1967, 1970a; Hall-Martin 1986; Mukinya 1973; Hitchins & Anderson 1983).

White rhinoceros

Censuses of the white rhino population in the Umfolozi–Hluhluwe complex were conducted annually or every other year by the Natal Parks Board from 1959 onwards. Initially these were carried out by fixed-wing aircraft, but from 1970 a helicopter was used. Prior counts were carried out on foot using large numbers of beaters, the first being organized in 1929 by a visiting American zoologist, Herbert Lang. Further ground counts were conducted by the veterinary department in 1932 and 1936. There were also two early aerial counts in 1948 and 1953.

The standard technique used in the fixed-wing aircraft surveys was to fly parallel strips 1100 m apart. A four man team was used, including a pilot, recorder and two observers. Complete coverage of the game reserve area and its immediate environs required about three days (Vincent 1969). Helicopter surveys entailed intensive low altitude searching of successive blocks. Ground counts used lines of beaters walking in parallel. This technique was retained for the northern hilly section of Hluhluwe Game Reserve, where aerial coverage was difficult.

All of these survey techniques have inherent errors, which must be reconciled before the rate of increase of the white rhino population can be established. In 1970 a survey by fixed-wing aircraft carried out a week after the first helicopter census recorded only 54% of the number of rhinos obtained from the helicopter. However, the total obtained from this fixed-wing aircraft survey was also lower than that obtained in the fixed-wing censuses over the previous two years. The totals recorded in 1968 and 1969 averaged 58% of the mean number of white rhinos counted by helicopter in 1970 and 1971. Allowing for some natural increase over this period, it will be assumed that censuses by fixed wing aircraft account on average for 60% of the number of white rhinos recorded in helicopter surveys. It will furthermore be assumed that any tendency to undercount in helicopter censuses is negligible. Melton (1978) found that flying speed in helicopter surveys influenced the number of medium-sized ungulates counted, but not the white rhino totals. Since ground counts were accomplished with large numbers of beaters and covered most of the area inhabited by white rhinos at the time, it is unlikely that many rhinos were missed, so they will be assumed to be 100% efficient.

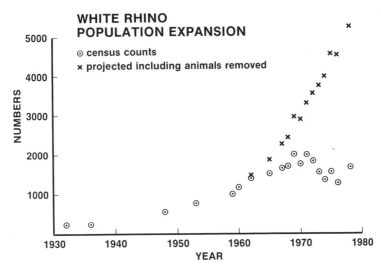

Fig. 11.1 Expansion of the white rhino population in the Umfolozi–Corridor–Hluhluwe complex, showing also the projected increase when animals removed from the population are included.

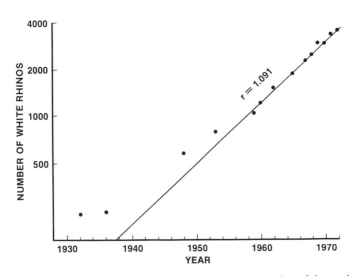

Fig. 11.2 Logarithmic plot of the projected expansion of the total white rhino population up to 1972.

Table 11.8. *Estimates of rates of increase by the white rhino population in the Umfolozi–Hluhluwe complex over different time periods*

Census method	Period	Instant. increase rate r	Std. error for r	Finite increase rate R (% p.a.)	95% confidence limits for R
All aerial censuses	1960–72	0.0917	0.0032	9.6	8.9–10.2
Early ground and air counts	1932–59	0.0611	0.0043	6.3	5.5–7.1
Fixed-wing aircraft counts	1965–69	0.1119	0.0126	11.8	9.4–14.3
Helicopter counts	1970–78	0.0622	0.0101	6.4	4.3–8.6

Allowance also needs to be made for the white rhinos translocated from the population from 1962 onwards, and for the potential recruitment to this segment. Details of the animals removed and the areas from which they were taken are documented in Natal Parks Board files. A higher proportion of subadults was removed than present in the overall population. Nevertheless it will be assumed that this segment would have increased at a rate similar to that shown by the overall population, i.e. at about 10% per annum.

The growth of the white rhino population in the Umfolozi–Hluhluwe complex is shown in Fig. 11.1, together with its projected expansion including animals removed. A logarithmic plot of the projected expansion shows a constant exponential rate of increase of 9.6% per annum over the period 1960–1972 (Fig. 11.2). Prior to 1959, the rate of increase was somewhat lower at about 6.3% per annum. Helicopter counts carried out over the period 1970–1978 also indicate a rate of increase of 6.4% per annum (Table 11.8). However, over this latter period there had been considerable disruption of the population by the greatly stepped-up capture and removal operations.

The proportion of calves less than a year old recorded in fixed-wing aircraft surveys averaged 9%, but very small calves are easily missed. Helicopter censuses over 1972–74 yielded between 10% and 14% subyearling calves, the higher figures being obtained after heavy disturbance of the population structure. A horseback survey in 1974 showed a proportion of 10% to be calves under a year old (Brooks 1974). These figures suggest a rate of increase of 8–9% per annum, allowing for a mortality rate among animals older than a year of about 2% per annum (Chapter 8).

Redistribution of the population as it expanded is evident from regional breakdowns of white rhinos recorded in different sections of the reserve complex and environs, with allowance for animals removed. Between 1953 and 1962, prior to any fencing, the core population between the two Umfolozi rivers increased by 58%, while the number of white rhinos south of the White Umfolozi River increased three-fold (Table 11.9). In 1967, 67 white rhinos were counted outside the southern boundary fence of the Umfolozi reserve, which after correction for undercounting bias indicates a total of about 120 animals. However, 274 white rhinos were removed from this region between 1967 and 1970. Two young males first appeared in the northern section of the Hluhluwe Reserve in 1961, while in 1967 ground counts indicated a total of 106 white rhinos in this region.

The magnitude of the dispersal rate can be estimated by comparing the rate of increase shown by the population core with that of the whole population, assuming no differences between regions in rates of birth or death. A population dispersal rate outwards from the between rivers section of Umfolozi of 2.4% per annum between 1953 and 1962 is indicated; and from the western section of 3.5% per annum between 1967 and 1970 (Table 11.9).

The age–sex categories of the animals dispersing is shown by the Natal Parks Board records of the animals removed from outside the boundary fence of the Umfolozi reserve between 1968 and 1973. Seventy-five percent of these were classified as adults or subadults over 4 years of age, 21% as immatures aged 1–4 years, and only 4% as calves under a year old. Of the adults plus subadults, 66% were males, while only 50% of the immatures plus calves were males. Of the 30 animals that I recorded in the north-eastern section of Hluhluwe in 1969, 19 were subadults aged 6–10 years, two-thirds of these male. There were two adolescents aged about 3 years, a male and a female. Also present were five adult males, and two females with calves. Both of the latter were primaparous, having given birth only after settling in the area (P. M. Hitchins personal communication).

If 80% of the animals dispersing are subadults (based on the Hluhluwe north sample), which make up about a third of the total population, the population dispersal rates calculated above represent a specific emigration rate of 7.5% per subadult per year.

The demographic parameters of the total white rhino population in the Umfolozi–Hluhluwe complex in 1969–70 are summarized in Table 11.10. These parameters may be used to formulate an age–class structured model of the population (see Appendix II). The simulation was started with a small nucleus of 29 animals, representing the number of white rhinos that supposedly occurred in the Umfolozi reserve at the time of its proclamation in

Table 11.9. *Rates of population expansion in different sections of the complex*

Region	1953 count × 1.67	1962 count × 1.67	Removals between counts[a]	Projected 1962 popul. incl. remov.	% increase over 9 years	Incr. rate % p.a.	1967 count × 1.67	1970 helic. count	Removals between counts[a]	Projected 1970 popul. incl. remov.	% increase over 3 years	Incr. rate % p.a.
UGR west	—	—	—	—	—	—	857	910	85	995	116	5.1
UGR between rivers	542	794	60	854	158	5.2	960	1103	86	1189	124	7.4
UGR south, inside fence	—	—	—	—	—	—	246	258	6	264	107	2.3
UGR south, outside fence	—	—	—	—	—	—	118	0	274	274	232	32.4
UGR south, combined	115	347	0	347	302	13.1	364	258	280	538	148	14.0
Corridor	72	222	0	222	308	13.3	248	313	0	313	126	8.0
HGR	33	52	0	52	158	5.2	92	90	0	90	98	—
Whole complex	762	1415	60	1477	194	7.6	1664	1764	366	2130	128	8.6

Notes: [a] Derived pro rata from data for numbers of animals removed per calendar year, from Brooks (1974).
Popul. incl. remov. = population including removals.
Incr. rate = rate of increase.
Helic. count. = helicopter count.
UGR = Umfolozi Game Reserve.
HGR = Hluhluwe Game Reserve.

Table 11.10. *Summary of the demographic parameters of the Umfolozi white rhino population as estimated for 1969–1970*

| | Adults | | Subadults | | | Calves | |
| | | | Post-pubertal | Pre-pubertal | | | |
	♂	♀	♂	♀		Juvenile	Infants
Age range (years)	10–45	7–45	6–10	6–7	2.5–6	0.5–2.5	0–0.5
Proportion of population (%)	20.5	25.6	7.7	3.0	21.4	21.9	
Annual mortality (%)	3.0	1.2		3.0		3.5	8.3
Annual natality (%)	—	40.0	—	0	—	—	—
Annual dispersal (%)	3.0	0		7.5		0	

1897. After 70 annual iterations, the size of the simulated population had grown to 12 500 animals, and its rate of increase had stabilized at 9.0% per annum (allowing no dispersal).

The model population stabilized in age structure with reproductively active adults constituting 43.5% of the total. This is slightly lower than the proportion of adults estimated for the real 1969 population (46%). The discrepancy can be accounted for on the basis that the real population had not attained a stable age distribution, since its rate of increase had been lower prior to 1959. This difference in age structure would explain half the difference between the rate of increase shown by the simulated population, and that recorded for the real population, over the period 1960–72.

In the simulated population, animals in the terminal age class of 45 years formed only 0.3% of the total population. This emphasizes one factor responsible for the low mortality rate recorded in the real population – the small number of animals reaching senility. Individuals dying of senescence in 1970 would have been born during the 1920s, when the total population numbered only about 200 animals. If 20 calves had been born in 1925, and had experienced the mortality schedule shown in Table 11.10, only 7 would remain to die of old age in 1970. This may be compared with the total mortality of 18 adults estimated for the real 1969 population. This suggests that the mortality rate of prime adult females was only about 0.5% per annum, and that of prime-aged males about three times as high.

Summary

Available data suggest the following maximum sustained rates of population growth for megaherbivores: (i) elephants – about 6.5% per annum; (ii) rhinoceroses – about 9% per annum; (iii) hippopotamus – about 11% per annum; (iv) giraffe – up to about 12% per annum, depending on predation levels. Higher rates of density increase may be shown where births are augmented by immigration, and where the population age distribution has not stabilized. The rate of population growth by white rhinos at Umfolozi apparently increased with increasing density, until disrupted by large-scale removals.

Population density and biomass

The population densities attained by different large herbivore species indicate their relative success in converting plant biomass into animal biomass in particular habitats. Regional densities assessed over geographic areas, including zones where the species may be absent, need to be differentiated from local (or ecological) densities, based on occupied habitats. Furthermore short-term concentrations should be distinguished from the year-round mean densities. To compare species of different size, numerical densities need to be converted into biomass units. If a wide range in body size is represented, it is more meaningful to make comparisons in terms of metabolic mass equivalents, i.e. body mass to the power three-quarters. The metabolic mass equivalent is an index of the total energy turned over through the population.

Elephants

Elephant population density varies widely in different parks in Africa (Table 11.11). Highest regional densities of over 3 elephants per km^2 (over 6000 kg km^{-2}) occurred formerly in the Luangwa Valley in Zambia and in the North Bunyoro region of Uganda. However, in the North Bunyoro region the density had been raised by compression from surrounding human settlement, and the elephant population was in a stage of decline. Local density was 5.5 per km^2 (about 11 000 kg km^{-2}) at Manyara in Tanzania, and over 4 per km^2 in the Addo Park in South Africa. Both of these populations were still increasing. On the alluvial soils bordering the Luangwa River, elephants attained densities of 7 per km^2 during the wet season. Dry season densities along the Chobe river frontage in Botswana also reach about 7 elephants per km^2. Highest elephant densities are

Table 11.11. *African elephant population density and biomass in different regions*

Area	Extent (km²)	Period	Density (km⁻²)	Biomass (kg km⁻²)	Reference
Luangwa Valley, Zambia	40 000	1973	2.15	4515	Caughley & Goddard 1975
South Luangwa Park	12 500	1973	3.35	7035	Caughley & Goddard 1975
Luangwa River alluvium	4170	1973	7.4	15 540	Caughley & Goddard 1975
Manyara Park, Tanzania	82	1970	5.5	11 550	Douglas-Hamilton 1972
Addo Park, South Africa	23	1979	4.1	8610	Hall-Martin 1980
North Bunyoro, Uganda	2727	1969	3.43	7203	Laws *et al.* 1970
Ruaha region, Tanzania	31 500	1977	1.39	2912	Barnes & Douglas-Hamilton 1982
Ruaha Park	10 300	1977	2.4	5040	Barnes & Douglas-Hamilton 1982
Msembe area, Ruaha River	130	1977	4.11	8631	Barnes & Douglas-Hamilton 1982
Queen Elizabeth Park, Uganda	1559	1972	1.4	2940	Eltringham 1977
Hwange, Zimbabwe	14 620	1980	1.33	2800	Cumming 1983
Tsavo Park, Kenya	20 000	1967	1.2	2520	Laws 1969b
Chobe region, Botswana	16 000	1985	1.1	2300	Work 1986
Sengwa Research Area, Zimbabwe	353	1984	0.9	1980	Cumming 1975
Kasungu Park, Malawi	2450	1978	0.5	1050	Jachmann 1980
Kruger Park, South Africa	19 000	1981	0.4	840	Hall-Martin 1984
Kafue Park, Zambia	20 000	1980	0.25	520	Bell 1982
Tai Park, Ivory Coast	3400	1983	0.23	480	Merz 1986
Amboseli region, Kenya	3000	1980	0.2	420	Moss 1983
Etosha Park, Namibia	17 000	1978	0.15	320	Berry 1980
Serengeti region, Tanzania	25 000	1972	0.1	210	Croze 1974

associated with well-wooded savannas or shrublands with intermediate rainfall (450–1200 mm). Elephant densities are relatively low in open grassy savanna, e.g. Serengeti in Tanzania, and in wooded savanna with higher rainfall, e.g. Kafue National Park in Zambia. Regional elephant densities in forest habitats in West Africa vary between 0.1 per km² and 0.33 per km², but have been depressed by past hunting.

Asian elephants exhibit densities of 0.8 per km² (1484 kg km^{-2}) in the Kaziranga Sanctuary in India, and about 0.5 per km² in Gal Oya in Sri Lanka. In Lobis-Bekok, Malaysia, elephant densities are 0.27 per km² in primary forest and 0.65 per km² in secondary forest. However, these populations have been disturbed by past hunting (Eisenberg & Seidensticker 1976; Ishwaran 1981; McKay 1973; Olivier, cited by Merz 1986a).

Hippopotamus

The effective ecological densities attained by hippo populations are generally calculated assuming that the grazing range extends 3.2 km from the river or lake margin. Hippo densities estimated in this way averaged 18 per km² (18 000 kg km^{-2}) along the shorelines of Lakes Edward and George in Uganda and Zaire during the early 1960s. Local concentrations reached effective densities of up to 31 hippos per km² (31 000 kg km^{-2}). Under such densities grassland degradation was occuring. Densities averaged 19 per km² along the 88 km section of the Nile River above Murchison Falls in Uganda, with a peak of 26 per km² shown over an 8 km section. Along the Semliki River in Zaire, the hippo density was over 10 per km². Along the Luangwa River floodplain in Zambia, the effective hippo density was about 8 per km² (8000 kg km^{-2}), without grassland deterioration. Along various rivers in the Kruger Park in South Africa, hippo densities averaged 1.1 per km² (Laws 1968b, 1981; Laws, Parker & Johnstone 1975; Pienaar *et al.* 1966a; Naylor *et al.* 1973; Verheyen 1954).

Giraffe

Giraffe in the Serengeti National Park in Tanzania attained a regional density of 1.2 per km², with local densities varying between 1.4 and 2.6 per km² (850–1600 kg km^{-2}). This population was still increasing. In the Timbavati Reserve in South Africa, the giraffe population had stabilized at a density of 2.6 per km². Census figures for the adjoining central district of the Kruger Park indicate a regional giraffe density of about 0.7 per km², but probably underestimate the true population. In Tsavo East in Kenya, the regional giraffe density is 0.2 per km² (Hirst 1975; Leuthold & Leuthold 1978; Pellew 1983a).

Rhinoceroses

The density of Indian rhinos in Chitwan National Park in Nepal averaged 0.3 per km² over the total area of 907 km², about 70% of which was unsuitable habitat. Highest local densities reached 4.8 per km² (5750 kg km⁻²). In the Kaziranga Park in India, Indian rhino density averaged 1.5 per km² over an area of 400 km² (Laurie 1978).

Black rhinos in Tsavo East in Kenya formerly had a mean density of 0.3 per km², with the highest local density reaching 1.6 per km² (1200 kg km⁻²) over an area of 544 km². The black rhino population in the Hluhluwe Reserve (215 km²) peaked at 1.4 per km² in 1961, but declined to 0.9 per km² by 1980. Elsewhere black rhino densities are considerably lower: 0.3 per km² at Ngorongoro, 0.17 per km² at Olduvai, 0.04 per km² at Serengeti (all in Tanzania), 0.14 per km² at Mara in Kenya, and 0.01 per km² at Etosha in Namibia (Berry 1980; Frame 1980; Goddard 1967, 1969; Hitchins & Anderson 1983).

White rhinoceros

Based on the results of helicopter censuses, the total white rhino population in the Umfolozi–Corridor–Hluhluwe complex in 1970–71 was 2000 animals over an area of 940 km², i.e. a regional density of 2.1 per km² (2800 kg km⁻²). In Umfolozi Game Reserve alone, 1530 white rhinos occurred at a mean density of 3.2 per km² (4300 kg km⁻²). Densities in the Corridor averaged 1.4 per km², and those in Hluhluwe 0.4 per km².

Within the study areas, local population densities and structure were estimated using the 'Territory Occupancy Index'. Since dominant territorial bulls were always present within the bounds of their own territories, they represented known animals occupying discrete areas. Populations of other animals within the territories were estimated from ratios of sightings compared with those of the territorial bull. Calculations are analogous to those made in the mark–recapture method of population estimation (see Caughley 1977). In the three study areas located in the western section of Umfolozi, the local population density of white rhinos averaged 5.7 animals per km², equivalent to a biomass of about 7600 kg km⁻² (Table 11.4).

Summary

African elephants and white rhinos attain local population densities exceeding 5 animals per km² in favorable habitats. These are below equilibrium densities, since populations were still increasing. Elephant biomass may exceed 6000 kg km^{-2} on a regional scale, and 10 000 kg km^{-2} on a local scale; while white rhinos at Umfolozi attained biomass levels of over 4000 kg km^{-2} regionally and 7500 kg km^{-2} locally. Indian rhinos show local densities approaching those of white rhinos. Hippos attain ecological densities as high as 10–20 per km^2, equivalent to a biomass of 10 000–20 000 kg km^{-2}, at which levels populations appeared to be stable. Population levels attained by browsing black rhinos and giraffe are considerably lower, up to 1500 kg km^{-2}.

Comparisons with smaller ungulates

The age composition of megaherbivore populations is similar to those typical of populations of medium-sized ungulates, if age classes are defined in functional terms. In stable populations, socio-sexually mature adults generally form 60% or more of the population; while in rapidly increasing populations adults make up 50% or less of the population. Most medium-sized ungulates show an adult sex ratio bias in favor of females. While local populations of elephants and rhinos may have an excess of females, the overall sex ratio calculated over the complete distribution range tends to be close to parity.

Megaherbivore populations can sustain maximum rates of growth of 6–12% per annum. For medium-sized ungulates breeding annually, population growth rates can be as high as 25–30% per annum, if predation is negligible. However, predation normally slows the population growth of medium-sized ungulates, and also giraffe, while other megaherbivores are limited only by food availability.

Under favorable conditions, medium-sized ungulates may attain biomass levels comparable to those of megaherbivores. For example, wildebeest attained a regional biomass exceeding 7000 kg km^{-2} in the Serengeti (Sinclair 1985), while the local biomass of African buffalo at Manyara is almost 10 000 kg km^{-2} (Mwalyosi 1977). However, biomass levels exceeding 2000 kg km^{-2} are exceptional for medium-sized ungulates, but typical of megaherbivore populations allowed to reach saturation densities. Differences in metabolic mass equivalent are less marked, due to the decrease in metabolic rate per unit of body mass with increasing body size. For equivalent biomass levels, megaherbivores are represented by far fewer animals per km^2 than is the case for medium-sized ungulates.

12

Community Interactions

Introduction

In this chapter I consider relations between megaherbivore populations and other species, both plant and animal. How different would the community be, in its species composition, habitat physiognomy, productivity and other ecosystem level features, if megaherbivores were absent? The kinds of interactions of importance include (i) the predatory and disturbing impacts of herbivores on vegetation, (ii) competitive and mutualistic interactions among herbivore species, and (iii) the consequences of changes in species representation for ecosystem structure and function.

Impact on vegetation

Larger herbivores exert a direct impact on vegetation by their consumption of plant parts, and by breaking or trampling plants. Plants are damaged by the removal of leaves, bark and other parts, by breakage of branches, which depresses growth, and through being felled or uprooted, causing whole plant mortality. Even low levels of leaf loss may make certain species less successful in competition with other plants, or hold them in the height range within which they are vulnerable to fires. Bark removal may allow attack by wood-boring beetles and pathogens, increasing fire susceptibility. Felling may be fatal to certain plants, but not to others able to regrow by coppice regrowth of the stump. Selective herbivory could suppress the environmental dominance of certain species, providing opportunities for other species to increase. Heavy utilization of all or most species may depress plant biomass and hence primary production. If only the herb layer is heavily impacted, this reduces fire frequency, thereby favoring woody plant invasion. Removal of the ground cover makes soils vulnerable to compaction and erosion, reducing resources for plant regrowth, and so may lead to progressive decline in overall plant abundance. Progressive denudation of the herbaceous cover is interpreted as a symptomatic of 'overgrazing'.

Elephants

African elephants feed on plants by plucking grasses, forbs and creepers, frequently uprooting them; by stripping leaves, fruits, twigs or bark from woody trees and shrubs; by breaking branches off to facilitate consumption of edible parts; and by pushing over or uprooting trees and shrubs, sometimes but not necessarily always to gain access either to shoots in higher levels of the canopy or to roots. The consequences of this utilization vary depending on the species selected, the levels of damage inflicted, season at which it occurs, elephant densities, the period over which impact is sustained, and the interactive effects of other factors such as fire, rainfall and soil properties.

Changes in the species composition of woody vegetation brought about by elephant feeding have occurred where direct or indirect mortality was inflicted on certain favored species. In the Sengwa Research Area in Zimbabwe, the dominant tree and shrub species in the riverine woodland were originally *Acacia tortilis* and *Grewia flavescens*. As elephants increased in numbers, these two species declined in representation, due to felling and debarking in the case of *A. tortilis*, and uprooting in the case of *G. flavescens*. At Lake Manyara in Tanzania, mature *A. tortilis* trees were being killed by elephants at a rate of 8% per annum, and saplings at a rate of 3% per annum, over 1975–79; but recruitment by surviving saplings was regarded as adequate to maintain the tree population. Heavy culling of elephants and impala in the Nuanetsi region of Gonarezhou National Park, also in Zimbabwe, resulted in abundant regeneration by *A. tortilis* in the riverine fringe. Plants grew to a height of 5–7 m in height within 7 years (Anderson & Walker 1974; Cumming 1981a; Mwalyosi 1987).

Populations of *Acacia* spp have been depressed due to felling, debarking or uprooting by elephants in other parts of Africa. In the Seronera region of the Serengeti Park in Tanzania, *Acacia tortilis* trees were declining at a rate of 6% per annum. However, regenerating plants less than 1 m in height were abundant and ignored by elephants. Other species declining as mature trees in this park due to elephant impact included *A. xanthophloea* (5.5% per annum) and *A. senegal* (2.6% per annum). At Tsavo East in Kenya, *Acacia tortilis* plants greater than 1 m tall declined in density by 65% between 1970 and 1974, but regenerating plants less than 1 m remained abundant. At Kidepo in Uganda, the predominant *Acacia gerrardii* suffered a 23% decline among large trees over a 3 year period, while *A. senegal* also decreased markedly. In this area small trees less than 1 m were uprooted and eaten by elephants. However, other factors besides elephants were responsible here for some of the mortality among mature trees. *Acacia nigrescens* has suffered heavy elephant impact by debarking and felling in Kruger Park in South Africa, Chobe and Tuli in Botswana, Sengwa in Zimbabwe and Luangwa in Zambia. Along the Ruaha and Magusi Rivers in the Ruaha Park in Tanzania, *Acacia albida* declined in density from 8–13 per ha to about 2 per ha over 11 years, largely due to bark removal by elephants. In the Luangwa valley in Zambia, elephants felled some stands of *A. albida*, while other stands remain only lightly affected. At Mana Pools in the Zambezi valley, tall *A. albida* tress have persisted with relatively little mortality for some 20 years since bark damage by elephants was first reported. However, regeneration of this species has been suppressed completely by elephant feeding on small plants. At Amboseli in Kenya, the loss of 90% of large *A. xanthophloea* trees over a 17 year period was due primarily to an upward shift in ground water levels resulting in increased soil salinity, rather than to elephant damage. However, in the Arusha Park in Tanzania moribund *A. xanthophloea* trees were attacked more frequently by elephants than healthy trees (Barnes 1985; Croze 1974b; Field & Ross 1976; Harrington & Ross 1974; Lamprey *et al.* 1967; Leuthold 1977d; Pellew 1983c; Vesey-Fitzgerald 1974; Western & van Praet 1973; personal observations).

The baobab *Adansonia digitata* is a tree species widely susceptible to elephant damage, due to its soft, pithy wood. Elephants strip off bark, then gouge deepening holes into the trunk, until eventually the tree collapses (Fig. 12.1). Declines in baobab populations have occurred widely where elephants have reached densities that have resulted in a shortage of food during the dry season. In the Tsavo East Park in Kenya, baobabs had been virtually eliminated by 1974, less than 20 years after first reports of damage by elephants. At Ruaha in Tanzania, baobab tree density declined from 72 per km^2 to 40 per km^2 over an 11 year period, a mean rate of decrease of 3% per annum. Few trees less than 2 m in girth survived. At Lake Manyara in Tanzania, damage to baobabs was light in 1969, but by 1981 only 13% of trees remained undamaged. Annual tree mortality there was about 1% per annum. However, extinction of baobabs was unlikely, as many trees oc-curred on steep slopes of the escarpment where they were inaccessible to elephants. Along the Zambezi River frontage at Mana Pools, 24% of baobabs had been killed by elephant damage (15% during one year), while away from the river 6% of trees were severely damaged although few were dead. *Sterculia* spp, which like baobabs have soft trunks, have suffered similar declines at Tsavo and Luangwa (Barnes 1980, 1985; Leuthold 1977d; Napier Bax & Sheldrick 1963; Swanepoel & Swanepoel 1986; Weyerhaeuser 1985).

Fig. 12.1 Baobab tree destroyed by elephants in the Luangwa Valley, Zambia.

At Tsavo East, mature *Commiphora* shrubs were reduced in density from a mean of 90 per ha in 1970 to 5 per ha in 1974. This period spanned a severe drought, during which one third of the elephant population died of starvation. In the rift valley stratum of the Ruaha Park in Tanzania, *Commiphora ugogoensis* shrubs declined in density, from about 30 per ha in 1971, to 15 per ha in 1977 and 1–2 per ha in 1982 (Barnes 1985; Leuthold 1977d).

Dominant woodland trees are also susceptible to population depression by elephants. In the Chizarira Game Reserve in Zimbabwe, *Brachystegia boehmii* trees were being killed by elephants at a rate of 20% per annum over 1971–73. Death was due to felling, ringbarking and uprooting. By 1980, this tree species had been all but eliminated, and the elephants were turning their attention to the remaining *Brachystegia glaucescens* woodland. In the nearby Sengwa Research area, elephants reduced woody plant biomass in the *Brachystegia* woodland by 45% between 1972 and 1976, largely by selective debarking of *B. boehmii*. *Pterocarpus angolensis*, also readily ringbarked by elephants, has been greatly reduced in *Brachystegia* woodland along the Chobe River frontage in Botswana. Mopane (*Colophospermum mopane*) woodland at Sengwa sufffered a reduction in woody plant biomass of about 6% between 1972 and 1976. However, mopane trees felled by elephants respond by coppice regrowth from the stumps. While 45% of miombo woodland trees and 80% of riverine

Fig. 12.2 Stand of *Colophospermum mopane* woodland largely felled by elephants. The mopane woodland in the background remains standing (Luangwa Valley, Zambia).

woodland trees pushed over by elephants were dead at Sengwa, two-thirds of mopane trees pushed over were regrowing. In the Luangwa Valley in Zambia, mopane was being felled at a rate of 4% per annum, with additional trees killed by ring-barking. However woodland destruction was patchy in its distribution (Fig. 12.2). Over large areas mopane persisted in the form of coppice regeneration 2–3 m high (Fig. 12.3). Extensive areas of low shrub mopane are a feature of the Kruger and Gonarezhou parks on basaltic flats, although soil depth and fire are other factors besides elephants preventing tall trees from emerging. At Kasungu in Malawi, areas of mixed *Brachystegia* woodland have been converted to stands of dense coppice regrowth by heavy elephant use (Anderson & Walker 1974; Bell 1981a; Caughley 1976c; Cumming 1981b; Guy 1981; Lewis in preparation; Thompson 1975; van Wyk & Fairall 1969; personal observations).

In the dense shrubby vegetation of the Addo Park in South Africa, woody biomass was reduced by half by the confined population of elephants. However, all shrub species persist, although in gnarled form. The less favored *Capparis sepiaria* has increased at the expense of more favored shrubs, but only the succulent *Aloe africana* has been reduced to near extinction. Stands of shrubby vegetation elsewhere likewise show little change despite heavy elephant use, including *Baikaea/Baphia* and

Fig. 12.3 **Stand** of *Colophospermum mopane* **converted to shrub coppice** form as a result of sustained elephant browsing (Luangwa Valley, Zambia).

Combretum/Commiphora thickets at Sengwa, *Lonchocarpus laxiflorus* shrublands at Murchison Falls, and *Combretum obovatum* thickets at Luangwa valley (Anderson & Walker 1974; Laws, Parker & Johnstone 1975; Penzhorn, Robbertse & Olivier 1974).

In evergreen forest patches in Uganda, elephants feed selectively on regenerating saplings of valuable timber species such as *Khaya* (mahogany), *Chrysophyllum, Cordea* and *Maesopsis*. This leads to dominance of the mature tree layer by *Cynometra* (Laws 1970; Laws, Parker & Johnstone 1975).

In some areas elephant impact has been such as to transform woodlands into open grassland. A striking example is the Rwindi–Rutshuru plain in the Virunga National Park in Zaire, as documented by paired photographs taken in 1934 and 1959 (see Bourliere, 1965, but fire was a contributory factor). In the Masai Mara Park in Kenya, fire was primarily responsible for the conversion of *Acacia* woodland into open grassland, but elephants exerted a secondary effect by browsing the *Acacia* regrowth. In the Murchison Falls Park in Uganda, the *Terminalia/Combretum* woodland was virtually eliminated in the central zone south of the Nile River. Tree density declined from between 430 and 1060 per km^2 in 1958 to 20 per km^2 in 1967. Tree death followed bark stripping by elephants, with damaged trees becoming infested with woodborers, and eventually killed by fire. At Chizarira in Zimbabwe, a *Brachystegia* woodland with a tree density of 1180 per km^2 was transformed into a lightly wooded grassland. At Tsavo East in Kenya, *Commiphora*-dominated thicket has been transformed into open savanna with scattered trees over an area of about 4400 km^2. However, these extreme habitat transformations have generally been associated with exceptional conditions. Both at Murchison Falls and at Chizarira, elephants were compressed into the sanctuary of the park by human occupation and hunting in their former range. Removal of grass by annual fires during the dry season forced elephants to concentrate their feeding on remaining woody vegetation. At Tsavo, human disturbance forced elephants to concentrate in an area with a mean annual rainfall of only 400 mm. Woodland destruction was exacerbated by severe droughts (Bell 1981a; Buechner & Dawkins 1961; Cumming 1981a & b; Dublin 1984; Laws, Parker & Johnstone 1975; Leuthold 1977d; Myers 1973; Parker 1983; Pellew 1983c; Phillipson 1975; Thompson 1975).

Nevertheless, elephants have induced declines in tree populations even in areas where they occur at relatively low densities. In the Kruger Park, with a mean density of 0.4 elephants per km^2, bark damage and felling of trees such as *Sclerocarya birrea* and *Acacia nigrescens* has been a source of concern. During two successive drought years over 1982–83, many trees suffered extensive damage, including *Combretum apiculatum, Acacia nigrescens, Adansonia digitata* and *Kirkia acuminata*. In the Seronera area of the Serengeti Park, *Acacia tortilis* trees were depressed from 48% to 3% of the total population of mature trees by a local density of elephant bulls of only 0.2 per km^2 (Coetzee *et al.* 1979; Pellew, 1983c; van Wyk & Fairall 1969; unpublished National Parks Board reports).

Large-scale reductions in elephants in Uganda resulted in a dramatic recovery by woody vegetation. At Murchison Falls, *Acacia sieberiana* trees grew to a height of 7–10 m over 24 years in plots from which elephants were excluded. One plot that had formerly been open grassland with widely scattered *Acacia* trees became transformed into close canopy *Acacia sieberiana* woodland. Associated with this change was an extreme build-up in soil nitrogen, and the herb layer became dominated by two forb species. With elephant browsing excluded, plant species richness declined from 45 to 22 species in the herb layer, and from 18 to 13 species in the tree layer. However, where *A. sieberiana* had replaced former *Combretum-Terminalia* woodland, plots showed overall increases in species richness. In the Queen

Elizabeth Park, *Euphorbia candelabrum* trees declined at a rate of 5.7% per annum over 1971–76 due to elephant damage, although low scrub regeneration was abundant. Following a drastic reduction in elephants, there was widespread regrowth by *E. candelabrum, Acacia sieberiana, A. gerrardii, A. hocki* and *Croton macrostachyus*, as well as by the shrubs *Securinega virosa* and *Turraea robusta*. In the Tsavo Park in Kenya, there was abundant regeneration by *Melia* and *Commiphora* in 1978, aided by above-average rainfall (Croze, Hillman & Lang 1981; Eltringham 1980; Hatton & Smart 1984; Laws, Parker & Johnstone 1975; Lock 1985; Smart, Hatton & Spence 1985).

Elephants serve as the prime dispersal agent of the seeds of many trees, especially in primary rain forest (Alexandre 1978; Bainbridge 1965).

Tree damage by Asian elephants is relatively light in the small parks in Sri Lanka where they are conserved. At Ruhunu, the main impact is in terms of distortion of the crown patterns of trees. At Gal Oya, about 25% of woody plants were damaged, but most of this (78%) took the form of stem or branch breaking or twisting. Only 8% of damage consisted of tree felling. Bark stripping from main trunks was rare, due to the lack of tusks of most of the elephants (Ishwaran 1983; Mueller-Dombois 1972).

Hippopotamus

Areas heavily grazed by hippos became converted from a medium-tall grass cover to a mosaic of tall and short grass areas, or even to extensive areas of short grass lawns. Terrace erosion may be evident, while bush density may increase due to the reduction in fires. On the Mweya peninsula in Queen Elizabeth Park, Uganda, palatable taller grass species such as *Themeda triandra, Heteropogon contortis* and *Cenchrus ciliaris* disappeared, except in grazing enclosures, and were replaced by the tussock-forming, relatively unpalatable *Sporobolus pyramidalis*, and by creeping *Chrysochloa orientalis*. On parts of the area erosion had removed 3–8 cm of sandy topsoil, and one erosion gully advanced 7 m over four years (Laws 1968b; Laws, Parker & Johnstone 1975; Naylor *et al.* 1973; Olivier & Laurie 1974; Petrides & Swank 1965; Thornton 1971).

Following the complete removal of hippos from the Mweya peninsula in 1958, *Sporobolus pyramidalis* increased, while *Chrysochloa orientalis* decreased. *Cynodon dactylon*, another creeping grass, almost vanished from the localities where it had formerly been common, but increased in other places. Another medium-tall grass, *Chloris gayana*, became the second-most common species, while *Cenchrus ciliaris* changed little in abundance. The basal cover of grasses declined over the first year from 15% to 5% (associated with rainfall that was only 60% of the mean), but then recovered partly to 11%. By 1975 the hippo population had recovered to a density of 25 per km², somewhat higher than the pre-1958 level. The grass community remained dominated by *Sporobolus pyramidalis*, with no further deterioration occurring. It seems that the earlier degradation was due largely to a seven year period of well below average rainfall. Nevertheless the more favorable species of taller grass, including *Hyparrhenia filipendula, Heteropogon contortis, Cenchus ciliaris* and *Themeda triandra*, predominated only in plots from which hippos were excluded (Lock 1972; Thornton 1971; Yoaciel 1981).

Giraffe

In the Serengeti Park in Tanzania, giraffe browsing reduced the growth of regenerating *Acacia tortilis* saplings to one third of that in adjacent exclosure plots. As a result, the time taken for *A. tortilis* plants to grow above 3 m in height, i.e. beyond fire-susceptible height classes, was retarded from 8 years to 21 years; while for *A. xanthophloea* it was retarded from 4 years to 19 years. At Serengeti, Kruger Park and other areas of high giraffe density, mature plants of favored *Acacia* species have their canopy shapes moulded into conical or hour-glass shapes by giraffe browsing pressure concentrated in the height range 2–4 m above ground. In the Hluhluwe-Umfolozi Reserve in South Africa, severe canopy distortion of *Ziziphus mucronata* trees was recorded, even though the density of giraffe was only about 0.2 per km² (Brooks & Macdonald 1983; Pellew 1983c).

Rhinoceroses

Indian rhinos seem to have little impact on the vegetation of Chitwan National Park in Nepal, at their prevailing density. Sumatran rhinos break down woody saplings up to 100 mm in diameter by walking over them; while males damage saplings up to 20 mm diameter by twisting them down with their horns prior to squirt-urination. However, at their low densities the impact on the vegetation is inconsequential.

Black rhinos concentrate their browsing on woody scrub under 1 m in height, especially acacias. Over half of the above-ground parts of plants in this height range may be consumed. Spindly *Acacia* saplings up to 3.8 m in height may be pushed over breaking the stem. The growth of browsed plants is retarded and some are killed. Black rhinos confined at Addo at a density varying between 1.3 and 5.2 per km² transformed a dense shrub thicket into open dwarf shrubs in a time span of a few years. In most areas black rhino densities appear insufficient to halt recruitment to taller size classes, at least in the absence of fire (Borner 1979; Hall-Martin, Erasmus & Botha 1982; Hitchins 1979; Hubback 1939; O'Regan in preparation; Vesey-Fitzgerald 1973).

White rhinoceros

When the Natal Parks Board assumed control of Umfolozi Game Reserve in 1952, *Themeda triandra* was the predominant grass species, growing in luxuriant medium–tall stands. In 1969, *Themeda*-dominated grass communities still covered 53% of the reserve area, but only 13% remained in 'climax' condition. Short grass communities characterized by creeping *Panicum coloratum* occurred over 17%, while other short or sparsely growing communities covered a further 30% of the reserve. Marked changes in grassland condition took place over the period of my study. Large sections that had been tall *Themeda* grassland in 1966 had become converted to short *Panicum–Urochloa* grassland by 1969 (Figs. 12.4 and 12.5). While the eastern slopes of the Madlozi valley had been well grassed in 1966, by 1969 they retained only a patchy grass cover with sheet erosion prominent (Figs. 12.6 and 12.7). The alluvial soils bordering the Madlozi stream has mostly tall grass in 1966, but by 1968 retained only a sparse cover of short grasses and forbs. While drought contributed to the grassland decline, the grazing impact of white rhinos and other larger herbivores was strikingly evident from a comparison of grassland condition inside and outside exclosure plots (Fig. 12.5). Since white rhinos formed over half of the total biomass of large herbivores, grassland changes could be ascribed largely to their grazing pressure. Furthermore, large-scale influxes of wildebeest and zebra into the Madlozi area took place only in 1968 (Downing 1972; Owen-Smith 1973).

In 1970, about one third of my Madlozi study area in the western section of Umfolozi was covered by short grass grassland, and much of the *Themeda*-dominated grassland was grazed down by the end of the dry season. Sheet erosion was prominent in sparsely-grassed areas, while erosion gullies had expanded in extent. Watercourses consigned chocolate-brown topsoil into the two Umfolozi rivers. The White Umfolozi, formerly a clear stream meandering over a sandy bed, had become as murky as the Black Umfolozi, and ceased surface flow for increasingly long periods during the dry season. The Black Umfolozi, formerly a fairly swift river with a rocky bed, had become silted up, and in 1970 ceased surface flow for the first time on record.

However, rainfall over the period 1966–70 was consistently below the

Fig. 12.4 Stand of *Themeda triandra* grassland grazed down by white rhinos, with patches of taller grass persisting around *Acacia* scrub (west of Madlozi, Umfolozi Game Reserve).

Fig. 12.5 A grazing exclosure plot south of Madlozi in Umfolozi Game Reserve, showing tall *Themeda* grassland within the exclosure, and short *Themeda–Panicum–Urochloa* grassland outside the exclosure (plot fence completed in 1967, photograph taken in October 1969).

Fig. 12.6 *Themeda triandra* grassland on the slopes of the upper Madlozi valley in Umfolozi Game Reserve in March 1966.

Fig. 12.7 The same view as in Fig. 12.6 photographed in August 1971, showing removal of the tall grass cover and sheet erosion.

long term average. The mid 1970s formed a period of well above average rainfall. Aided by a 30% reduction in the biomass of all grazing ungulates through culling, including white rhinos, tall grass cover improved markedly. Over 1978–83 rainfall was exceptionally low, creating perhaps 'once in a century' drought conditions. By late 1980, all tall grass reserves in Umfolozi had been grazed down, even as far as the tops of the highest hills. Short grassland areas retained only a sparse stubble, with much bare soil. Following dry season rain in 1981, an annual forb spread over much of the bare soil. Culling reduced the total biomass of grazing ungulates in Umfolozi to 60% of the 1971 level, except in a 63 km² control block left unculled. White rhino numbers were reduced to 40% of their 1971 total. In the control block where animal densities remained high, the grass cover declined markedly. An outbreak of harvester termites (*Hodotermes mossambicus*) removed remaining grass stubble during 1982–83, resulting in further grass mortality. Soil erosion rates increased drastically, and *Acacia* scrub 1–2 m tall became prominent in formerly open patches. The control block thus showed all the signs of extreme overgrazing, while conditions in the rest of the reserve where animal densities had been reduced were somewhat better.

During the 1983/84 season, torrential cyclonic rains resulted in a precipitation 60% above average, and rainfall remained high through 1984/85. By March 1985, tall *Themeda* grassland had reappeared over most of Umfolozi, including parts of the control block; while *Panicum coloratum* had spread into the spaces between surviving grass tufts in short grass areas. White rhinos and wildebeest moved into the control block, due to the predominance of unfavorable tall grass elsewhere (Brooks & Macdonald 1983; Emslie in preparation; Walker *et al.* 1987).

The history of grassland dynamics in Umfolozi Game Reserve over this period demonstrates a strong interactive effect between grazing pressure and rainfall. Grasslands deteriorated everywhere during periods of low rainfall, but the change was less extreme in areas protected from grazing, or where grazer densities were markedly reduced, than in areas where animal densities were left unmanaged. The major grazing impact was due to white rhinos, although other grazing ungulates such as wildebeest, zebra, impala and warthog contributed. Nevertheless, grasslands recovered following high rainfall even in areas that had appeared badly degraded, at least to the extent that they became attractive to short grass grazers. In areas where animal densities had been reduced, an excess of tall grass made conditions unfavorable for many of the ungulate species.

Summary

African elephants can exert a negative influence on populations of mature trees of sensitive species. These include species easily uprooted by felling (e.g. *Acacia* spp, *Commiphora* spp), those with bark that is easily stripped (*Acacia* spp, *Pterocarpus angolensis, Brachystegia boehmi*), and trees with soft pithy trunks (*Adansonia digitata, Sterculia* spp). Other species that respond to trunk breakage by coppice regrowth of the stump, such as *Colophospermum mopane*, persist despite severe elephant impact, although their growth form is changed. Elephants uproot and consume small woody plants, at least those taller than 0.5 m, thereby suppressing replacement of the mature trees that they kill. Under extreme conditions where elephants are crowded into small areas, they can transform wooded savanna into open grassy savanna. Woody plant damage is exacerbated following grass removal by fire or drought. In forests, elephants of both species may suppress shade-tolerant 'climax' trees by creating open gaps colonized by faster-growing species. Elephants commonly uproot grass plants while feeding, but despite this their impact on the grass layer appears negligible.

Sustained grazing pressure by white rhinos and hippos can transform medium–tall grassland areas into communities dominated by short or decumbent grass species. With reduced canopy cover, soil erosion may be accelerated, and due to reduced fire frequency woody scrub may invade grassland. Deterioration in the grass cover generally takes place in association with below-average rainfall.

Browsing pressure by black rhino and giraffe can slow or suppress the growth of regenerating woody plants. By keeping plants within the fire susceptible zone for longer, browsing may retard recruitment to mature height classes.

Effects on other large herbivores

Different species of large herbivore are potential competitors for food resources. However, competition is effective only where a food resource shared in common by two species is limiting, so that a reduction of the available food by one species depresses the population density that the other species can attain. For herbivores food quality is generally more limiting than food quantity (Sinclair 1975; Rhoades 1985). A reduction in plant biomass due to consumption by one species may improve food quality for another, by stimulating regrowth of nutrient-rich new leaves, or by making foliage more accessible (McNaughton 1976). Megaherbivores are the prime species in terms of biomass in 13 out of 16 African parks for which good census data are available (Chapter 14), as well as in those Asian parks where they remain effectively conserved (Eisenberg & Seidensticker 1976). Although the food intake of megaherbivores per unit of biomass is only about two-thirds that of medium-sized ungulates, one or other megaherbivore species is generally pre-eminent in the larger herbivore community in terms of consumptive demand.

Populations may also interact less directly. A change in the structure or species composition of the vegetation may make certain species vulnerable to predation (Sinclair 1985). Water availability may be improved, or reduced. The subject of competition remains controversial, because it is difficult to distinguish direct interactions between populations from independent responses to changing habitat conditions, which may be induced by weather fluctuations or other external causes. Experiments are difficult to perform on large herbivores, so that in general the only information available to suggest competitive or facilitatory effects consists of relative population changes over time. Such evidence must be treated with due caution. Where two populations change in antisynchrony, this suggests the possibility of a competitive interaction, but the population changes could be due to independent responses to some other environmental factor (Owen-Smith 1988 in press).

Elephants

African elephants form the greatest component of large herbivore biomass in 10 out of 16 parks, while they are second to hippos in two parks, and fall just below African buffalo in one park. Only in the grassy savanna of the East African plateau, e.g. the Serengeti and Nairobi parks, are they a minor faunal constituent.

At Tsavo East in Kenya, elephant density reached high levels before the population crash associated with the 1971 drought. The opening up of the woody vegetation that they caused was associated with declines in the populations of other browsing ungulates. Although data on population changes are unavailable for Tsavo National Park, there are figures for the neighboring Galana Ranch, where very similar habitat changes occurred. Between 1963 and 1981–82, lesser kudu declined by 90%, gerenuk by 80% and giraffe by 40%. Black rhinos decreased to very low numbers, but poaching was a contributory factor in this case. In compensation, grazing ungulates such as Grevy's zebra and oryx, as well as Grant's gazelle, increased markedly (Parker 1983). These population changes are probably a response to the opening up of *Commiphora* shrublands and riverine thickets by elephants, rather than to competition by elephants for browse. For example, black rhino show relatively little overlap with elephants in their food preferences at Tsavo (Chapter 3).

In association with a 40% increase in elephant numbers in the Ruaha Park and environs in Tanzania between 1973 and 1977, there were apparent declines by zebra, impala, eland, kudu and black rhino, while buffalo and giraffe increased. While the overall diversity of large herbivores appeared to have dropped, little reliance can be placed on these differences, which are based on only two aerial censuses (Barnes 1983a; Barnes & Douglas-Hamilton 1982; Norton-Griffiths 1975). Along the Chobe River in Botswana, opening up of *Combretum* thickets by elephants has resulted in far fewer bushbuck being seen than was the case earlier (D. Work personal communication). At Sengwa in Zimbabwe, although elephants reduced total woody biomass, there was more browse available within the height reach of medium-sized ungulates (Guy 1981). In the Kruger Park in South Africa, trees pushed over by elephants were sought out by kudus during a drought (personal observations). In the Addo Park in South Africa, numbers of browsers, in particular kudu and eland, increased following the expansion of elephants into a new section of the dense valley bushveld (A. Hall-Martin personal communication).

While elephants may derive half their food intake from the grass layer, competitive influences on grazing ungulates have not been reported. Instead, elephants may facilitate grazing by other ungulates by opening up tall grass in valley grasslands (Vesey-Fitzgerald 1960). On the other hand, by creating dense stands of woody coppice that shade out the grass layer, elephants may create unsuitable conditions for grazers such as sable and Lichtenstein's hartebeest (Bell 1981a).

Hippopotamus

Hippos are the major species in Queen Elizabeth National Park in Uganda, forming 40% of the large herbivore biomass (Field & Laws 1970; Eltringham & Din 1977). In 1958 hippos were eliminated from the 4.4 km² area of the Mweya Peninsula. By 1968, the total biomass formed by other large herbivores had increased three-fold. Elephant, buffalo and waterbuck showed increases, while warthog and bushbuck declined. By 1973 hippos had recovered to more than their former density. Numbers of elephant, buffalo and waterbuck using the area had dropped, while warthog numbers were unchanged. However, population changes in such a small area are unreliable. The drop in buffalo numbers was due to the breakup of one large herd using the area (Eltringham 1974, 1980).

Giraffe

Giraffe derive most of their food from levels out of reach to other browsers, and thus do not compete directly, except perhaps for fruits and pods that may otherwise fall to the ground. By retarding growth by regenerating *Acacias*, giraffe are likely to facilitate browsing by smaller ungulates (Pellew 1983c; J.T. du Toit personal communication).

Rhinoceroses

Indian rhinos are pre-eminent among wild herbivore species in biomass at Chitwan in Nepal, and are second to elephants at Kaziranga in India (Eisenberg & Seidensticker 1976). Their effects on other ungulates are unreported.

Black rhinos generally feed at a lower level than other browsing ungulates like kudu and giraffe, reducing the potential for competition with other large browsers. Since woody plants sprout soft new shoots following browsing, feeding by black rhinos may increase food quality for smaller browsers. However, they may compete for forbs with kudu, nyala and perhaps small antelope like steenbok and dikdik (Joubert & Eloff 1971; B. P. O'Regan personal communication).

White rhinoceros

When the Natal Parks Board assumed control of Umfolozi Game Reserve in 1952, other large ungulates besides rhinos had been virtually eliminated by the shooting campaign carried out by the veterinary authorities to eliminate hosts for tsetse fly. Numbers of wildebeest, zebra, buffalo, waterbuck, impala and nyala thereafter increased steadily through colonization from Hluhluwe Game Reserve via the Corridor. By 1970, the total biomass of large herbivores in Umfolozi exceeded 8000 kg km^{-2}, of which white rhino formed 50% (Table 12.1). Hluhluwe Game Reserve supported a similar biomass, but with white rhino comprising only 6%. At about this time, the Natal Parks Board vastly increased their culling quotas for all grazing ungulates, in an attempt to alleviate perceived overgrazing of grasslands. White rhino numbers in Umfolozi were reduced from 1550 in 1972 to 1070 in 1976, and 600 by 1982. Wildebeest had declined by 1982 to 30% of their 1970 peak, while numbers of buffalo, zebra and impala remained unchanged or increased despite the removals.

Table 12.1. *Ungulate populations in the Umfolozi–Corridor–Hluhluwe Complex*

Areas: Umfolozi Game Reserve (UGR) (south of Black Umfolozi River) – 456 km²; Hluhluwe Game Reserve (HGR) – 215 km²; Corridor (Cor) – 270 km²; Complex – 940 km².

Species	estim. mean weight[a] (kg)	Estim. undercount factor	Numbers				Biomass (kg km^{-2})			
			UGR	Cor	HGR	Total	UGR	Cor	HGR	Total
White rhinoceros	1350	1.0	1361	313	90	1764	4028	1565	568	2532
Wildebeest	160	0.9	3745	2170	1159	7074	1310	1290	862	1200
Zebra	210	0.8	1027	1308	1096	3431	472	1017	1072	765
Buffalo	410	1.0	426	640	651	1717	384	974	1240	750
Impala	40	0.5	2506	4114	9850	16470	220	610	1836	700
Nyala	55	0.5	2634	696	4360	7690	318	142	1114	448
Waterbuck	155	0.67	1567	396	39	2002	532	228	28	330
Kudu	165	0.67	1260	85	331	1677	457	52	253	295
Warthog	40	0.5	3346	1138	1762	6246	294	170	328	264
Black rhinoceros[b]	700	—	60	69	199	328	91	179	647	246
Giraffe	800	1.0	17	0	17	34	30	0	63	29
Reedbuck	45	0.5	946	76	10	1032	92	6	—	48
Mountain reedbuck	30	0.5	218	8	20	246	14	—	—	8
Total							8242	6239	8011	7615

Notes: Not censused: bushbuck, bushpig, grey duiker, red duiker, eland, klipspringer, steenbok, blue duiker.
[a] Equals ¾ of average adult weight for most species; for white rhinoceros based on known age structure; weights from Hitchins (1968), Wilson (1968), Smithers (1971).
[b] Not censused, populations from Hitchins (personal communication 1971).

Sources: Derived from the Natal Parks Board helicopter census of August 1970 (Vincent 1970), with corrections for undercounting bias (modified from Melton 1978).

Only half of the decrease by wildebeest could be accounted for in terms of removals. Three other ungulate species declined in numbers in Umfolozi subsequent to 1970, although not subjected to culling. By 1982 common reedbuck numbers had dropped to about 15% of their 1967 abundance, bushbuck to about 25% of their 1967 abundance, and waterbuck to about a quarter of their 1970 numbers. Wildebeest may have been influenced adversely by the increase in tall grass during the 1970s, but an increase by lions was probably also a contributory factor. Waterbuck may have been under stress from the earlier reduction in tall grass cover caused by the increasing grazing impact of white rhino, although other factors such as higher tick loads and greater predator numbers were probably contributory factors. Reedbuck almost certainly suffered from the opening up by white rhino grazing of the tall grass cover that they depend on as a refuge from predation. In the case of bushbuck, competition from the expanding nyala population has been invoked; but they were probably also influenced adversely by the opening up of vegetation cover in bottomland regions, to which white rhinos contributed. Thus the only species that suffered adversely from the vegetation impact of the high white rhino biomass were reedbuck and, more equivocally, waterbuck and bushbuck (Brooks & Macdonald 1983).

Summary

Megaherbivores are generally pre-eminent in terms of biomass in large herbivore communities occupying African savanna. Their consumptive demands therefore have a greater impact on vegetation than that of other large herbivore species in the community, while they exert additional vegetation impact through breakage and trampling. Megaherbivores do not compete directly to any significant extent with other larger herbivore species. However, the vegetation changes that they induce may affect other herbivore species. By opening up thickets, elephants reduce cover for species such as bushbuck, but increase browse availability for other species. By transforming stands of tall grass into short grass, white rhinos and hippos may exert a negative influence on ungulate species dependent upon taller grass for either food or cover; but such grassland changes may be beneficial for species favoring short grass. However, by generally promoting the replacement of tall mature woodlands or grasslands by rapidly growing shrubs or short grasses, high megaherbivore densities are more likely to favor than be to the detriment of smaller species of herbivore.

Comparisons with smaller ungulates

Wildebeest in large concentrations, such as occur in the Serengeti Park in Tanzania, can transform stands of medium-tall *Themeda* grassland into short grass. By trampling and horning, they also damage and suppress small woody plants in the grass layer. However, their grazing pressure on taller grasses is restricted mostly to the dry season, so that changes in grassland composition generally do not result (Dublin 1984; McNaughton 1984). Short grass grazers like wildebeest and impala may, by their sustained grazing pressure on short grass areas, cause progressive denudation of the grass cover. Domestic cattle at high numbers, such as are commonly stocked in African tribal areas, can certainly convert medium–tall grasslands into degraded areas of sparse short grass and encroaching bush (Lamprey 1983). Wild populations of African buffalo do not have this effect; the culling of buffalo in the Kruger Park is justified primarily in terms of their competition with other ungulate species for water (Joubert 1983). Small browsers may have a significant impact on the survival of woody plant seedlings, but effects on woody plant communities remain to be documented. Populations of wild ungulates, apart from megaherbivores, have caused major changes in African savanna vegetation only in circumstances where predators have been eliminated, or movements confined by fencing to small areas, or in the vicinity of waterpoints (Cumming 1982).

13

Body size and population regulation

Introduction

For an expanding population to be transformed into a stable one, density dependent changes must occur either in rates of recruitment, in population losses, or in both (Caughley 1977). Recruitment can decline due to (i) a decline in female fecundity, brought about by reduced litter sizes, or increased intervals between births; (ii) higher post-natal losses; or (iii) ages at first parturition being retarded. Losses can increase due to (i) increased mortality, whether as the direct result of nutritional deficiencies, or as a result of predation or disease; or (ii) to increased emigration from the area.

Caughley & Krebs (1983) suggested that there is a fundamental dichotomy between the processes of population regulation in small mammals (under about 30 kg in body mass), and those operating in larger mammals. The former are regulated mainly by intrinsic mechanisms, i.e. by behavioral or physiological responses acting before food becomes limiting. In contrast, large mammals are regulated largely by extrinsic factors, such as the direct effects of food limitations on survival and reproduction.

Goodman (1981) emphasized that the population dynamics of large, long-lived mammals are much more sensitive to variations in annual survival rates than to corresponding variations in fecundity. These circumstances favor deferred reproduction as an adaptive response to adverse conditions.

Riney (1964) maintained that populations of large herbivores expanding from low densities in favorable habitats inevitably overshoot the carrying capacity of the vegetation. He suggested that the typical pattern was an eruptive oscillation, which could be subdivided into four stages: (i) an initial phase of rapid expansion with high fecundity and low mortality, with vegetation condition beginning to deteriorate towards the end of this period; (ii) a phase of temporary stabilization as the population reaches, then exceeds, the carrying capacity of the habitat; vegetation degradation is accelerated, and animal condition declines; (iii) a phase of rapid decline in numbers as the population adjusts to a lowered carrying capacity brought about by its impact on food resources; (iv) a final phase of stabilization in the degraded habitat at a density very much lower than that reached during earlier stages. Riney proposed that the overshoot of carrying capacity was an inevitable consequence of the upward momentum built into the growing population due to the high proportion of animals in younger age classes. Thus even if age-specific fecundity and mortality rates were to adjust instantaneously to those values characterizing a stable population, the breeding segment would continue to expand for a period, due to continuing recruitment into it of those young animals already alive.

Caughley (1976a, 1977) generalized Riney's description into an analytic model of plant-herbivore systems. For plants the rate of renewal of primary resources, such as sunlight and water, is largely independent of the existing plant biomass. Under these conditions a simple logistic model of population growth is applicable. In contrast, there is an interaction between herbivore populations and the plants that form their resource base. Interactive systems tend to display oscillations in both plant and herbivore biomass, because of the lag in the effect of herbivore feeding on future vegetation growth. If in the model vegetation biomass is reduced to less than half of its potential in the ungrazed state, persistent oscillations develop. These take the form either of a stable limit cycle, or of divergent oscillations leading to the extinction of the herbivores alone, or of both herbivores and plants.

Fowler (1981) suggested that large mammals are liable to overshoot their ultimate mean population levels due to the non-linear dependence of demographic variables on population density: i.e. changes in birth rates and mortality are initiated only at population levels close to carrying capacity.

Caughley (1976b) suggested, specifically with regard to elephant populations, that no stable equilibrium with vegetation is reached. Elephants inevitably increase to densities sufficient to exert a major impact on tree populations, resulting in the death or emigration of most of the elephants. The pattern of change generated is one of a stable limit cycle with a period of about 200 years.

Other hypotheses have been advanced to explain why elephant populations fail to stabilize before severe damage to woody vegetation has been inflicted. These are:

1. Range compression. Laws (1969b, 1970) and others suggested that elephants have been forced into national parks by hunting, settlements and other human disturbances in surrounding areas.

Human activities furthermore inhibit movement by elephants out of these sanctuaries. As a result population densities reach atypically high levels.

2. Population eruption. Elephant numbers were depressed over much of Africa at the end of the last century as a result of ivory exploitation. Following the cessation of ivory hunting, populations entered a phase of eruptive expansion, which has caused them to temporarily overshoot carrying capacity (Spinage 1973).

3. Elimination of human predation. Kortlandt (1976) argued that humans have been major predators of proboscideans since the Pleistocene, possibly displacing the extinct sabretooths from this role. Pienaar (1983) suggested that the rapid expansion of elephant, hippo and buffalo populations in the Kruger Park was related to the elimination of hunting.

A pattern of herbivore populations increasing inevitably to densities at which they overexploit and depress their own food resources is inherently unstable, and seems incompatible with large, K-selected mammals like megaherbivores. To be considered in this chapter are these questions: (i) Do expanding megaherbivore populations inevitably show an eruptive oscillation? (ii) Is vegetation degradation an inevitable consequence of megaherbivore population expansion? (iii) Can megaherbivores attain a stable equilibrium with vegetation? (iv) What difference does dispersal make? (v) How do megaherbivores differ from small ungulates in processes of population regulation?

I will use simulation modelling as an analytic aid. Simulation models can be used to explore possible behaviors of a system given certain assumptions about its structure and the interactions taking place within it. Caution is advisable in extending conclusions derived from any model to the real system that the model is intended to represent. Modelling is best approached as a heuristic thought-provoking exercise.

Demographic models

I developed a computer-based simulation model intended to be sufficiently detailed to capture the essential features of the population dynamics of a large, long-lived megaherbivore like white rhino. The basis of the model is a population structured into age classes differing in their mortality, natality and emigration rates (see Appendix II). A simulation incorporating the demographic characteristics of the Umfolozi white rhino population in 1970 closely replicated the observed features of this expanding population.

Some indication of possible changes that could bring about stabilization is given by the demographic characteristics of the black rhino population in the Hluhluwe Reserve. This population had declined from a total of about 300 animals in 1961, to 199 individuals in 1972. In contrast, the black rhinos occupying the adjoining Corridor and Umfolozi Reserve seemed to be in a phase of population expansion. A comparison between the population statistics from these two regions suggests that the following changes had occurred at Hluhluwe: (i) mean calving interval lengthened from 2.3 to 2.7 years; (ii) age at first parturition retarded from 6.5 to 12 years; (iii) infant mortality increased from 9% to 59% per annum, with hyena predation largely responsible; (iv) adult mortality increased to 3.5% per annum in females and 7.3% per annum in males; (v) mortality of immature plus juvenile animals increased to 5.7% per annum (the population sample for the Corridor plus Umfolozi was too small for mortality rates of adults and subadults to be estimated) (Hitchins & Anderson 1983).

Another possible example is given by the African elephant population in the Murchison Falls National Park in Uganda. This showed a decline from an estimated 22 000 animals in 1946 to 9400 in 1966, with about half of the reduction due to control or sport hunting. Demographic changes included an increase in mean calving interval from 4 to 7 years, a retardation of female puberty from 12 to 18 years, and an increase in calf mortality from 28% to 43% (Laws, Parker & Johnstone 1975).

For white rhinos it is anticipated that the main demographic changes would occur in age at first parturition and in birth interval, and to a lesser extent in infant survival. By trial and error with the computer simulation, it was found that the following combination of changes would achieve a stationary white rhino population: (i) natality rate for prime aged females reduced from 45% to 25% per annum; (ii) natality of old females reduced from 25% to 15% per annum; (iii) age at first calving retarded from 6 years to 10 years; (iv) infant losses tripled from 8% to 24%; (v) juvenile mortality increased from 3.5% to 8% per annum; (vi) subadult mortality increased from 3% to 6% per annum; (vii) mean mortality of prime adults doubled from 1.5% to 3.0% per annum; (viii) mortality of old animals (over 35 years) increased to 8% per annum. In the model stationary population, 58% of the animals are adults (> 10 y), 27% are subadults (3–10 y), and 14.5% are calves (0–3 y). Infants under one year of age form 5.2% of the population, compared with 11% in the real population in 1972. The terminal mortality of adults at the age of 45 years adds an additional amount of 1% per annum to the overall mortality rate of adults, so that this becomes 5% per annum. These characteristics suggest a white rhino population experiencing fairly severe nutritional limitations.

To make the model dynamic, density-dependent functions must be formulated for rates of natality, mortality and emigration. Furthermore, the density level at which the population stabilizes must be decided.

Since the white rhino population was still increasing at near its maximal rate in 1970, biomass levels prevailing then were evidently some way below the upper limit based on food resources. This upper limit I will term the saturation density. Considering that retrogressive habitat changes had been initiated, the saturation density cannot lie too far above the 1970 density. Arbitrarily it will be assumed that the white rhino population in 1970 was at two-thirds of its saturation density. The mean density of white rhinos over the 450 km² area of Umfolozi in 1970 was 3.2 per km², equivalent to a biomass of 4300 kg km⁻². Thus the assumed saturation biomass is 6500 kg km⁻², or a mean white rhino density over the whole extent of Umfolozi nearly as high as that prevailing in the western high density zone in 1970.

To make the model more realistic, physiological lags in responses to changing biomass levels will be allowed (supposedly related to the stored body reserves carried through from previous years). Thus the mortality and natality rates of adults and subadults become functions of mean biomass averaged over the preceding three years, juvenile mortality becomes a function of mean biomass over the preceding two years, and infant mortality of the prevailing biomass over the preceding year. Age at first parturition, because of its dependence on growth rate, will be taken to depend on mean biomass over a five year period.

In the model, dispersal is an option that can either be allowed or prevented. Other modifications will be introduced below. Results of the simulation runs were as follows:

1.   Expanding population with no vegetation interaction

If the model population is started at a density similar to that of the Umfolozi white rhino population in 1925, it increases rapidly to reach the saturation biomass after 25 years (Fig. 13.1(*a*)). The population overshoots the saturation biomass by 13%, and undergoes a small oscillation before stabilizing. Thus even with the multiple time lags built into the model, the oscillation is fairly minor. Nevertheless, the peak biomass reached is 65% greater than that prevailing for the real population in 1970, when retrogressive changes in grasslands were strikingly evident.

Dispersal was prevented in the above simulation. If dispersal is allowed, at rates similar to those estimated to be occurring in the real white rhino population, the model population levels off below the saturation biomass and barely oscillates (Fig. 13.1(*b*)). The equilibrium biomass level attained is only 25% greater than that prevailing for the real population in 1970. Thus dispersal dampens oscillations and causes the population to equilibrate at a lower density than that reached in the absence of dispersal.

2. Increasing population with vegetation impact

For this simulation, the peak wet season biomass of grass is assumed to be 200 g m^{-2} y^{-1} (based on the measurements of P. Dye, 1983, in a savanna with similar rainfall at the Matopos Research Station in Zimbabwe). Of this total, 50% is assumed to be unavailable or inaccessible (e.g. on steep hillslopes); and 50% of the remainder is assumed to be eaten or otherwise removed by other herbivores, including insects such as harvester termites as well as other ungulates. Thus the amount of edible and accessible food that is available for consumption by white rhinos amounts to 50 g m^{-2} y^{-1}. From the data presented in Chapter 5, the eating rate of a white rhino is about 1.5% of body mass (dry mass/livemass) per day. Hence the total grass consumption (in dry mass) by white rhinos over the course of a year amounts to five times the standing live biomass of white rhinos.

For the grass population, a simple logistic model will be assumed. The maximum between-year growth rate of the grass population (as distinct from the within season accumulation rate of grass biomass) is taken to be 50% per annum. Agriculturalists commonly assume that the threshold for overutilization of grasslands lies at a consumption level removing about 50% of annual above-ground grass production. McNaughton (1979) showed that short grass grasslands in the Serengeti can sustain somewhat higher grazing pressures; but the graminoids in this region seem to be exceptional in their tolerance of grazing, due partly to the extremely nutrient-rich volcanic soils. In the model it will be assumed that, if white rhinos remove less than 50% of the available grass production, they have no impact on the grass population; they simply remove part of the annual growth, which is replenished completely the next season. If consumption levels exceed the 50% threshold, the grass population the following year is depressed, due to the death of some tillers or tufts (see Appendix II for the formulation of this effect). The model parameters are adjusted such that, at the saturation biomass of white rhinos, grass consumption exceeds the overgrazing threshold.

The model was re-run with identical starting conditions to those used for the model with no vegetation impact. With dispersal prevented, the output shows slowly fading oscillations by both white rhino and grass populations, with a period of 35 years (Fig. 13.1(c)). If emigration is allowed, the

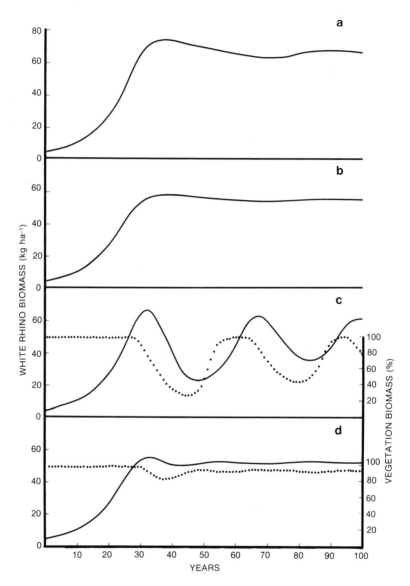

Fig. 13.1 Model output showing the pattern of increase by the simulated white
rhino population under different conditions. Parameter values given in
Appendix II, rainfall constant. The solid line indicates white rhino biomass;
dots indicate vegetation biomass, as a percent of the maximum. (a) no
vegetation interaction, no dispersal, (b) no vegetation interaction, with dispersal
allowed, (c) with vegetation interaction, no dispersal, (d) with vegetation
interaction, with dispersal allowed.

oscillations are dampened, because the peak density attained by the white rhino population only marginally exceeds the threshold biomass for overgrazing to be induced (Fig. 13.1(d)).

In general the precise behavior of the model is determined by the specific values selected for the vegetation parameters. Only very minor oscillations develop if grass production is such that the overgrazing threshold is not reached. If the potential rate of increase of the grass population is raised, the amplitude of the oscillations is reduced.

3. Fluctuating rainfall regime

This simulation investigates how an initially stabilized population of white rhinos might respond to year to year differences in rainfall. Grass production, and hence the amount of food available for herbivore consumption, can differ widely between years in association with rainfall fluctuations. Hence if the herbivore population is food-limited, its potential saturation density changes between years. For regions receiving annual rainfall totals of less than about 700 mm per year, grass production is almost linearly related to the current season's rainfall (Dye 1983). Thus in the model the saturation biomass of white rhinos becomes a variable whose value is dependent upon the seasonal rainfall total. In addition, the extent to which the grassland is overgrazed depends on the relation between the current season's grass production, which is linearly related to rainfall, and the consumptive demand of the white rhinos, which is dependent upon the prevailing white rhino biomass.

The available rainfall data for Umfolozi Game Reserve spanned a 22 year period from 1959 to 1981. In order to generate a sufficiently long time period for the model population to move away from the initial conditions, the available rainfall figures were replicated four times to yield an 88 year sequence, the last 22 years of which correspond to the real rainfall received over the period 1959–81.

When run under conditions of no dispersal, the model white rhino population fluctuates widely in biomass, rising to peaks during phases of high rainfall, such as prevailed over 1959–64 and 1973–77, and declining rapidly during phases of low rainfall, such as occurred over 1964–73 and 1977–81 (Fig. 13.2(a)). Because of the lag in population response, an overgrazing effect depresses grass populations at the start of each dry phase.

If dispersal is allowed, the population fluctuations are greatly reduced, although the mean population biomass is closely similar in the two simulations (4733 kg km^{-2} with no dispersal, and 4582 kg km^{-2} with dispersal

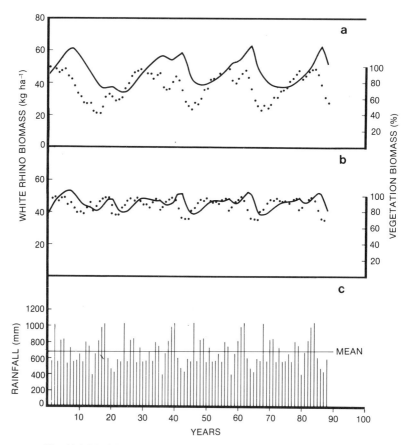

Fig. 13.2 Model output showing biomass changes by a simulated white rhino population under a variable rainfall regime, based on the Umfolozi records. The solid line indicates white rhino biomass; dots indicate simulated changes in the population of plants serving as a food resource. Parameter values given in Appendix II. (a) vegetation interaction, no dispersal, (b) vegetation interaction, with dispersal allowed, (c) rainfall pattern used in the simulation.

(Fig. 13.2(*b*)). The depression of the grass population due to overgrazing is greatly reduced.

If a model population expanding under conditions of rainfall variability is simulated, the regular pattern of the population oscillations depicted in Fig. 13.1 is disrupted. Should the initial population peak happen to coincide with a drought period, the depression of plant populations is more severe than it would otherwise be.

4.   Effects of body size

In order to compare processes of population regulation in mega-herbivores with those operating in smaller ungulates, the parameters of the simulation model were adjusted to represent a medium-sized antelope, such as a wildebeest. The differences in parameter values between the two versions of the model are listed in Appendix II, together with details of the changes made in formulation.

If the saturation biomass of 'wildebeest' is set at the same level as that of white rhinos, the degree of overgrazing that occurs at saturation levels is much more severe than was generated in the white rhino simulation. This is because medium-sized ungulates exhibit a higher feeding rate per unit of biomass than do megaherbivores. Such a situation simply generates a crash by the 'wildebeest' population, associated with a reduction in grass biomass to baseline levels. Hence the saturation biomass of 'wildebeest' was adjusted in the model so that at saturation they consume the same fraction of grass production as is eaten by 6500 kg km^{-2} of white rhinos in an average rainfall year.

If the model is run so as to simulate an expanding population of wildebeest with no emigration, the outcome is basically the same as that of the comparable white rhino simulation: the population increases to exceed its saturation biomass by 13%, and then undergoes dampened oscillations with a period of 16 years (Fig. 13.3(a)). The equilibrium biomass towards which the population tends is 25% less than the saturation biomass, because of the effect of overgrazing on the grass population. If emigration is allowed, the initial overshoot of carrying capacity is reduced to 2%, and the oscillations fade away more rapidly (Fig. 13.3(b)). The eventual equilibrium density is the same as that attained with dispersal prevented; but the population appears healthier in terms of the mortality rates prevailing at this equilibrium biomass.

In the model formulation, the mortality rates at saturation densities of 'wildebeest' are nearly three times as great as they are for white rhinos. The higher mortality of medium-sized ungulates can be interpreted as due largely to predation. Thus to simulate a situation without predators, a constant 5% was subtracted from the density-dependent mortality levels in each age class. With this change, the model 'wildebeest' population becomes highly unstable in the absence of emigration. It increases rapidly to a high biomass, then crashes due to grassland depletion, with oscillations tending to be sustained (Fig. 13.3(c)).

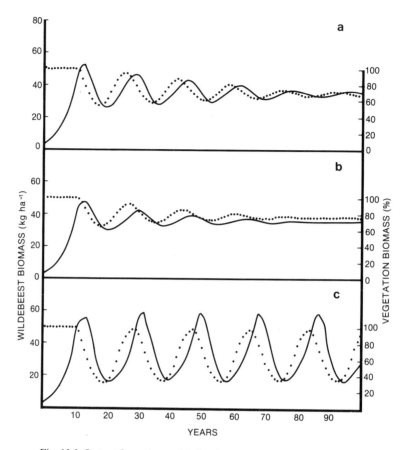

Fig. 13.3 Output from the model showing pattern of increase by a simulated wildebeest population under different conditions. Parameter valued given in Appendix II, rainfall constant, with vegetation interaction. Solid line indicates wildebeest biomass; dots indicate vegetation biomass, as a percent of the maximum. (a) no dispersal, (b) with dispersal, (c) no dispersal, and predation eliminated.

Conclusions for the modelling exercise

The output of the simulation model suggests the following conclusions, which might have some validity for real populations of large herbivores:

1. Population oscillations result mainly from the interaction between herbivores and vegetation, rather than from age structure effects. They develop where levels of consumption by the herbivore at peak biomass are such that plant populations are depressed.

2. Dispersal dampens population oscillations, lowers the peak

biomass level attained, and hence reduces the impact of herbivores on the vegetation. Dispersal also avoids or alleviates the vegetation overutilization that might otherwise develop when drought follows a period of favorable rainfall.

3. Predation and dispersal have basically similar effects on populations. They remove animals before starvation levels are reached, and thereby depress population density below the saturation level set by nutritional limitations. Even if density-independent, and hence strictly not regulatory in their operation, they still have this effect.

4. Smaller ungulates are not intrinsically more or less likely to exhibit populations oscillations than are megaherbivores. This depends on the specific details of their interaction with vegetation, rather than on their demographic characteristics. However, the vulnerability of medium-sized ungulates to predation means that their populations are less likely to reach saturation densities in the absence of dispersal than are megaherbivore populations.

Interactions with vegetation

Is it inevitable that megaherbivore populations increase to densities at which they depress their own food resources, before crashing to a lower density in an impoverished environment, as postulated by Riney (1964)? Megaherbivores – at least elephants, white rhinos and hippos – certainly do modify vegetation structure and species composition, at densities somewhat below their saturation levels. The critical question is, are these vegetation changes detrimental to future food resources?

Jachmann & Bell (1985) present evidence that African elephants respond differently to palatable and unpalatable species of woody plant. Unpalatable species are pushed over independently of their height, while palatable species are pushed over mostly in taller height classes where their canopy is out of reach. Palatable species commonly respond to trunk breakage by coppice regrowth from stumps. The result is the conversion of stands of tall trees offering little accessible forage to dense shrubby regrowth with an abundance of branches within easy foraging reach. Repeated browsing of such patches maintains them in this favorable growth stage for an indefinite period. Examples of woodlands having undergone such beneficial transformations of structure include stands of mixed *Brachystegia* in the Kasungu Park in Malawi (Bell 1981a), and of *Colophospermum mopane* in Luangwa Valley, Zambia. In forest areas elephants of both species encourage, by breaking canopy trees, the growth of gap-colonizing early seral species, which are generally more palatable than slow-growing climax trees (Laws, Parker & Johnstone 1975; Mueller-Dombois 1972).

A further possible benefit is suggested by recent research on plant anti-herbivore chemistry. Repeated heavy browsing of plants may stress them to an extent sufficient to depress their contents of deterrent chemicals (Bryant, Chapin & Klein 1983). By concentrating their browsing on stands of coppice regrowth, elephants may not only make more food available, but also produce better quality food.

Species of *Acacia*, widely favored by African elephants, generally show no seedling establishment under the canopies of mature trees. By pushing over trees, elephants may promote the replacement of mature trees by a much greater density of regenerating saplings, as has happened in the Serengeti (Pellow 1983c). However, other factors such as fire and giraffe browsing can interact to suppress the growth and recruitment to reproductive height classes of these saplings. In many areas feeding by elephants on small plants is sufficient to prevent their recruitment to mature size. Browsing pressure on coppice regrowth of mopane and *Brachystegia* may also be heavy enough to prevent any plants growing through to tree height; but notably mopane at least can produce seed within shrub height classes (R.J. Scholes personal communication). By eliminating baobab, *Commiphora* and other species serving as food sources during drought periods, elephants reduce the ability of future elephant populations to survive droughts. However, if such food sources were not utilized, the population would suffer higher mortality during the current drought.

Under extreme conditions elephants may eliminate woody plants, being forced then to subsist mostly on grass through the dry season. Since dormant grass is a less nutritious food source for elephants than woody browse, the carrying capacity of the habitat for elephants is reduced. The only clear example of such a situation is the Murchison Falls Park in Uganda (although there is suggestive evidence that similar conditions might have been developing in other areas). At Tsavo East in Kenya, woodlands were opened up by the near elimination of *Commiphora* over extensive areas; but other woody species, apart from baobab and *Sterculia*, were less affected, and abundant regeneration of woody scrub has occurred (Croze, Hillman & Lang 1981).

In the Ruaha Park in Tanzania, elimination of *Commiphora ugogoensis* and depression of *Acacia albida* and baobab populations has not yet exerted any notable effect on elephant population trend. However, a reduced ability of elephants to survive droughts can be anticipated. At Luangwa, the demographic parameters reported by Lewis (1984) suggest that the elephant population has stabilized. While vegetation measurements are lacking, depression of food resources appears limited to destruction of baobabs and local elimination of *Acacia* species (personal observations 1982). Since grasses and mopane shrubbery provide the staple food sources, depression of food availability would seem to be relatively minor.

White rhinos and hippos transform stands of tussock grasses to lawnlike expanses of low-growing or creeping species. Decumbent grasses have a higher leaf to stem ratio, and commonly also higher protein and lower fiber contents in their leaves, than taller grasses. Thus expanses of short grass offer a higher concentration of leaf material and a higher food quality than tall grass stands (McNaughton 1985). However, white rhinos are dependent upon reserves of taller grass for subsistence through the dry season when short grass has been grazed down to stubble. If all areas of tall grass were converted to short grass, white rhinos would become more vulnerable to starvation-induced mortality during drought periods. Trampling by white rhinos furthermore compacts soil structure, and the impact of raindrops on bared soil surfaces also reduces water infiltration. Such effects can lead to reduced grass production from short grass areas.

Hippos at Queen Elizabeth Park in Uganda stabilized during the 1950s at ecological densities three times as high than those attained by white rhinos in the western part of Umfolozi in 1970. Despite such intense grazing pressure, short grass swards persisted, apart from a temporary, drought-related decline. Soil erosion rate increased strikingly, but without worsening effects on food production for hippos (Eltringham 1980; Yoaciel 1981).

Most park populations of elephants are still in a phase of increase from the low densities that persisted at the end of the ivory hunting era at the turn of the century (Spinage 1973). The vegetation structure and composition has developed largely in the absence of elephants over a 70–100 year period. While the state of the vegetation is now being considerably modified by expanding elephant numbers, it is not yet obvious whether long term food production for elephants is being increased or depressed. Only in the Luangwa Valley and Murchison Falls Parks have elephant populations ended their phase of expansion. At Luangwa, the effects on food resources appear minor, at least superficially. At Murchison, the woody browse component was almost eliminated over a large area; but the situation was greatly exacerbated by the compression of elephants into a restricted area by surrounding human settlements.

Caughley (1976b) suggested that the elephant–woodland interaction might generate a stable limit cycle rather than quasi-equilibrium. This outcome seems unlikely, for several reasons. Firstly, even the stability of a limit cycle is unlikely to persist in African savanna environments where rainfall and hence vegetation production vary so widely between years. Secondly, the effects of elephant feeding and related damage on vegetation do not necessarily result in depression of food production. The main effect seems to be a reduction in reserve food sources important during droughts. This could make populations more susceptible to episodic mortality during drought periods. Thus populations may exhibit fluctuations in response to climatic cycles.

Dispersal

For megaherbivores, population stability (in the sense of a dynamic equilibrium with varying food production) can result from lengthened birth intervals and delays in attainment of sexual maturity. However, these population responses are too slow-acting to avert severe vegetation damage and resulting starvation mortality during drought periods. The only adjustment that can prevent or reduce episodic overexploitation of food resources during droughts is dispersal. Caughley & Krebs (1983) suggested that, while dispersal was an important population regulatory mechanism among small mammals, it was unimportant among large mammals. Are megaherbivores thus different from medium-sized ungulates in their dependence upon dispersal?

Among small mammalian herbivores such as voles and hares, there is widespread evidence that populations grow to higher levels, with associated overexploitation of food resources, where dispersal is prevented (Krebs *et al.* 1973). Emigration is an important process in the regulation of snowshoe hare populations, with overutilization of shrubs occurring at peak densities (Keith & Wyndberg 1978; Wolff 1980). Lidicker (1975) distinguished two forms of dispersal among small mammals: (i) presaturation dispersal taking place during the phase of population increase, generally by prime individuals; (ii) saturation dispersal from high density populations, mostly by juveniles and social outcasts. Among small mammals, most dispersal occurs during the presaturation phase, and may be largely density-independent (Gaines & McLenaghan 1980).

Among white rhinos, the territorial system has little influence on dispersal because it affects only a segment of adult males, and not breeding females or immatures. Presaturation dispersal, mainly by subadults, evidently occurs during the phase of population expansion (see Chapter 11). Dispersal movements seem to be partly a direct response to the depletion of habitat resources, and partly an innate tendency to wander during the subadult period. Likewise, African elephants extended their range southwards through the Kruger Park well before they reached saturation densities in the northern district of the park (Pienaar, van Wyk & Fairall 1966b).

Dispersal movements have also been documented for medium-sized ungulates (Owen-Smith 1983). The thar populations studied by Caughley (1970) expanded from nuclei of animals dispersing from the original center of release. Impalas ear-tagged as juveniles in Hluhluwe were subsequently found up to 30 km away in Umfolozi, the initial colonizers being adult males together with immature animals of both sexes (Hitchins & Vincent 1972). For reindeer on South Georgia, signs of overgrazing were followed by the emigration of 400 animals across a glacier that had formerly presented a barrier (Leader-Williams 1980). For white-tailed deer, dispersal rates of 13% per annum were shown by yearling females, and 7% per annum by adult females, with male dispersal rates even higher (Hawkins, Klimstra & Autry 1971). For kudus, in contrast, my own (unpublished) observations indicate dispersal rates of only about 1% per female per year. Dispersal movements certainly occur among most, if not all, ungulates. Rates of dispersal may be lower than they are in small mammals, but so are all other demographic rates.

Small rodents have the potential to more than double their density between years. Thus if mortality and natality rates are functions of the density levels prevailing the previous year, the degree of overshoot of saturation densities will be more extreme than it might be for populations with a lower potential rate of increase. In such circumstances it is advantageous for individuals to move elsewhere, rather than remaining on in a locality where food resources are inadequate to meet demands.

For medium-sized ungulates predation may replace or mask dispersal. Situations of severe resource depletion tend to arise only in circumstances where predator populations have been severely depressed; or where predators are completely absent, as in New Zealand and on other islands.

Megaherbivores are characterized by low mortality and natality rates, and therefore by a high degree of population inertia. They tend to attain biomass levels close to the threshold for vegetation over-exploitation. Furthermore they depress slowly renewing components of the vegetation such as trees or climax grassland. Predation, except in some circumstances by humans in recent times, is a negligible factor for adults. Laws (1969b) stated, with regard to elephant overpopulation, that the important impact of humans in recent times has been to compress populations and block dispersal movements. This conclusion can be generalized to apply to other megaherbivores as well.

Taylor & Taylor (1977) point out that dispersal does not avoid mortality; it merely shifts it elsewhere. Death rates must on average be equal to the sum of *in situ* mortality and net dispersal. Thus emigrants move into areas where their survival chances are reduced relative to source populations. It remains to be explained why dispersal tendencies persist over evolutionary time.

Taylor & Taylor argue that species are confronted repeatedly by hostile situations that are transient and spatially variable in their intensity. Under such conditions, emigrants have a chance of finding more favorable localities, whereas if they remained they would certainly experience starvation induced mortality, or at least lowered reproductive output. Theoretical aspects of this problem have been discussed by Gadgil (1971), Roff (1974, 1975) Lomnicki (1978, 1982) and Taylor (1981a,b). Lomnicki (1978) suggested that population outbreaks are most likely under homogeneous conditions where there are no vacant areas available to absorb emigrants. Hamilton & May (1977) demonstrated theoretically that, even when the mortality of emigrants was extremely high, and the environment offered no vacant sites for colonization, it could still be adaptive for parents to commit more than half of their offspring to be migrants.

Laws (1981a) suggested that in the past elephants experienced a regional mosaic of areas that fluctuated in favorability, with such fluctuations tending to be out of phase in different areas. Nevertheless, the problem remains that long term mortality rates must generally be higher than those measured in local areas in the short term. Thus either emigrants must experience chronically high mortality rates, or catastrophically high mortality must intervene at intervals in established populations.

Major animal die-offs associated with severe droughts are an acknowledged but poorly documented phenomenon. In Botswana, an estimated 15 000 wildebeest died during the 1964 drought, and at least as many again in 1970 (Child 1972). However, the situation was exacerbated by the consequences of the fences erected to control cattle movements, which restricted access to waterholes. In Kenya, the total ungulate biomass on the Athi–Kapiti plains was reduced by 44% during the severe drought of 1961 (Stewart & Zaphiro 1963; Talbot & Talbot 1963b). However, a contributory influence was competition for water with domestic livestock belonging to the Masai people. At Klaserie in the eastern Transvaal, the total biomass of grazing ungulates was reduced to 18% of pre-existing levels by two successive drought years over 1981–83. The main species affected were wildebeest, zebra, buffalo and impala and warthog, with browsers such as giraffe and kudu suffering much less. An exacerbating factor here was the widespread distribution of waterholes, so that no reserve grazing areas remained. In the neighbouring Kruger Park, population declines by susceptible species did not exceed 30–35% (Walker *et al.* 1987).

Population crashes on a similar scale have been documented for elephants at Tsavo East, and for hippos in the Kruger Park. However, elephants and rhinos have survived other equally severe droughts with only minor increases in mortality among the adult segment. The major impact of these droughts falls on juvenile recruitment, both through reduced calf survival and deferred conceptions. It may be that the populations of megaherbivores surviving droughts with little mortality had not yet saturated the carrying capacity of their habitats, so that with time these populations are likely to experience mortality on a similar scale to that recorded for elephants at Tsavo East. On the other hand, there is reason to suspect that the Tsavo East population crash may be an exceptional occurrence, which resulted only because elephants were compressed into a region of very low rainfall by surrounding human disturbance.

Where there are no fence or other restrictions, elephants and white rhinos have responded to food limitations during droughts by wide-scale movements. These may allow animals to locate regions where conditions are somewhat more favorable than elsewhere. Whether this occurs depends on the relation between the scale of spatial variability in the intensity of the drought in relation to the extent of the movements undertaken by the animals. Elephants may move over a distance of 100 km or more, white rhinos over 20 km or more. By such opportunistic movements, animals may escape the full severity of the local conditions reached during drought episodes. Hippos, on the other hand, are somewhat more restricted in their movements; and notably their demographic features tend towards more *r*-selected characteristics than those of other megaherbivores.

As pointed out by Goodman (1981), for large mammalian herbivores comparatively small increases in mortality may have quite a large influence on population trends. Combined with deferred reproduction, regional populations may sag somewhat during drought episodes, then surge back when conditions become more favorable again. However, coupled with the effects of the drought are the impact that populations have on vegetation when animals become forced to subsist on structural components such as large trees or taller grasses. Thus a mosaic effect might result, with the localities that were temporarily more favorable during droughts being severely impacted, then abandoned for a period to recover while the population shifts its centers of concentration elsewhere. In this scenario, fluctuations in regional populations appears as low amplitude, long period ripples, rather than as the more robust ebb and flo experienced by smaller species. However, at the local level plants may experience episodes of extreme concentrations of megaherbivores, followed by recovery periods after centers of animal distribution have shifted elsewhere.

Summary

Megaherbivore populations tend to reach saturation densities at which nutritional limitations restrict further increase. The major regulatory responses occur through changes in fecundity, i.e. in birth intervals and in age at sexual maturity. These may be sufficient to halt population growth without much increase in mortality. Megaherbivores may initiate changes in vegetation structure and composition somewhat below saturation biomass levels. Vegetation changes may result in increased food availability, due to the replacements of mature woodlands or grasslands by faster-growing, more nutritious pioneer plants. However, reserve sources of food used during droughts may be depressed. Dispersal movements allow population densities to adjust in the short term to regional variations on food production related to rainfall variability. Where dispersal is prevented or populations are compressed by human disturbance, local vegetation impact may become severe and precipitate a collapse by the herbivore population. However, in circumstances where large-scale movements can be undertaken, populations may avoid the most severe effects of local droughts, although not without severe impacts on vegetation structure. Regional fluctuations in megaherbivore population levels are likely to be less extreme, and take place over a somewhat longer time scale, than those of smaller ungulates.

14

Body size and ecosystem processes

Introduction

In this chapter I consider how the contribution of large herbivores to community and ecosystem processes varies with increasing body size. The ecosystem features to be covered include the biomass levels sustained, energy fluxes and nutrient cycling through this biomass, and the stability of these features over time. The basic question is, how different would these patterns and processes be if megaherbivores were absent from the system?

Biomass levels
Population biomass

The biomass level that a species population sustains represents a relation between the production of food in the environment, and the ability of animals of the species to transform the food into animal biomass. In African savanna regions, vegetation production is proportional to land surface modified by rainfall, while the resting metabolic requirements of an animal per unit of mass are proportional to its body mass raised to the power minus one-quarter. Therefore, if the amount of food available in the vegetation were independent of body size, the population biomass supported per unit of land area should vary in relation to $M^{0.25}$, i.e. larger species should tend to sustain somewhat higher biomass levels than smaller species.

However, two factors modify the simple relationship developed above. Firstly, the mass-specific metabolic requirements of free-ranging animals, allowing for activity costs, may be scaled in relation to a body mass exponent slightly different from -0.25. For herbivorous mammals, the best available estimate of the scaling exponent is -0.27 (from Nagy 1987, see Chapter 5), i.e. field metabolic requirements scale almost identically to basal metabolic requirements. Secondly, larger herbivores are able to tolerate a lower quality diet than smaller herbivores (see Chapter 5). This means that, for larger animals, a higher proportion of the available herbage becomes acceptable food. Taking into account these considerations, the population biomass levels sustained by large herbivore species should increase with increasing size according to a body mass exponent somewhat greater than 0.27.

Damuth (1981b) examined the relation between local (ecological) population densities and body mass for mammalian primary consumers. For his complete data set, covering some 307 species from mice to elephants, density was related to $M^{-0.75}$. Since biomass equals density multiplied by unit body mass, population biomass levels vary correspondingly in relation to $M^{0.25}$. Damuth also considered relations between density and body mass for local mammalian communities occupying particular habitats in different parts of the world. For none of these communities did the slope of the log–log regression between density and body mass differ significantly from -0.75. Pooling the data from these individual communities yielded an estimate for the power coefficient b relating density to body mass of -0.70 ± 0.08. Damuth concluded that energy metabolism at the population level was independent of body size.

Peters & Wassenberg (1983) examined the relation between abundance and body size for a wider range of animals than was covered by Damuth. For mammalian herbivores of north temperate regions, they obtained a value for the power coefficient b of -0.61 ± 0.14. Thus biomass levels are proportional to $M^{0.39}$. This exponent is somewhat higher than that predicted from the scaling of field metabolic requirements. Peters & Raelson (1984) subsequently found numerical density and body size among mammalian herbivores to be related to $M^{-0.88}$ on a global basis; but the overall slope is biased by the fact that small mammals in North America exhibit densities considerably higher than those shown by small mammals in the tropics. For North American herbivores, ecological density is proportional to $M^{-0.66}$, which is closely similar to the general relation reported earlier by Peters and Wassenberg. However, for 'larger-tropical' herbivores, which includes mainly animals occupying non-forest habitats in Africa and Asia, crude densities are proportional to $M^{-0.30}$, i.e. biomass proportional to $M^{0.70}$. Thus for herbivores in tropical savanna and woodland regions, larger species exhibit somewhat higher biomass levels relative to smaller species than can be explained on the basis of metabolic requirements alone, at least on a regional basis. A similar finding has been reported for local communities of birds, fishes and granivorous rodents (Brown & Maurer 1986). In other words, large species appear to use a disproportionately large share of food resources in these communities.

Herbivore census figures have been published for a number of African national parks and equivalent reserves. However, population estimates are notoriously unreliable, due to limitations in the census methods that can be applied to extensive areas. Aerial surveys tend to undercount small species that are less visible from the air, and are unsuited to woodland habitats and broken terrain. With care, and photographic back-up for large groups, aerial counts can provide reasonably unbiased estimates for populations of species from the size of buffalo upwards. For medium- and small-sized ungulates, aerial surveys need to be supplemented by ground counts, whether by vehicle or on foot, in order to work out correction factors.

Besides the Hluhluwe–Umfolozi Reserve, population estimates for a further ten national parks were selected from the literature as being reasonably unbiased for a range of small to large ungulates. Despite the wide scatter in the population biomasses shown by particular species in different areas, there is a significant positive correlation between biomass and body mass (Fig. 14.1). The weighted linear regression yields an estimate for the power coefficient b of 0.65 ± 0.33, which is close to the value obtained by Peters & Raelson (1984) for larger tropical herbivores. The data for Umfolozi are more reliable than those for any other park, being based on foot estimates from my study areas for smaller species such as duiker and steenbok, and on corrected helicopter counts for larger species. The Umfolozi biomass figures (Table 12.1) yield as estimate for b of 0.76 ± 0.46.

A tendency to undercount smaller species to a greater extent than larger species could be incorporated in even the best census data available, and this effect would tend to bias regression coefficients on the high side. To circumvent this effect, rather than considering regional population totals, one can look at the highest ecological densities reached by each species within local areas of a few square kilometers in extent, preferably estimated from specific studies on the species concerned. For the range of African large herbivores for which such data were obtainable, local population biomass varies in relation to body mass raised to the power 0.71 ± 0.28 (Fig. 14.2). However, most of the variation in local biomass levels occurs among species under 100 kg in body mass. Wildebeest in the Serengeti and waterbuck at Nakuru attain population biomasses of about 9000 kg km^{-2}, closely similar to the maximum biomass levels shown by elephant, hippo, white rhino and buffalo. Black rhino, giraffe and other mainly browsing species exhibit biomasses considerably lower than those of grazers of similar body mass. For grazing ruminants alone, biomass levels appear to increase even more steeply with body mass: for these species, an estimate for b of 1.36 ± 0.50 is obtained.

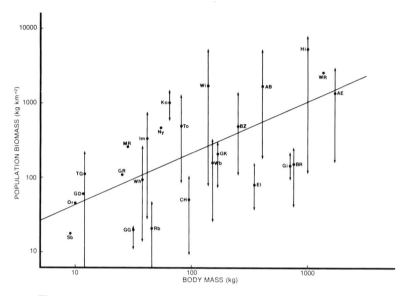

Fig. 14.1 Population biomasses for African large herbivores in relation to body mass. Key to species labels given in Appendix I. Arrows indicate the range, dots the geometric means, for the following National Parks: Amboseli, Kenya (Western 1975); Bouba Njida, Cameroon (Van Lavieren & Esser 1980); Kruger, South Africa (Pienaar 1982); Manyara, Tanzania (Mwalyosi 1977); Nairobi, Kenya (Foster & Coe 1968); Queen Elizabeth, Uganda (Field & Laws 1970; Eltringham & Din 1977); Sengwa, Zimbabwe (Cumming 1975); Serengeti, Tanzania (Sinclair & Norton-Griffiths 1979); Tsavo, Kenya (Leuthold & Leuthold 1976); Umfolozi–Hluhluwe, South Africa (Natal Parks Board unpublished records for 1970 and Table 12.1); Virunga, Zaire (Bourliere 1965). Regression line (based on weighted linear regression): $BIOM$ (kg km^{-2}) = 6.9 $M^{0.65}$ (SE(b) = 0.016, R^2 = 0.42, N = 113, P = < 0.0001).

Thus for African large herbivores, population biomass levels increase more steeply with body mass than is predicted on the basis of a simple relation between metabolic requirements and body mass. This implies that the effective food density available for consumption increased with increasing body mass, and that over a wide range in body size this increase occurs approximately in relation to $M^{0.70}/M^{0.27} = M^{0.43}$. This means that the abundance of food resources available to white rhino, with a unit body mass of 1350 kg, is over 7 times that available to Thomson's gazelle, with a unit mass of 13 kg. Among grazing ruminants up to wildebeest size, the increase in resource density occurs in almost direct relation to body mass, implying that a 200 kg wildebeest has 20 times as much food available to it as a 10 kg oribi.

Four possible hypotheses may be advanced to explain why large animals

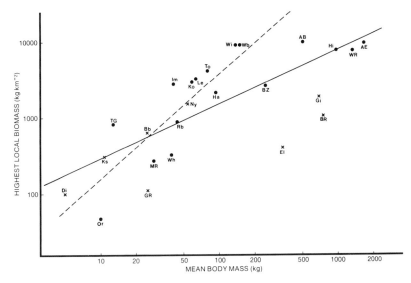

Fig. 14.2 Highest local biomasses reported for African large herbivores in relation to body mass. Dots represent grazers, crosses browsers. Data for each species as follows: African buffalo–Manyara (Mwalyosi 1977); African elephant–Manyara (Douglas-Hamilton 1972); black rhino–Tsavo (Goddard 1970a); bushbuck–Queen Elizabeth (Waser 1974); dikdik–Serengeti (Hendrichs & Hendrichs 1971); eland–Giant's Castle (Scotcher 1982); giraffe–Timbavati (Hirst 1975); greater kudu–Andries Vosloo (Allen-Rowlandson 1980); grey rhebuck–Highmoor (Oliver *et al.* 1978); hartebeest–Nairobi (Gosling 1974); hippopotamus–Virunga (Bourliere 1965); impala–Sengwa (Murray 1980); klipspringer–Semien (Dunbar & Dunbar 1974); kob–Torro (Buechner & Roth 1974); lechwe–Kafue (Sayer & Van Lavieren 1975); mountain reedbuck–Highmoor (Oliver *et al.* 1978); nyala–Lengwe (Bell 1980); oribi–Highmoor (Oliver *et al.* 1978); reedbuck–St Lucia (Venter 1979); sable antelope–Matopos (Grobler 1974); Thomson's gazelle–Ngorongoro (Estes 1967); topi–Queen Elizabeth (Jewell 1972); warthog–Sengwa (Cumming 1975); waterbuck–Nakuru (Wirtz 1981); wildebeest–Serengeti (Sinclair & Norton-Griffiths 1979); white rhino–Umfolozi (this study); zebra–Ngorongoro (Turner & Watson 1964). Regression line (all species); $LBIOM$ (kg km^{-2}) $= 56\ M^{0.71}$ (SE(b) $= 0.014$, $R^2 = 0.53$, $N = 25$, $P = <0.0001$); for grazing ruminants only (dotted): $LBIOM = 9.2\ M^{1.36}$ (SE(b) $= 0.26$, $R^2 = 0.73$, $N = 12$, $P = <0.0004$).

utilize a disproportionately large fraction of food resources relative to smaller animals.

1. Large animals can tolerate a lower quality food intake than can smaller animals, as a consequence of the gut capacity–metabolic rate relation (the Jarman–Bell Principle, Chapter 5).
2. Acceptable food patches are more continuously distributed for larger species than for smaller species, with fewer lacunae of unsuitable habitat (Peters & Wassenberg 1983).

3. Being influenced less by predation, larger animals attain densities closer to the saturation capacity of the vegetation than do smaller species (Chapter 13).
4. Larger animals dominate smaller animals in competition for food (Brown & Maurer 1986).

Let me first dispose of the latter two hypotheses. Medium-sized antelopes like wildebeest and waterbuck appear no less susceptible to predation than small antelope like reedbuck and gazelles (see Schaller 1972); yet still achieve a vastly higher metabolic biomass than the latter. Relative invulnerability to predation is a feature only of species in the megaherbivore size range, but biomass levels do not increase significantly between wildebeest size and rhino size.

While interspecific competition between large and small species has been documented for the granivorous rodents studied by Brown *et al.* (1986), it is not a feature of large ungulate communities. Instead the grazing and browsing effects of large species tend more often to increase the availability of herbage that is nutritionally acceptable to small species than the reverse (McNaughton 1976; Chapter 12).

As documented in Chapter 6, large ungulates accept a greater dilution of nutrients by indigestible fiber than small ungulates. This allows them to utilize vegetation components that would be submaintenance in quality for small species, particularly during the dry season period of food restriction; for example the mature dry leaves and stems of grasses, and the roots, bark and twigs of woody plants. There are small mammals with cecalid digestion, including hares and certain rodents, which also consume fibrous plant tissues such as bark and roots, compensating for low nutrient concentrations through increased food intake. Nevertheless, the extent of removal of plant biomass by large herbivores vastly exceeds that documented for small mammals (although not all of the material consumed is digested).

When conditions are favorable, i.e. high soil nutrients relative to rainfall, medium-sized antelope like wildebeest or impala (and domestic cattle) may remove as high a fraction of plant biomass as megaherbivores such as white rhino or hippo. Hence the maximum biomass levels attained for example by wildebeest on the Serengeti Plains are metabolically at least equivalent to the highest biomass levels recorded for megaherbivore species (Fig. 14.2). Below about 100 kg in body mass, ungulates becomes increasingly restricted to the green leaf fraction of the vegetation (Chapter 6).

Very large herbivores furthermore have the ability to utilize vegetation growing in regions of higher rainfall where medium-sized ungulates become restricted in their distribution largely to zones of nutrient concentration, such as floodplains. Although less favorable in terms of nutrient:fiber or secondary metabolite ratios, overall vegetation production in such regions is greater than it is in semi-arid savannas. A notable example is the grasslands of Uganda and Zaire bordering the lakes of the western rift valley, where the biomass is dominated by elephant, hippo and buffalo. However, African elephants are not adapted anatomically to exploit grasses efficiently, while white rhinos at Umfolozi have not been allowed to reach their equilibrium density. An elephant-sized grazer, such as the extinct mammoths or *Elephas recki*, would probably have outperformed any extant species in terms of ability to transform plant biomass into animal biomass.

The extent of the increase in resource use with body size among herbivores in the size range 5–200 kg appears far greater than can be explained by the Jarman–Bell Principle alone. Furthermore, this principle is less applicable to omnivores feeding on seeds and animal matter, which vary relatively little in nutritional quality, than it is to folivorous herbivores. Hence it does not explain the disproportionate use of food resources by large species in communities of birds and rodents, as reported by Brown & Maurer (1986).

On a regional scale, small antelope species like duiker and steenbok appear no less widely distributed than megaherbivores such as elephant and giraffe (for example as shown by species distribution maps for the Kruger Park, Pienaar 1963). However, at the scale of habitat patches as represented by plant communities, steenbok are notably more restricted in their occurrence than are larger browsers like kudu and giraffe (J. T. du Toit in preparation). Thus, large herbivores appear to be habitat generalists, or at least to respond to a larger scale of habitat patchiness, relative to smaller herbivores. This pattern is likely to be typical of other trophic categories besides large herbivores. The patch dispersion hypothesis thus provides the most general explanation for the monopolization of resources by large species at habitat scales equivalent to home range size or larger.

Community biomass

Coe, Cumming & Phillipson (1976) demonstrated a significant relation between the total biomass of the large herbivore community in different African savanna ecosystems and mean annual rainfall. They related this to the controlling influence of rainfall on primary production (Rosenzweig 1968; Rutherford 1980). However, certain areas exhibited a large herbivore biomass about twice as great as that predicted on the basis of the overall mean regression. Coe *et al.* (1976) explained the exceptionally high biomass of large herbivores in the Manyara and Amboseli parks as due to the additional influence of ground water on primary production; and the high biomass in the Virunga and Queen Elizabeth parks to the eutrophic volcanic ash soils prevalent. Notably, however, in all four of the exceptional parks the biomass is dominated by megaherbivores – either elephants or hippos, or both. Umfolozi Game Reserve also exhibits a large herbivore biomass about one third higher than is predicted by the regression derived by Coe *et al.* (1976) for a rainfall of 700 mm (the data for Umfolozi used by Coe *et al.* were derived from earlier, inaccurate censuses by fixed-wing aircraft).

Bell (1982) pointed out that areas underlaid by basement granite were poorly represented in the data set used by Coe *et al.* (1976), and that regions of basement geology exhibit a lower biomass of large herbivores for a given rainfall than areas with volcanic soils or nutrient rich sedimentary deposits. The effect of basement geology is most marked where rainfall exceeds 700 mm per annum, due to leaching effects on soil nutrients. Furthermore, Bell suggests that the predominance of large animals, in particular elephant and buffalo, tends to be higher in savanna regions with basement geology and relatively high rainfall than in nutrient-rich savannas. This effect he related to the tolerance of very large herbivores for the lower quality, more fibrous vegetation that predominates under conditions of low soil nutrient status.

East (1984) divided African large herbivores into two groups depending on how their biomass levels responded to soil nutrient status. Species associated with arid/eutrophic savannas decline in biomass in regions underlaid by basement geology where annual rainfall exceeds about 800 mm; while species associated with moist/dystrophic savannas show biomass levels that increase steadily with rainfall, up to rainfalls about 1000 mm per annum, independently of geological substrate. All megaherbivore species were allied with the arid/eutrophic division of savannas. Elephants in particular showed considerably lower biomass levels, for a given rainfall, in regions underlaid by basement geology than in areas of volcanic material or karroo sediments. However, elephant, hippo and black rhino, together with buffalo and eland, are widely distributed in areas of basement geology with annual rainfalls exceeding 1000 mm; whereas smaller ungulates associated with arid/eutrophic savannas have a restricted occurrence in such regions.

For many African national parks, megaherbivore species together make up from 40% to 70% of the total large herbivore biomass (Fig. 14.3). The exceptions include (i) areas where elephants have been exterminated, or reduced to low numbers, such as Hluhluwe (South Africa), Bouba Njida (Cameroons) and Akagera (Rwanda); (ii) areas of open grassy savanna

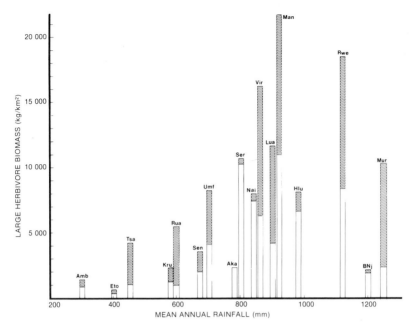

Fig. 14.3 The total biomass of large herbivores in different African ecosystems in relation to rainfall. Shaded segments represent the fraction of the total biomass formed by megaherbivores. Key to species labels given in Appendix I. Based on data reported for the following National Parks: Aka–Akagera, Rwanda (Spinage *et al.* 1972); Amb–Amboseli, Kenya (Western 1975); BNj–Bouba Njida, Cameroon (Van Lavieren & Esser 1980); Eto–Etosha, Namibia (Berry 1980); Kas–Kasungu, Malawi (Bell 1981b); Kru–Kruger, South Africa (Pienaar 1982); Lua–Luangwa, Zambia (Naylor *et al.* 1973); Man–Manyara, Tanzania (Mwalyosi 1977); Mur–Murchison Falls South, Uganda (Laws *et al.* 1975); Nai–Nairobi, Kenya, in 1961 (Foster & Coe 1968); Rua–Ruaha, Tanzania (Barnes & Douglas-Hamilton 1982); Rwe–Rwenzori (Queen Elizabeth), Uganda (Field & Laws 1970; Eltringham & Din 1977); Sen–Sengwa, Zimbabwe (Cumming 1975); Ser–Serengeti, Tanzania (Sinclair & Norton-Griffiths 1979); Tsa–Tsavo, Kenya (Leuthold & Leuthold 1976); Umf–Umfolozi; Hlu–Hluhluwe, South Africa (Table 12.1); Vir–Virunga, Zaire (Bourlière 1965).

where elephants are uncommon, such as Serengeti (Tanzania), and Nairobi (Kenya). In other areas where open grasslands are prevalent, such as Virunga (Zaire) and Rwenzori (Uganda), the abundance of hippos compensates for the moderate densities of elephants; while at Umfolozi white rhinos are the dominant large herbivore.

No relation between the proportion of the total biomass made up by megaherbivores and prevailing rainfall is apparent in Fig. 14.3. The most extreme predominance by megaherbivores occurs in Tsavo (Kenya), under

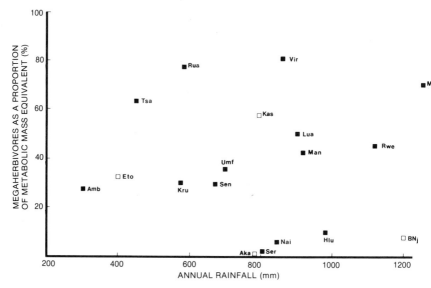

Fig. 14.4 Megaherbivores as a proportion of the metabolic mass equivalent ($M^{0.75}$) of the total large herbivore community in different African ecosystems, in relation to mean annual rainfall. Key to labels and references as in Figure 14.3. Closed symbols represent eutrophic geological substrates, open symbols represent oligotrophic geological substrates.

fairly arid conditions. There are insufficient data for regions of basement substrates to examine the effect of underlying geology on megaherbivore predominance. It appears that megaherbivores of one or more species form the major biomass component in a wide variety of African savanna ecosystems, irrespective of rainfall or geology.

Energy flux

Because of the metabolic weight–body size relation, the fraction that the megaherbivore component contributes to energy turnover by the large herbivore community is somewhat less than is suggested by relative population biomass levels. Proportional contributions to energy turnover may be estimated by transforming body mass into the metabolic mass equivalent, i.e. $M^{0.75}$. On this basis, megaherbivores are responsible for between 30% and 60% of the total energy flux through the large herbivore community for a range of African ecosystems (Fig. 14.4). Exceptions to this pattern are the East African grassy savannas; but this discrepancy may be due simply to the absence of white rhinos, which were formerly abundant in this region, as judged by their prevalence in Pleistocene deposits at Olduvai in Tanzania.

Where megaherbivores are absent, energy flux through the large herbivore component would be reduced correspondingly, unless there were compensatory increases in the density of smaller ungulates. The available evidence suggests that, if compensation occurs, it is only partial. Parks containing megaherbivores exhibit a higher total biomass per unit area of large herbivores than parks under similar rainfall conditions where megaherbivores are absent, or greatly reduced in numbers (Fig. 14.3). The very high elephant biomass at Manyara and Luangwa is not associated with any obvious impoverishment in other ungulate species. In the Mweya Peninsula in Queen Elizabeth Park, a five-fold increase in hippos between 1967 and 1973 was associated with declines by buffalo and elephant; but nevertheless energy flux through the large herbivore community increased to a record 160 000 joules h^{-1} km^{-2} in 1973 (Eltringham 1980). In Umfolozi, overall large herbivore biomass and energy flux increased as the white rhino population expanded, until populations were reduced by culling after 1970.

It is widely claimed that ruminant artiodactyls tended to replace cecalid perissodactyls and proboscideans during the course of the Plio-Pleistocene, owing to the superior efficiency of foregut fermentation over hindgut fermentation. This conclusion was challenged by Cifelli (1981) on the basis of the fossil record. The main decline in perissodactyls took place fifty million years ago during the late Eocene, and the radiation of artiodactyls during the Pliocene and Pleistocene was not accompanied by any detectable effect on perissodactyl diversity. In fact periods of increase in generic diversity have tended to coincide in the two orders, suggesting that ruminants and cecalids responded independently to the same environmental factors.

Data for a number of African conservation areas show that cecalids account on average for 35% of the energy metabolism among large herbivores, and that the proportion can be 50% or more in some areas (Fig. 14.5). If *Elephas recki* or white rhino had persisted in many of these faunas, this proportion would be higher still. Thus the amount of energy transformed by hindgut fermentation matches that taking place among the array of foregut fermenters. There is no basis for the contention that hindgut fermentation is less efficient than foregut fermentation, in terms of the primary production metabolized.

The main difference is that ruminants exhibit a much greater diversity of species than cecalids. Typical African ecosystems contain no more than 2–4 species of cecalid, compared with some 12–15 species of ruminant. This contrast in species richness can be explained on the basis of fiber tolerance.

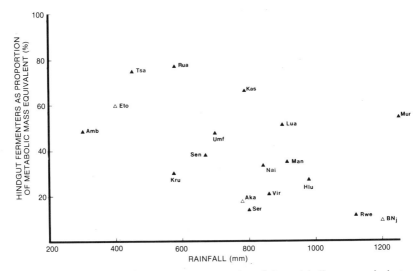

Fig. 14.5 Hindgut fermenters as a proportion of the metabolic mass equivalent ($M^{0.75}$) of the total large herbivore community in different African ecosystems, in relation to mean annual rainfall. Key to labels and references as in Figure 14.3.

Being more fiber-tolerant, cecalid species can exploit a wider range of plant parts and forms than can any one species of ruminant; although, as is typical of ecological generalists, they utilize these somewhat inefficiently. A ruminant species is a specialist for a narrow range of fiber contents, restricted not only by the effects of fiber levels on nutrient concentrations, but also by the consequences for rates of digestive throughput. Bell (1969) noted that zebras have a significantly wider distribution in East Africa than any species of large ruminant, and a similar pattern is evident elsewhere. These relationships have been modelled, and their consequences for niche separation among African ungulates discussed, by Owen-Smith (1985).

The two extant species of elephant are supreme generalists. They consume a spectrum of plant parts from carbohydrate-rich fruits to tall grasses, bark and even the trunks of certain trees. In ecological terms, success is to be measured not in terms of species richness, but rather in terms of the fraction of primary production transformed into animal biomass per unit of land area. Individual species of ruminants may outperform megaherbivores in localized habitats; but if trophic efficiency is assesed within an ecosystem-wide context, African elephants and white rhinos metabolize more plant energy into animal biomass than the most successful species of ruminant, except in extremely eutrophic regions.

Nutrient cycling

Soil nutrient status generally declines with increasing rainfall due to leaching effects. However, Botkin, Mellilo & Wu (1981) presented evidence that, in African savanna ecosystems, soil nutrient status (as assessed by percent base saturation) tends to increase over the annual rainfall range 700–1200 mm, declining at higher rainfall. They related this pattern to the trend shown by large herbivore biomass, which peaks in the range 900–1100 mm annual rainfall. These authors concluded that the large herbivore community makes an important contribution to soil nutrient status by maintaining a pool of nutrients in animal biomass against the forces of leaching. Even though the animal pool is small relative to the nutrient pools in vegetation and soil, its contribution to nutrient cycling is disproportionate, since the turnover rate of animal biomass is vastly greater than that of other pools. Furthermore, nutrients released in faeces and urine are immediately available for quick uptake by plants. In the rainfall range emphasized by Botkin *et al.* (1981), megaherbivores typically contribute 40–80% of the metabolic flux through the large herbivore component (Fig. 14.4). Furthermore, by trampling or pushing over tall grasslands and mature woodlands, megaherbivores accelerate the release of nutrients from these structural tissues via the decomposer chain. By transforming mature phases of vegetation with a slow turnover rate to pioneer phases with more rapid growth rates and higher nutrient concentrations, megaherbivores also increase the amount of acceptable forage available for consumption by smaller ungulates. Changes in soil nutrient status moreover favor faster-growing, more nutrient-rich plant species.

Where herbivore densities are low, the nutrients contained in grass biomass and in tree trunks are released through the agency of fire. However, a large fraction of the nitrogen content is volatilized, while mineral nutrients deposited on the bared soil are susceptible to leaching. By reducing grass biomass, grazing megaherbivores such as white rhino and hippo reduce the incidence of fire. The tendency of woody plants to invade grasslands in the absence of fire may be suppressed by simultaneous browsing pressure from elephant and black rhino.

Jachman & Bell (1985) document the conversion of less productive mature woodlands to more productive stands of shrubby regeneration in the Kasungu Park in Malawi. They propose that this beneficial consequence is effective only in moist dystrophic savannas; and that, even in these savanna forms, increased browse production may fail to materialize where grassy conditions supporting intense fires develop following the opening up of the woodland canopy. In eutrophic savannas, where the woodland is generally dominated by *Acacia* species, open grassy conditions tend to replace savanna woodlands, due to the combined effects of fire and sustained heavy browsing on the regenerating *Acacia* saplings (Bell 1984).

However, in eutrophic savannas most of the grass biomass is of an acceptable quality to be grazed down by white rhinos, and by medium-sized ungulates like wildebeest if they occur in sufficient concentrations. The result could be a herbivore controlled savanna with an open shrubby woody component and predominantly short grass cover, with fire in consequence an insignificant factor. The energy and nutrient fluxes occurring where such conditions prevailed in the past would have been vastly greater than anything recorded in existing ecosystems today, at least where water was not limiting. Nevertheless, there would still have been large areas of Africa far from surface water, where grazing ungulates and elephants would have been sparse, and fire the ultimate non-selective consumer of vegetation biomass.

Ecosystem stability and disturbance

Much of developing ecological theory at the population, community and ecosystem levels has been based on notions that conditions tend towards stasis if undisturbed by human intervention, and thereafter resist further perturbations. However, as the period over which scientists have monitored African savanna ecosystems has been extended (currently up to three decades), stability has become more elusive. The wildebeest population in the Serengeti has increased six-fold, and the buffalo population 2.5-fold, following the elimination of the disease rinderpest from the region. In the Kruger Park, kudu increased 2.5-fold over the decade that I monitored population dynamics, in response to changing rainfall (Owen-Smith 1984). Elephant densities have changed dramatically in a number of parks, due to immigration, poaching or drought. During the 1890s, populations of buffalo, wildebeest and many other ruminants crashed to low levels through most of Africa as the rinderpest pandemic swept across the continent. During the 19th century, elephant and white rhino were reduced to low numbers everywhere, except for elephants in the equatorial forest region, by hunters with firearms.

An alternative ecological perspective, currently taking hold, considers that equilibrium conditions are transient, soon disturbed by events ranging in frequency and scale from fire and rainfall fluctuations to tectonic events and global climatic shifts (Pickett & White 1985; Hansen in preparation).

The pages of this book have revealed little evidence of stasis by megaherbivore populations in recent years. What historic or prehistoric evidence is available further reinforces the picture of disequilibrium and transformation. Archeological evidence from the Tsavo region of Kenya indicates that the vegetation was considerably more open some 100 or more years ago than it was prior to the elephant impact of the 1960s (Thorbahn 1984: Tyrrell 1985). The *Commiphora* shrub steppe prevalent there until recently probably arose following the near elimination of elephants during the ivory hunting era of the 19th century. The dense shrubbery favored tsetse fly, making the region unsuitable for cattle-owning peoples; and it was the lack of a settled human population in the area that led to Tsavo being proclaimed a national park in 1948, rather than any great abundance of animals (Parker 1983).

In recent history, periods of concern about overgrazing by white rhinos, hippos and other ungulates, and damage by elephants to woody plants, have invariably coincided with episodes of below average rainfall (Brooks & Macdonald 1983; Phillipson 1975: Walker & Goodman 1983). Annual vegetation production may decline to 50% of the mean level, and catastrophic mortality among medium-sized ungulates may result if such conditions persist for two consecutive years. Megaherbivores are not immune to large-scale mortality under such conditions, as shown for elephants at Tsavo East and hippos in the northern Kruger Park (Chapter 13). Nevertheless, megaherbivores are more resilient than smaller ungulates to the effects of drought, generally responding by a decline in conceptions rather than by much increase in adult mortality. To subsist through the crisis periods, megaherbivores turn to vegetation components such as woody stems and less nutritious grasslands. The damage that they cause to structural features of vegetation during such episodes of low rainfall may leave a lasting impact on vegetation structure, and drastically alter habitat conditions for a range of other animal and plant species.

In the absence of megaherbivores, fire becomes the major influence on savanna structure. It is this situation that prevailed over much of Africa, especially in the southern part of the continent, until reversed in a few protected parks during the last quarter-century. For the rest of biological time, prior to the invasion by western man and his weapons, megaherbivores were likely to have been a dominant influence on ecosystem structure and dynamics over much of the savanna region of Africa and, as will be shown in the next chapter, the rest of the world.

15

Late Pleistocene extinctions

Introduction

Prior to the late Pleistocene, megaherbivores were represented by a wider variety of taxa than occur today, and were present on all continents. Their disappearance from Europe and the Americas took place at the end of the last glacial period of the Pleistocene, around 11 000 years ago, and was synchronous with the extinction of numerous other large mammal forms. The extinctions occurred during a time of rapid climatic change, with associated transformations in habitat conditions. Another important event took place at about the same time: the entry of humans into the Americas, following their expansion through the furthest corners of the Old World. The relative importance of climate and associated habitat changes versus human predation as causal agents in the late Pleistocene extinctions remains an unresolved problem (Martin & Wright 1967; Remmert 1982; Martin & Klein 1984).

Since climatic change and human range expansion are so closely interwoven in time, wider patterns need to be considered in order to understand the causal links in these extinctions. These include the geographic distribution of extinctions, and variations in the incidence of extinctions among genera of differing body size. In the following analysis I focus specifically on large mammalian herbivores, since it is generally accepted that extinctions of carnivores and of large scavenging birds were related to their dependence upon the herbivores as a food source. It is the herbivores that are likely to be most responsive to changing habitats; and it is also such species that were the prime targets as prey for the expanding human population. By large herbivores, I mean those species weighing more than about 5 kg (the minimum ungulate body mass).

The weight of circumstantial evidence will be evaluated in relation to these questions: (a) Is the climatic shift that occurred around the end of the Pleistocene, and associated habitat changes, a sufficient explanation of the pattern of extinctions among large herbivores? (b) Can human predation alone account for the extinctions? (c) Were both of these factors acting in concert an essential requirement for the extinctions? (d) What were the causal mechanisms leading to the extinctions?

Pattern of extinctions

In North America, terminal dates for extinct species span a narrow range in time between 12 000 and 10 000 years ago. Thirty out of the 40 genera of large herbivores that were reportedly extant during the earlier part of the late Pleistocene (Rancholabrean) disappeared around this time. Extinctions included four genera of megaherbivore: *Mammuthus* (mammoth), *Mammut* (mastodont), *Cuvieronius* (gomphothere) and *Eremotherium* (ground sloth), together with other genera of ground sloths, muskox, deer, camelid and equid. Interestingly, most of the surviving species are from genera that had immigrated into North America from Eurasia earlier in the Pleistocene (Gilbert & Martin 1984; Martin 1984a).

South America suffered an even higher number of extinctions, involving 41 genera of large herbivore. Among these were six genera of megaherbivore: the gomphotheres *Cuvieronius, Haplomastodon* and *Stegomastodon; Toxodon*, a hippo-like notoungulate; and the giant ground sloths *Eremotherium* and *Megatherium*. Also disappearing were genera of equid, camelid, litoptern, edentate and cervid. Terminal dates span a similar time range to those in North America (E. Anderson, 1984; Martin 1984a).

In Europe, six species of megaherbivore belonging to five genera disappeared during the course of the late Pleistocene: *Mammuthus primigenius* (woolly mammoth), *Elephas namadicus* (straight-tusked elephant), *Coelodonta antiquitatis* (woolly rhino), *Dicerorhinus kirchbergensis* (forest rhino), *D. hemitoechus* (steppe rhino) and *Hippopotamus antiquus* (which was somewhat larger than the living *H. amphibius*). Medium-sized ungulates disappearing included the Irish elk *Megaloceros*, the muskox *Ovibos* (which survives in North America), and *Saiga* (which still occurs in temperate Asia). Mastodonts (*Mammut*) became extinct in Europe early in the Pleistocene. *Elephas, Dicerorhinus* and *Hippopotamus* disappeared from continental Europe around the end of the last interglacial some 100 000 years BP, although all of these genera persisted in tropical Asia or Africa. Mammoths existed in Russia until about 11 000 years BP, but disappeared from central and southern Europe somewhat earlier. Dwarf elephants (*Elephas falconeri*) standing only a meter high at the shoulder, and similarly dwarfed hippos, survived in Sicily and other Mediterranean islands until as recently as 8000 years BP. In northern Asia, woolly mammoth and woolly rhino persisted in Sibera until around 11 000 years BP. The giant grazing rhinoceros *Elasmotherium*, bearing a massive single horn on its forehead, survived on the steppes of Manchuria until some time in the late Pleistocene, although precise dates are unavailable (E. Anderson 1984; Davis 1985; Fortelius 1982; Kurtén 1968; Martin 1984a; Vereshchagin & Baryshnikov 1984).

In Australia, a comparable episode of extinctions occurred between 26 000 and 15 000 years BP. This included the rhino-sized *Diprotodon*, giant wombats, and several species of giant kangaroo (Horton 1984).

In southern Africa, three genera of large herbivore disappeared at the end of the Pleistocene between 12 000 and 9500 years ago: the giant buffalo *Pelorovis*, the giant hartebeest *Megalotragus*, and a giant warthog *Metridiochoerus*. Other ungulate species becoming extinct in southern Africa at about the same time included the giant Cape horse *Equus capensis*, and two species of springbok (*Antidorcas*). *Pelorovis* was absent from East Africa after the late Pleistocene, but persisted in the Saharan region until about 4000 years ago. Its extinction from the latter area coincided with advancing desert conditions. An earlier pulse of extinctions occurred between the early and Middle Stone Age periods some time between 200 000 and 130 000 years BP. This included *Elephas recki* (= *iolensis*), *Hippopotamus gorgops, Sivatherium*, the hartebeest genus *Parmularius*, and several large suids (Klein 1984a, 1984b; Martin 1984a).

Further waves of extinction occurred in Madagascar and New Zealand less than 1000 years ago. In Madagascar these included a dwarf hippopotamus as well as giant lemurs and the elephant bird *Aepyornis*. In New Zealand, where mammals were completely lacking apart from bats, extinctions involved particularly the giant herbivorous birds known as moas (Dinornithidae) (A. Anderson 1984; E. Anderson 1984; Martin 1984a).

The wave of extinctions was thus especially severe in both North and South America, involving proportionally about 75% of the genera of large mammalian herbivores on both subcontinents (Fig. 15.1). In Europe and Australia, the episode of extinctions was less severe, accounting for about 45% of large herbivore genera, while the timing of extinctions was somewhat more spread out than in the Americas. In Africa, large herbivore extinctions were minor and no different from earlier pulses during the Pleistocene.

In those continents suffering severe extinctions, the incidence of generic extinction was positively correlated with body size. All megaherbivore genera disappeared, compared with 76% of genera in the size range 100–1000 kg, and 41% of genera between 5 kg and 100 kg. Among small mammalian herbivores weighing less than 5 kg, extinctions involved less than 2% of genera (Fig. 15.2).

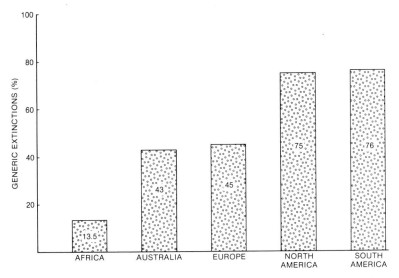

Fig. 15.1 Geographic patterning of late Pleistocene extinctions of mammalian large herbivore genera (this includes all extinctions occurring during the course of the late Pleistocene, i.e. the last 130 000 years, of genera with body masses > 5kg). Data from E. Anderson (1984), Klein (1984a, 1984b), Martin (1984a).

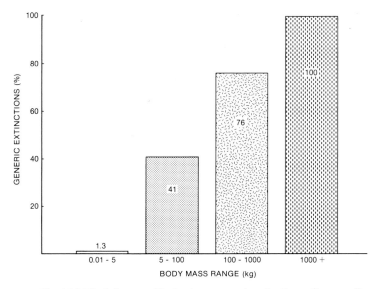

Fig. 15.2 The influence of body size on generic extinctions of mammalian herbivores in North America, South America, Europe and Australia. Sources as in Fig. 15.1.

Climatic change

The paleontological record shows that episodes of extinction commonly coincide with periods of rapid climatic shift. One such period was the end of the Miocene some five million years BP, when global temperatures dropped 5–10 °C and conditions became somewhat drier than they had been earlier (Brain 1985). At about that time more than 60 genera of land mammals (of which 35 weighed more than 5 kg) went extinct in North America, including the last North American rhinoceros (*Teleoceras*) and several genera of gomphothere. Another wave of extinctions occurred around the end of the first glacial cycle of the Pleistocene around two million years ago, when 35 genera of mammal (20 weighing over 5 kg) disappeared from North America. The next major extinction event was that at the end of the Pleistocene, when a total of 43 mammalian genera (39 of which exceeded 5 kg) went extinct. Nevertheless, the total generic richness of the mammalian fauna of North America did not decline until the Holocene, as prior to this time extinctions were balanced by originations and immigrations of new genera (Gingerich 1984; Webb 1984). In Africa, periods of turnover in bovid lineages, including both extinctions and speciations, appear to be associated in time with climatic shifts between wet (pluvial) and dry (interpluvial) periods during the course of the Pleistocene (Vrba 1985).

The habitat changes accompanying the end of the last Ice Age have been documented in some detail for the far north of Eurasia and North America (Hopkins *et al.* 1982). During the glacial period, an open grassy steppe extended from Alaska and north-western Canada through Siberia to northern Europe. The evidence for the grassland rests largely on the associated fauna. Large primarily grazing herbivores predominated, including woolly mammoth, woolly rhino (in Eurasia only), bison, equids, sheep and camelids, together with mixed feeders such as caribou, muskoxen and pronghorn. Other grassland species included badgers, ferrets, ground squirrels and voles. Also present were large carnivores including sabertooths, lions and canids. Browsing ungulates such as moose, elk and smaller deer were rare. How grassy or barren this arctic steppe was is still a source of contention. The pollen record shows only a sparse presence of grasses, together with low shrubs such as *Artemisia* and, in Siberia only, *Selaginella* mosses. Belts of woodland possibly occurred along river valleys. Winter temperatures were frigid, with strong winds redistributing soils in the form of fertile loess deposits. Precipitation was low and probably mainly in summer, so that snow cover was sparse. Winds probably drifted snow into valleys, promoting localized areas of higher plant biomass. Continental glaciers occurred to the east on the Canadian shield, and in the west on the Scandinavian shield.

Global warming began around 14 000 years BP, initiating a rise in sea levels. The Beringian land bridge connecting Alaska to Siberia became submerged, allowing warmer water to penetrate the Arctic Ocean. Winds diminished in intensity as continental glaciers retreated, and more precipitation fell during autumn and winter. The pollen record shows that tree birch (*Betula*) extended its range steadily northwards in North America between 14 000 and 8000 years BP, followed shortly by spruce (*Picea*). The present vegetation in much of Alaska and Siberia is low shrub tundra, with graminoids restricted to waterlogged meadows or stony upland sites. Soils are waterlogged due to underlying permafrost, and leached of nutrients. Large herbivore densities are low, with caribou, muskoxen and moose predominating (Ritchie & Cwyner 1982; Guthrie 1982, 1984; Hopkins 1982; Matthews 1982; Schweger 1982).

The environmental changes that occurred in the far north were extreme, involving a radical transformation of the entire ecosystem – climate, soils, vegetation and fauna. It is difficult to imagine a grazing ungulate fauna surviving in the habitats occupying this region today. Mountain sheep remain in mountainous areas, and bison persist to the south; but mammoth, woolly rhino, equids, camelids and antilocaprids have vanished.

In the conterminous United States, the mid-continental grasslands and southern deserts date only from the early Holocene. During the last (Wisconsin) glacial, the prevailing vegetation over much of the continent was a coniferous parkland or open forest, with an admixture of hardwoods and an understorey of shrubs, grasses and forbs. This parkland occurred south of the steppe-tundra in a broad band from the east coast to Wyoming and as far south as Georgia and Kansas. Open herbaceous vegetation similar to the modern prairie occupied southern Illinois until 35 000 years BP. Wetter conditions than those of today prevailed in the region of the Great Salt Lake in Utah. Communities of animals and plants were generally more diverse than they are at present. For example, in the Appalachian region mammoth, mastodont, ground sloths, bison, equids, muskoxen, caribou, stag-moose (*Cervalces*), deer and tapir coexisted during the late Pleistocene.

The border between steppe and forest vegetation shifted steadily northwards as conditions warmed at the end of the Pleistocene, reaching northern Minnesota about 11 000 years BP and Quebec about 10 000 years BP. In South America, the most pronounced change was a large-scale contraction of open steppe habitats about 10 000 years BP, with forests thereafter becoming much more extensive (Gilbert & Martin 1984; Guilday 1984; Markgraf 1985; Wright 1977, 1984)

In Europe, open steppe vegetation extended through the British Isles and into northern France during the height of the last (Weichselian) glaciation. During this period, around 20 000 years ago, mammoth, bison, horse and reindeer were abundant in France, as documented in cave paintings. To the south of the herbaceous steppe, a shrub steppe graded into forest composed of spruce, pine, larch, birch and alder. As the climate warmed, shrubs extended northwards, with tree birch reaching Denmark and Poland about 12 500 years BP. By 11 000 years BP, a birch–poplar–pine forest covered most of north-western Europe, although the forest was open and parklike on its northern fringe. Areas of open larch parkland still persist today in places in Siberia (Wright 1977; Yurtsev 1982).

In Australia, hyperarid conditions expanded over most of the continent between 26 000 and 15 000 years BP. During this period, woodlands became restricted to a narrow band along the east coast (Hope 1984; Horton 1984).

In Africa, the Pleistocene was characterized by alternating wetter and drier conditions. Drier conditions favored a predominance of grazing ungulates, and wetter ones a predominance of browsers. A mainly grazing ungulate fauna prevailed in the Cape between 30 000 and 12 000 years BP, since which time browsing ungulates have predominated. At a site in the northern Transvaal where the local vegetation is currently mixed deciduous savanna, vegetation over the past 35 000 years varied between an open forb-rich grassland with forest patches, and semi-arid Kalahari thornveld. At one time in the late Pleistocene, open country grazers like blesbok and springbok extended their ranges into Zimbabwe where *Brachystegia* wood-lands now prevail; while at other times lechwe, a grazer of seasonally flooded grasslands, occurred in the southern Kalahari region. On the East African plateau, conditions were drier than at present during the latter part of the last glacial period and open grassy vegetation was prevalent. At the beginning of the Holocene, there was an expansion of forests, although not sufficient to displace the grassy savanna on most of the plateau (Klein 1984b; Livingstone 1975; Scott 1984; Vrba 1985).

The habitat changes that occurred at the end of the Pleistocene involved more than just latitudinal shifts in vegetation biomes. Species of plant and animal responded individually to environmental changes, so that the eco-logical communities we find today are very different from those existing during the Pleistocene. Graham & Lundelius (1984) proposed that 'coevolutionary disequilibrium' between plants and herbivores was a factor in the extinctions. Guthrie (1984) suggested that the mosaic interspersion of communities that prevailed during the Pleistocene was transformed into a broader scale zonation of vegetation in the Holocene, and that the resulting decline in local habitat diversity made it more difficult for large herbivores to secure their seasonal habitat requirements.

However, to account for the late Pleistocene extinctions, evidence is needed that the climatic and habitat changes between the last glacial period and the Holocene were more extreme in rate or magnitude than was the case during the numerous previous transitions between glacial and interglacial conditions. Guthrie (1984) proposed that there is a trend of intensifying seasonality, reaching greater extremes during the Holocene than during previous glacial or interglacial periods. The evidence for this is drawn largely from the differences in faunal community structure between the Pleistocene and the Holocene. Species that today exist only in the central and southern United States coexisted in Pleistocene faunal assemblages alongside species now occurring only in the far north of Alaska and Canada. Some evidence of increasing resource restrictions comes from the decline in body size shown by many lineages, including mammoths, towards the end of the Pleistocene. However, the record of temperature changes, as indexed by oxygen isotope ratios both in marine foraminifera and in Antarctic ice, shows no significant difference in rate between the Illinois–Sangamon and Wisconsin–Holocene transitions. Furthermore, during the Sangamon interglacial temperatures evidently rose to 2–3 °C warmer than those experienced during Recent times (Shackleton *et al.* 1983; Lorius *et al.* 1985). Hence the reorganization of habitats relative to glacial conditions was likely to have been more extreme then than at present.

The extreme habitat transformations that occurred in the far north are sufficient to account for the extinctions of grazing ungulates in this region. However, the almost synchronous disappearance of a variety of large herbivore genera at lower latitudes in North America, and from tropical as well as temperate South America, remains unexplained. All of these lineages had survived numerous previous transitions between glacial and interglacial conditions, albeit as new species or with evolutionary modifications. In Europe, grazers, including horse, ass, camel, yak and saiga, survived on the steppes of central Asia; only megaherbivores and muskox disappeared. Extinctions were equally severe in both North and South America; while in Africa and tropical Asia, at similar latitudes to South America, extinctions were few.

No explanations have been advanced as to why habitat changes were especially inimical to the survival of megaherbivores. As was documented in previous chapters, extant elephants and rhinos are tolerant of a wide variety of food sources and habitat conditions. During the Pleistocene, woolly mammoth had an enormous geographic range, extending through Eurasia from Spain, Italy and China in the south to Russia and Siberia, and in North America from Alaska to the Great Lakes region. Further south it was replaced by the Columbian mammoth, which extended from the United States through Mexico as far south as El Salvador. Judging from their fossil remains, mammoths of various species were especially abundant over most of this range throughout the Pleistocene. Woolly rhino were prominent in Pleistocene faunas through much of Eurasia, especially in Mongolia where grassy steppe still persists (Agenbroad 1984; Vereshchagin & Baryshnikov 1984).

From what is known about the ecology of extant species, the following kinds of habitat changes would have been especially detrimental for Pleistocene large herbivores: (i) reduced habitat diversity, particularly a lack of adjacent food resources for both the periods of summer growth and reproduction, and winter subsistence; (ii) a marked lowering of soil nutrient status, and hence of forage quality both of grasses and woody browse; (iii) a general reduction in the standing biomass of available food, or of its accessibility, due for example to a persistent cover of deep or crusted snow. Nevertheless, it is difficult to understand why the Holocene grasslands of the United States and Central Asia were inadequate for the continued existence of such tolerant grazers as mammoths must have been. Even the most uniform regions of vegetation have local zones of higher diversity along rivers or alluvial fans or in mountainous areas. Bison persisted in vast numbers on the mid-continental grasslands; while the Columbian mammoth, which had been abundant throughout this region during the Pleistocene, fell by the wayside. Extinctions included not only grazers, but also browsing mastodont and gomphothere, and semidesert inhabitants such as some of the ground sloths and camelids.

While there are numerous examples of megaherbivore extinctions prior to human presence, these were generally associated with ecological replacements. For example, in Europe mastodonts were replaced by *Elephas namadicus*, which was likewise a woodland browser. In contrast, the late Pleistocene extinctions involved the termination of particular ecological types in the fauna.

Thus hypotheses based on climatic change fail to explain either the geographic distribution of extinctions, or the fact that likelihood of extinction increased with body size among herbivores. Extinctions in the Americas and Europe coincided in time with a period of rapid climatic and habitat change, but in Australia they were associated with a period of climatic extreme. There is no independent evidence, apart from biotic changes, to show that the climatic change between the last glacial period and the Holocene was more extreme than during previous glacial/inter-glacial transitions. However, reductions in body size indicate that many herbivore species were under nutritional stress, and distributional changes of species suggest that seasonality may be more extreme today than it was in the past.

Human predation

In Europe, humans gradually extended their range northwards during the late Pleistocene, reaching 50 °N latitude by about 40 000 years BP. The earliest evidence of human hunting of mammoths dates from around 70 000 years BP. Specialized hunting of mammoths is associated with narrow flint speartips capable of penetrating the thick hide. Human interest in the Pleistocene megafauna, including mammoth, rhino, bison and reindeer, is graphically illustrated in cave paintings in France dating from about 20 000 years BP. Mammoths are absent from fossil assemblages after about this time in south-western Europe, southern Russia and China; while other species formerly associated with mammoths in steppe faunas, including horse, bison and reindeer, persisted. In Russia and Siberia, there is abundant evidence of humans hunting mammoths, horses, bison and reindeer, and in some places woolly rhinos. Some encampments in Siberia have houses constructed from the accumulation of mammoth bones. The most recent European dates for mammoth, from the Ukraine and Switzer-land, are from around 12 000–13 000 years BP. At the end of the paleolithic period between 12 000 and 10 000 years BP, *Mammuthus primigenius, Coelodonta antiquitatis* and *Ovibos* disappeared finally from their last refuges in Siberia (Liu & Li 1984; Martin 1982; Müller-Beck 1982; Vereshchagin 1967; Vershchagin & Baryshnikov 1984).

In North America, the earliest unequivocal evidence of human occupa-tion dates from around 13 000 years BP. Claims for earlier human presence are based on rather dubious artefacts and remain controversial, especially in comparison with the cultural stage reached by source populations in Eurasia by that time. At about 11 500 years BP, the characteristic stone spear-heads of the Clovis culture became abundant throughout the conterminous United States. After 11 000 years BP, the Clovis points were abruptly replaced by the somewhat different stone points of the Folsom culture. While Clovis points commonly accompany mammoth remains, Folsom points are associated only with bison and smaller species. There is also evidence of the hunting of mastodonts during Clovis times. In the conterminous United States, the most precise terminal dates for extinct species, from Racholabrea, California, cluster between 11 200 and 10 800 years BP, i.e. at the end of the Clovis period. The disappearance of ground sloths from the Grand Canyon in Arizona occurred close to 11 000 years BP. Elsewhere, the youngest reliable dates for mammoth, mastodont, ground sloth, equid, camelid and stag-moose are around 10 500 years BP. In Alaska, last dated records of mammoth are around 13 500 years BP (Fisher 1984; Haynes 1982; Marcus & Berger 1984; Mead & Meltzer 1984; Phillips 1984; West 1984).

In South America, there is evidence both of human presence and of the hunting of gomphotheres in Venezuela from at least 13 000 years BP, and from Brazil perhaps as far back as 32 000 years BP. Humans had evidently reached southern Patagonia by 12 600 years BP. By about 11 800 years BP gomphotheres had disappeared in Venezuela, although other now extinct species persisted until about 10 000 years BP. In Patagonia, ground sloths were not recorded after 10 500 years BP. Records elsewhere suggest that some of the extinct forms may have persisted until as recently as 9000 years BP (Gruhn & Bryan 1984; Guidon & Delibrias 1986; Markgraf 1985; Martin 1984a).

In Australia, the entry of aboriginal people from south-east Asia occurred prior to 30 000 years BP, and humans had spread as far south as Tasmania by 25 000 years BP. However, associations between human artefacts and large vertebrate fossils are rare. The main wave of large mammal extinctions occurred around 18 000 years BP, the coldest period of the last glacial, with a few species persisting at the southern fringe until 15 000 years BP (Horton 1984; Murray 1984; Reed 1970).

In Madagascar, there is close coincidence between the arrival of the first human occupants around 1000 years BP, and the disappearance of such large animals as dwarfed hippopotamus, several genera of lemurs, two species of giant tortoise, and two genera of giant flightless birds. In New Zealand, the extinction of moas followed closely in time the arrival there of the Maori immigrants from Polynesia. Evidence of human hunting of moas is plentiful (A. Anderson 1984; Martin 1984a; Trotter & McCulloch 1984).

Hominid presence in Africa has been continuous and continent-wide since the time of the Pliocene australopithecines. *Homo erectus* spread from Africa to tropical Asia in the mid-Pleistocene perhaps 700 000 years ago. Both these regions have thus had a long association between the evolving human lineage and the large mammal fauna. The Middle Stone Age people occupying southern Africa between 130 000 and 70 000 years ago hunted mainly eland and other medium-sized ungulates, and evidently avoided elephants and rhinos. Late Stone Age people occupying this region after 40 000 years BP hunted especially warthogs and bushpigs. Human predation may have been a factor in the extinctions of giant buffalo and giant hartebeest at the end of the Pleistocene (Klein 1977, 1984a, 1984b; Volman 1984).

Martin (1967, 1973, 1984a and b) has been the chief protagonist for human overkill as the prime cause of the large vertebrate extinctions of the late Pleistocene. Support for this hypothesis rests on the disproportionate extinctions among larger mammals relative to other animal groups, particularly in the Americas where animal populations would have been especially vulnerable through lack of previous contact with humans. The close synchrony between the appearance of humans and animal extinctions at different latitudes in Europe lends further support. Recent extinctions of large mammals and birds in Madagascar and New Zealand are unequivocally related to human colonization of these islands. However, in Australia the aboriginal colonists coexisted with the megafauna for some 10 000 years before the main wave of extinctions occurred.

Martin (1984b) envisaged a 'blitzkrieg wave' of humans spreading southwards through the Americas. Vulnerable animal species were quickly eliminated, forcing the colonists to keep moving on. This unsettled pattern is advanced to explain the paucity of associations between human cultural artefacts and the extinct fauna in North America, in contrast with the situation in Eurasia. Martin claims further support for this hypothesis from the close synchrony between the terminal dates for large mammal species in particular regions, and the putative time of arrival of human hunters, as they dispersed southwards.

The frequent associations between mammoth remains and Clovis spearpoints indicates that human predation on mammoths was relatively common. However, dates for mammoth fossils in North America do not support Martin's notion of a blitzkrieg wave rolling southwards (Agenbroad 1984). A further challenge comes from reported dates of human presence and hunting of large mammals in South America predating the Clovis period in North America.

However, the main problems confronting the overkill hypothesis are (i) to explain the synchronous disappearance not only of mammoths, mastodonts and ground sloths, but also of a wide range of medium-sized mammals, many of which are not obvious prey species for humans; and (ii) to account for the extinctions of certain species of bird, and of some small mammal species, that were not directly dependent upon the large mammal fauna for prey or carrion (Grayson 1977; Stedman & Martin 1984).

The role of megaherbivores

Human predation was undoubtedly a major factor in the extinctions of megaherbivores. There is abundant evidence of the importance of mammoth, and to a lesser extent woolly rhino, as a prey of humans in northern Eurasia. There is also clear evidence of human hunting of mammoths, mastodonts and gomphotheres in the New World shortly before their disappearance. The responses that modern elephants and white rhinos adopt against carnivores is to stand ground, protecting young animals behind the bodies of adults. However, such tactics are inappropriate against an organized band of humans armed with projectile weapons. Very large animals that had evolved with no contact with humans would be especially vulnerable to this novel predator. Even after aeons of coevolution, white rhinos still vacillate between standing their ground and fleeing when confronted by humans approaching from upwind. In contrast, African elephants and black rhinos react aggressively to any close approach by humans, in a way sufficient to inhibit hunting by humans not armed with guns. In Africa and tropical Asia, large mammals were able to improve their defenses as humans gradually improved their skills as hunters. In northern Europe, the confrontation was more sudden as humans spread northwards. In the Americas, it was particularly cataclysmal, as hunters who had perfected their techniques over centuries of practice in northern Eurasia met a particularly naive fauna.

The extinction of *Elephas recki* in Africa between the Early and Middle Stone Age periods cannot readily be ascribed to human predation, since archeological sites show that African hunters of the time avoided very large mammals. Nevertheless, indirect effects of human hunting may have played a part. The developing reliance of humans over this period on hunting larger animals rather than scavenging probably entailed increased control over fire as a tool in attracting animals and removing screening vegetation. From its grazing dentition, *E. recki* would have been dependent more on grass and less on woody browse than the modern *Loxodonta africana*. Thus regular removal of the tall grass cover by burning would have been especially detrimental to its subsistence through the dry season.

Burning is, however, beneficial to a short grass grazer like the white rhino, able to feed in the green flush following the fire. The disappearance of white rhino from East African savannas early in the Holocene could well be related to increasing hunting skills by humans occupying this region. White rhinos persisted into modern times only in the southern and north-central regions of Africa. The former area was occupied only by stone age hunters until perhaps 2000 years ago; and among the Iron Age colonists, the Zulus at least did not regard white rhino meat as edible. The latter area remained relatively uninhabited even in the present century.

For a predator to drive its prey to extinction, there are just two requirements: (i) a rate of harvest exceeding the maximum sustained recruitment rate of the prey population; (ii) the possibility for predation to be sustained even when prey become rare. For modern megaherbivores, maximum rates of population increase vary from 6–7% for elephants to about 10% for hippos. If mortality due to predation were to exceed such levels, populations would decline inexorably towards extinction. A possible scenario is (i) a regional prey density of about 2 per km^2; (ii) a regional human density of 0.05 per km^2 (similar to that estimated for the European paleolithic); (iii) a kill rate of one animal per band of 25 people per week sustained over the year.

Megaherbivore populations under nutritional stress due to adverse habitat changes would be less able to sustain the harvest quotas reported above. Populations restricted to localized pockets of acceptable habitat by the climatic reorganization of the terminal Pleistocene would have been especially vulnerable to being hunted down to the last animal; in particular, animals as large as megaherbivores have nowhere to hide. But what about the smaller ungulates, with higher potential rates of population increase, that also went extinct? For them a human overkill scenario is far more difficult to sustain.

We must turn next to a consideration of the consequences of the elimination of megaherbivores from ecosystems. Prior to human arrival, populations of mammoths, mastodont and ground sloth would have existed at saturation levels where further increase was prevented by food limitations. Vegetation would undoubtedly have been in a severely disturbed state. Grazers such as *Mammuthus*, *Stegomastodon* and *Mylodon* would have consumed or uprooted a large fraction of grass biomass, creating areas for colonization by low-growing pioneer grasses. In such disturbed grasslands, fires would have been less severe, allowing establishment by woody trees and shrubs. However, these large beasts may have prevented the woody component from growing into a mature woodland by incidental behaviors such as rubbing or exercising strength. Browsers such as *Mammut*, *Cuvieronius*, *Haplomastodon*, *Megalotherium* and *Eremotherium* would have created open gaps by felling trees in more heavily wooded habitats. These gaps would have been colonized by faster-growing, and hence more palatable, species of tree and shrub. A better-developed herb layer in these gaps may have allowed fires to penetrate, promoting further opening of the tree canopy. Such activities by both grazing and browsing megaherbivores acting in concert would have produced the kinds of savanna parklands, with a diverse mixture of different species of tree and herb, as have been identified from the pollen record. From this perspective, the change to less diverse zonal vegetation at the end of the Pleistocene would not be a causal factor in the megaherbivore extinctions, but rather a consequence of the removal of their disturbing influence on plant communities.

What would the consequences of these habitat changes have been for smaller herbivores? Such species had coexisted with megaherbivores through all of their evolutionary time. The kinds of vegetation changes that megaherbivores induced would have been beneficial in terms of stimulating growth by the more nutritious pioneer species of grass and shrub, and by creating a small-scale mosaic of plant communities. Furthermore, the activities of megaherbivores would have facilitated nutrient cycling, leading to further improvements in nutritional quality. Many of the medium-sized ungulates may have been dependent over a large part of their range on the kinds of habitats created by the presence of associated species of megaherbivore.

With the demise of mammoths, mastodonts and giant ground sloths, forests would have tended to close in, and taller grasses to expand so as to support fierce fires. The distributions of smaller species of herbivore must have contracted to localized sites of edaphic disturbance where a diversity of more nutritious species of plant still persisted. Discontinuities in the distribution of suitable habitat following the removal of megaherbivore disturbance may have made it more difficult for these species to follow shifting vegetation zones through the climatic changes of the late Pleistocene. Whether these habitat changes alone were sufficient to cause the extinctions of other species of herbivore, or whether human predation on the localized and vulnerable populations was necessary, must remain conjectural.

In summary, the scenario that I propose follows this sequence:
1. Sustained human predation causes naive populations of megaherbivores to decline as harvests exceed replacement potentials.
2. The climatic and habitat changes associated with global warming compress megaherbivore distributions and places these species under increased food stress, making them more vulnerable to being hunted to extinction.
3. The disturbing influence of megaherbivores on vegetation is thereby removed, leading to trends towards dominance by late successional plant species and more uniform zonal habitats.

4. Vegetation becomes less favorable for smaller herbivores, inducing population declines by these species, with predation by humans deprived of megaherbivores probably promoting their slide towards extinction; a crucial factor was perhaps the inability of the island population remnants to follow shifting vegetation zones.
5. Mammalian carnivores and scavenging birds decline due to the disappearance of much of their food resource.
6. Birds and other small vertebrate species dependent upon the disturbed habitats created by megaherbivores disappear.

It has been frequently claimed that climatic change was the major factor leading to the extinctions of large herbivores in the late Pleistocene, with human predation a secondary factor placing additional stress on populations. My conclusion is that extinctions were inevitable for those megaherbivore species with a maximum rate of population increase inadequate to support sustained human predation. The timing of the extinctions may have been controlled largely by habitat change, which confined populations to localized areas, thereby making them more vulnerable to sustained human hunting until the last pair had been killed. Thus the hyperarid period of the late Pleistocene in Australia may have made populations of *Diprotodon* and giant kangaroos such as *Sthenurus* especially sensitive to being hunted to extinction by the early Aborigines.

Whatever the interwoven effects of habitat change and human predation were in the late Pleistocene extinctions, there are certain basic points to be emphasized. The one certain factor distinguishing the terminal Pleistocene from the end of the previous glacial periods was the presence of humans as sophisticated hunters. Human hunting must inevitably have had a great impact on the populations of very large species, which had been largely free of predation by carnivores except on juveniles. The elimination of the impact of megaherbivores must inevitably have caused habitat changes, apart from those brought about by climatic shift. By hunting to extinction the vulnerable herbivores, humans initiated habitat transformations that may have played an important role in the extinctions of lesser herbivores, apart from their direct hunting impact on these species. In other words, if it had not been for human presence, most of these species would have survived into the Holocene. Thus I conclude that human predation on megaherbivores was the necessary and sufficient cause of the wider spread of extinctions among large mammals at the end of the Pleistocene.

Summary

The disappearance of numerous genera of large mammals and birds at the end of the Pleistocene has been related both to the major climatic and habitat changes occurring at this time, and to human overkill. Neither explanation is adequate in isolation. Climatic hypotheses fail to explain the geographic concentration of extinctions in the New World, and the positive correlation between extinctions of mammalian herbivores and large body size. Human predation readily explains the elimination of megaherbivores, but not of numerous smaller species.

The demise of megaherbivore populations through their inability to sustain organized human predation resulted in the removal of their positive influence on vegetation diversity and nutritional content. This precipitated cascading effects on other species dependent upon disturbed habitats. Human predation on megaherbivores was thus the prime factor in the late Pleistocene extinctions, with the climatic changes and their effects on vegetation aiding the timing of the extinctions.

16

Conservation

Introduction
The decline of megaherbivores did not end with the termination of the Pleistocene. During the nineteenth century, expanding human settlements and continued hunting reduced Asian species to isolated populations, and ivory exploitation led to African elephant becoming rare over most of southern, eastern and western Africa. Following the advent of firearms, white rhino declined in southern Africa from a widespread and abundant species to the brink of extinction over the course of 60 years. In north-east Africa, white rhino recently suffered an even more dramatic decrease, from several thousand animals distributed through three countries in the early 1960s, to a remnant of about 15 restricted to one park in Zaire at the time of writing. Over much of Africa, remaining populations of elephant and black rhino are suffering steady attrition due to continuing human exploitation for ivory and horn. Javan rhino and Sumatran rhino were listed by the IUCN among the world's twelve most threatened animal species; Indian rhino, Asian elephant and black rhino are listed as endangered; and African elephant and white rhino, while currently safe numerically, remain vulnerable to poaching pressures.

Where populations of megaherbivores have been effectively protected, a contrasting conservation problem has arisen. Populations have increased to levels where they have induced vegetation changes such as to threaten the survival of other animal and plant species in these areas. As a result elephant, hippo and white rhino have been culled in the sanctuaries set aside for their protection. Besides these three species, the only other ungulate that has been culled routinely in a large national park (one of over 2000 km²) is buffalo (in Kruger and Hwange). Cumming (1982) pointed out that, apart from elephant and hippo, no large herbivore species has been responsible for major vegetation degradation unrelated to human interference through actions such as provision of artificial water supplies, predator reduction or confinement to relatively small areas.

The choice of action to combat or alleviate both of these problems continues to generate much argument among conservationists. Is intervention in the form of culling really necessary in order to safeguard habitat conditions for other species? Should megaherbivore populations be exploited for economic gain? Can rapidly diminishing populations be effectively protected in the wild, given the limited extent of most conservation areas? Is the effort in terms of finance and manpower required to protect these species justified? Much of the debate revolves around philosophical differences in attitudes and objectives among conservationists. Thus I will first outline the various objectives of conservation. From this perspective, I will assess the merits and likely consequences of different actions, making recommendations for the most effective choices of action in relation to problems both of overabundance and of overexploitation.

Conservation objectives

In terms of the 'World Conservation Strategy' formulated by the International Union for the Conservation of Nature and Natural Resources (IUCN), the objectives of conservation are (i) to sustain life support processes; (ii) to maintain biotic diversity; (iii) to retain those species, or ecosystems, of particular benefit or interest; (iv) to keep future options open (IUCN 1980; Miller 1983). In the context of national parks and other designated conservation areas, these broad objectives tend to get translated into the more practical goal of retaining the full historic diversity of habitats and species in the region (Leopold *et al.* 1963; Leopold 1968; Brooks & Macdonald 1983; Pienaar 1983).

Conflicting with this goal is the fact that habitats and populations are not static. Thus the question becomes, how much change is permissible? Walker & Goodman (1983) suggest that preventing change causes ecosystems to lose resilience. This means that their biotic communities become less capable of absorbing climatic stresses or other disturbances, which inevitably occur sooner or later, resulting in more severe losses of species than would otherwise have occurred. There is increasing awareness among ecologists that disturbances of varying severity, frequency and extent (including fire, drought, floods, hurricanes, disease, and climatic shifts) have been an integral component of ecosystem functioning in the past. Observations both on coral reefs and in tropical forests suggest that species diversity is highest at intermediate levels of disturbance (Connell 1978).

Another conservation viewpoint places emphasis on conserving a representative suite of natural ecosystems, to serve as benchmarks against which to compare adjoining man-modified regions. This may involve the establishment of so-called biosphere reserves, encompassing conservation areas as integral units alongside adjoining developed areas. For national parks to function as benchmark ecosystems in this context, natural processes must be allowed to proceed with minimal human intervention. Nevertheless, the effects of surrounding developments, for example the ingress of alien species, may intrude into parks and require combatting (Caughley 1981; Ferrar 1983; Houston 1971; Siegfried & Davies 1982; Sinclair 1983).

More pragmatic concerns relate to the cost-effectiveness of conservation. If the future existence of national parks is to be safeguarded in developing African countries, they should not be a financial drain on the limited economic resources of these countries. Their persistence would be more secure if they generated material benefits for these countries, and more particularly for surrounding rural people. From this perspective, controlled hunting of species like elephants and rhinos, together with the sale of animal products like ivory, could make a particularly valuable economic contribution (Anderson 1983; Konigkramer 1983; Martin & Taylor 1983; Myers 1972).

Problems of overabundance
Elephants
African elephants have been responsible for changes in vegetation in a number of conservation areas, as outlined in Chapter 12. Such changes are generally regarded as threatening to the continued survival of the plant species damaged, and of habitat conditions for other animal species, as well as detrimental to aesthetic qualities of landscapes. Nevertheless, reductions in species diversity as a result of elephant-induced changes in vegetation are undocumented. Declines in certain tree or animal species have undoubtedly occurred; but whether these decreases were balanced by increases in other formerly rare species remains uncertain. Reports from Tsavo in Kenya suggest that, while woodland ungulates declined following the elimination of *Commiphora* thickets by elephants, open country grazers increased (Parker 1983).

In Kruger Park in South Africa and Hwange Park in Zimbabwe, elephants have been culled routinely to contain their populations within pre-assigned limits, with the aim of restricting vegetation changes. In other parks in Namibia, Zimbabwe and Uganda, temporary cropping operations have been carried out to reduce elephant populations. A cropping operation was also started at Luangwa Valley in Zambia, but abandoned due to its inefficiency. In Kenya and Tanzania, intervention in the form of culling has been resisted, except for a few limited operations for scientific purposes. Extensive habitat changes have occurred in some parks where laissez-faire management was followed, for example at Tsavo in Kenya and Ruaha in Tanzania. However, vegetation changes perceived as detrimental have not been avoided in parks where culling has been carried out.

Barnes (1983a) analyzed the factors involved in deciding whether to initiate elephant culling in the Ruaha Park in Tanzania, with the aim of halting the rapid and progressive decline in certain tree species in the rift valley section. To not only stop the woodland decline, but also allow the potential for recovery, about three-quarters of the total population of 24 000 elephants would need to be removed. The logistics of such an operation lay beyond the capabilities of the Tanzanian wildlife department. The most cost-effective time for culling to be initiated would have been before woodland damage became notable.

In the Nuanetsi region of Gonarezhou Park in Zimbabwe, a reduction of the elephant production by 40% (from 2 per km^2 to 1.3 per km^2), combined with culling of impala, resulted in vastly improved regeneration by *Acacia* trees in the riverine fringe. However, a 20% reduction of elephants in the Chizarira Game Reserve, also in Zimbabwe, alleviated woodland damage for only one year. In the Kruger National Park, where pre-emptive culling is carried out to prevent elephants from increasing above a ceiling density of about 0.4 per km^2, between 300 and 700 elephants are removed annually. Ivory, hides and meat resulting from this operation are marketed at a considerable profit. However, this policy did not prevent severe damage to tree populations from occurring during the 1982–84 drought. Furthermore, concern has been expressed about the cruelty to the elephants and the disturbance of their behavior resulting from chasing by helicopter and killing by paralysing drugs (de Vos *et al.* 1983).

White rhino and hippo

White rhino and hippo exert their modifying impact on the grass layer. The result is not only changed grassland conditions for other animals, but also increased soil erosion. Accelerated rates of sheet and gully erosion threaten the overall productivity and stability of the ecosystem. Soil is a slowly renewing resource, so that losses cannot be remedied in a short time span. Furthermore, with fires precluded, grassland areas tended to become invaded by *Acacia* scrub. These are the classical symptoms of the 'overgrazing' problem that is widespread in areas of human subsistence pastoralism over much of Africa.

In the Umfolozi Reserve, the management response to the grassland changes brought about by expanding populations of white rhino and other grazing ungulates was to institute culling programs aimed at reducing the densities of these species. The target densities were based on agricultural assessments of permissible stocking densities for cattle in the region. The white rhino population was reduced from 1550 in 1970 to 1070 in 1976 and 600 in 1982. All rhinos were caught and removed live, animals being transported to other wildlife areas, or sold to private farms and zoos. The wildebeest population was reduced by about one third over this period, with other ungulate species affected somewhat less. Overall a 50% reduction in large herbivore biomass was achieved between 1972 and 1983. As a consequence, very little increase in animal mortality occurred during the severe drought experienced over 1982–83. Following exceptionally high cyclonic rainfall in early 1984, grass recovery was striking not only in the culled area, but also in a control block left unculled (Brooks & Macdonald 1983; Owen-Smith 1973, 1981; Walker *et al.* 1987).

Hippos at high densities cause similar changes in grasslands and soils, although only a 3 km wide zone adjoining rivers or lakeshores is affected. To arrest the perceived habitat deterioration, seven thousand hippos were removed from the Queen Elizabeth National Park in Uganda between 1957 and 1967, the meat being made available to local people. Another four thousand hippos were cropped in the Murchison Falls Park. After cropping ended, hippos regained their former densities; but the grassland remained in an improved condition, perhaps due to higher rainfall (Eltringham 1980; Laws, 1968b, 1981b).

General problems

In general terms, managers are concerned about the following consequences of the vegetation changes induced by expanding megaherbivore populations: (i) radical modification of certain habitat types, leading perhaps to the loss of species dependent upon them; (ii) elimination of certain sensitive plant species; (iii) reduced vegetation cover leading to accelerated erosion and a decline in the overall productivity of the ecosystem; (iv) depression of the resource base for megaherbivore populations themselves; (v) loss of aesthetic features of landscapes, such as mature trees or lush grasslands.

As outlined in Chapters 13 and 14, vegetation changes are inevitable as megaherbivore populations increase towards saturation levels at which they become limited directly by food resources. At the time that national parks and game reserves were established, the prevailing vegetation communities had developed under low megaherbivore densities, due to past hunting. To some degree, vegetation is now merely reverting to its state in prehistoric times, when megaherbivore populations were unaffected by human hunting. Nevertheless, the rate and extent of vegetation change may be greatly exacerbated where animal populations are compressed into the confines of national parks by surrounding human disturbance.

The term 'overabundance' used in the heading to this section is a value judgement, implying that populations exceeding certain levels are detrimental to the objectives of conservation. However, some level of vegetation disturbance is probably beneficial to both habitat diversity and productivity. The vegetation impact of megaherbivores is only likely to become detrimental if it exceeds the disturbance regimes that prevailed over the evolutionary times of associated species, in either magnitude or frequency (Hansen & Owen-Smith in preparation). However, there is no *a priori* basis for specifying what these prehistoric regimes were. The only way to discover what they might have been is to create conditions as close as possible to those prevailing before intervention by modern man.

In Chapter 13, I argued that dispersal was the major short term process adjusting megaherbivore populations to the changing carrying capacity of their habitats. As a result of the suppression of dispersal movements by fences, settlements, hunting and other boundary restrictions, megaherbivore densities are likely to be higher and less responsive to rainfall-related variations in vegetation growth than would have been the case in the evolutionary past. The impact of elephants on trees, and white rhinos or hippos on grasses, may hence reach such extreme severities, and persist for sufficiently extended periods, as to lead to lowered species diversity. Sensitive tree species, like baobab, may be virtually exterminated, except in inaccessible localities such as steep hillslopes. If reserve grazing in the form of tall grass stands are eliminated, populations of white rhinos and other grazers become vulnerable to large-scale population crashes during droughts, with increased risks of local species extinctions during such events (Owen-Smith 1983). Hence laissez-faire inaction cannot be justified where ecological processes have been distorted by boundary restrictions. Some form of population manipulation is needed to counteract the effects of fences or other inhibitions on movements.

Population culling as currently applied has generally been aimed at achieving some arbitrarily set population level. For example, the target elephant density in the Kruger Park simply represents one elephant per square mile, a figure originally suggested on the basis of the elephant densities that were believed to exist in both the Kruger and Hwange National Parks at the time that problems of vegetation damage were first confronted. Subsequently, more accurate counts showed that the elephant density in Hwange at the time was about 2.5 times this figure. At Luangwa Valley in Zambia, the elephant density reached was three times that in Hwange, although rainfall was only 40% higher (poaching has subsequently reduced this density). The relatively low density of elephants maintained in Kruger Park has not prevented severe damage to certain tree species and woodland types. In the Serengeti, damage to mature *Acacia* woodland occurred at an elephant density of under 0.2 per km².

Similar considerations apply to short grass grazers such as white rhino and hippo. Even at low densities, white rhinos concentrate their grazing in existing short grass patches, and by removing the grass cover accelerate soil erosion from these areas. At higher densities, larger areas become affected. The choice as to the most desirable ratio of short grass to tall grass grassland is debateable.

Somewhat different concerns arise from the effects of culling operations on behavior and population structure. Adverse behavioral responses to humans, whether in vehicles or on foot, can degrade the experience of visitors to parks. In the Kruger Park, elephant culling is organized so as to take out complete family units, the aim being to minimize the disturbing effect on the rest of the population. Nevertheless, the mean distance travelled over a 12 hour period by surviving family units in clans from which other family units had been culled was 19.2 km, compared with 6.1 km for undisturbed family units (Hall-Martin, quoted by de Vos *et al.* 1983). The long range communication among elephants at sound frequencies below the audible range of humans could provide a means for transmitting panic responses through the population. Despite supposedly random selection procedures, buffalo culling in the Kruger Park is strongly biased towards subgroups of non-breeding individuals within herds (Mason & van der Walt 1984). Disruption of population structures reduces the potential for future scientific studies. Furthermore, pre-emptive culling may also pre-empt the operation of the processes of natural selection that have shaped the characteristics of the species.

Dispersal sinks

The conditions prevailing in national parks and other conservation areas today differ in one major respect from the situation in prehistoric times: the presence of boundaries restricting dispersal. Even if park boundaries are not fenced, movements beyond borders are generally inhibited by settlement and hunting in the surrounding area. Parks in effect tend to become island ecosystems, at least so far as larger mammals are concerned, and thus become subject to the kinds of changes that island communities experience. Populations of some species tend towards higher densities than they would attain under conditions where free dispersal was possible, while other populations become vulnerable to extinction. If the area is too small, the genetic diversity of populations may decline through inbreeding (East 1981; Lomolino 1985; Soule 1980).

If national parks and similar conservation areas are to replicate the conditions that prevailed within their borders in pristine times, opportunities for dispersal need to be created to reduce the island effect. Where emigration is prevented or inhibited by boundary fences or adjoining human disturbance, dispersal sinks must be created within the borders of the conserved area. This can be done by designating certain areas as vacuum zones for particular species. Culling would be carried out periodically so as to remove all animals of that particular species settling within the vacuum zone. No culling would be carried out in the remainder of the park, where populations would be allowed to attain their own equilibria, buffered from extreme change by the dispersal option allowed by nearby vacuum zones (Owen-Smith 1974a, 1981, 1983; Petrides 1974).

Vacuum zone culling offers these potential advantages:

1. The surplus individuals to be culled are selected by natural mechanisms operating within the population.
2. Population densities in the rest of the area can adjust to the supply rates of food resources, and the fluctuations therein associated with rainfall variations.
3. There is no artificial distortion of population structure within the bulk of the population.
4. Aesthetic qualities in the remainder of the park remain untarnished by the side-effects of culling operations.
5. Spatial diversity is enhanced. In particular the vacuum zones serve as refuges for those species sensitive to the impact of megaherbivores on vegetation.

Several practical issues arise with regard to the implementation of vacuum zone culling. I answer each of these in turn.

1. Will rates of dispersal be adequate to halt population increase before vegetation damage is incurred? The answer to this is clearly no. Vegetation damage in the form of tree breakage, or suppression of tall grasses, occurs even at low densities of elephants or white rhinos. The term damage is moreover a subjective one. It should be rephrased as damage to conservation objectives, i.e. detriment to overall species diversity. From this perspective, if dispersal is unimpeded then species diversity must be closely similar to that which would have occurred in prehistoric times. The vacuum zones

ensure the retention of vulnerable species – for example baobabs in the case of elephant, or reedbuck in the case of white rhino.

2. Will animals not become conditioned to avoid vacuum areas as a result of the culling carried out in them? Elephants are well known to be sensitive to hunting disturbances. For example, in the Luangwa Valley where controlled hunting areas adjoin the national park, elephants cross the river into the hunting areas in the evening and move back into the security of the park at dawn (Lewis 1986). If culling were carried out regularly in the vacuum zones, this would inhibit their colonization. The solution is to ensure that culling is carried out irregularly in the form of brief blitzes. Thus animals within the vacuum zones are killed or captured before they have a chance to react to the culling operation. After such an operation, any animals remaining in the region would be left unmolested for a time sufficient for any fear responses to fade, due to lack of reinforcement.

3. Would not the large number of animals removed from vacuum zones in a short period saturate marketing outlets? This is an inherent problem with reduction cropping operations. However, the number of animals to be removed from any one vacuum zone would not be large – at the most, perhaps 100 elephants or white rhinos – and culling can be rotated among different vacuum zones. Moreover, since population densities within the vacuum zones are kept well below saturation levels, there is no special time urgency in scheduling: culling can be carried out when it is most convenient in terms of markets, manpower and facilities. The culling team, whether part of the wildlife department or a private contractor, could move from one vacuum zone to another according to a pre-arranged and planned schedule. Operations could furthermore be carried out with maximal efficiency, since there is no need for concern about the age or sex classes of social units to be removed.

4. How many vacuum areas would be needed, and how large would each need to be? Vacuum zones need to be distributed so that they are readily encountered by animals moving out of saturated zones. They should be selected initially to be somewhat smaller than the home ranges covered by population units, so as to avoid depleting neighboring populations.

For example, for white rhinos in the Umfolozi-Hluhluwe complex, I recommended that four vacuum zones be established, each covering about 50 km² (Owen–Smith 1973). The total area encompassed by the vacuum zones would thus represent 20% of the total conservation area. Vacuum zones situated around the periphery of the reserve would facilitate the removal of animals with least blemish to the rest of the reserve. Where feasible, it would be an even better arrangement for vacuum zones to be located in adjoining tribal areas, so that harvesting could be carried out by neighboring people to their direct financial benefit.

In the case of white rhinos, demographic adjustments in the population core caused by nutritional limitations could halve the annual rate of increase from nearly 10%, as prevailed at the time of my study, to about 5%. With a total population of 1500 white rhinos in the 750 km² of core area, about 75 animals would need to be removed annually. However, the number to be removed from each vacuum zone would not be a constant figure; it would vary in relation to conditions of food availability within the core area, which in turn will vary in relation to rainfall. There would be no need to calculate any quota. The procedure would simply be to remove all the white rhinos found within the limits of the vacuum zones about once every four years.

In the case of African elephants, each vacuum zone would need to cover 250–500 km², the approximate extent of clan home ranges. Since elephants have a slower rate of population increase than white rhinos, each vacuum zone could be culled only about once every 10 years. This low frequency of culling would probably be adequate to avoid conditioning elephants against entering the vacuum areas.

5. Should the siting of vacuum zones be rotated? It would be preferable to retain vacuum zones in fixed locations. This allows a contrast in habitat conditions between these zones and the rest of the conservation area to build up, with beneficial consequences for overall spatial heterogeneity and species diversity. Vacuum zones thus become permanent fixtures sited most conveniently for removals. It is unlikely that habitat conditions would become so extreme as to discourage immigrants, at least during crucial drought periods.

6. Is vacuum zone culling applicable to ungulates besides megaherbivores? Intervention in the form of culling is necessary only for those species that (i) can exert a major impact on habitats, and (ii) for which natural population regulatory mechanisms are too slow-acting to alleviate such impacts. For ungulates from the size of buffalo downwards, predation can act to depress populations below the level at which vegetation impact becomes severe and persistent. Though there can be time lags before predator populations respond to prey increases, r_{max} values are higher for carnivores than they are for ungulates (Western 1979). Thus I consider that culling is justified for smaller ungulates only where predator populations are artifically depressed – as they are in many small conservation areas. However, vacuum zone culling would hardly be practical in such small parks, though this depends on the home range extent of the species causing problems in relation to the size of the area.

Problems of overexploitation

At the time of writing, problems of megaherbivore overabundance persist only in South Africa, Zimbabwe, Botswana and Namibia. Elsewhere in Africa, situations of perceived overabundance of elephants have been transformed within a few years into problems of rapidly dwindling populations. Rhino populations have undergone particularly rapid attrition throughout the continent north of the Zambezi. The cause is organized illicit hunting, spurred by the vastly inflated prices fetched by rhino horn and ivory.

Such patterns demonstrate dramatically the sensitivity of megaherbivore populations to human predation. The maximum population growth rates that can be sustained by such large animals are slow: a harvest of 10% in the case of rhinos, and 7% in the case of elephants, would be sufficient to push populations into steady decline. In a moderate-sized national park containing say a thousand rhinos or elephants, a rate of loss to poaching of 6–8 animals a month would initiate a downward plunge in numbers that would accelerate over time. Outside of equatorial forests, these huge beasts can find no seclusion from determined hunters.

Megaherbivore populations can sustain only a very limited degree of human exploitation for subsistence or profit. The elephant culling routinely undertaken in the Kruger Park in South Africa, and at Hwange in Zimbabwe, has been commercially viable in the absence, until recently, of any losses to poachers in these areas. Unless the control of such illegal hunting is very tight, the population surplus left for any legitimate economic exploitation can easily be reduced to zero. Moreover, modelling demonstrates that the economically most profitable time to harvest elephants for ivory is upon their death from old age, due to the fact that tusk growth continues throughout life (Pilgram & Western 1986). This would not apply to rhinos, from which horns can be removed without killing animals. However, proposals to remove horns from rhinos on a commercial basis would make minimal contribution to the conservation of these species, except in the form of a few mutilated specimens in certain small parks.

Effective responses to situations of overexploitation lie firmly within the socio-political realm. The need is to restrict rates of harvest to the low levels that populations can sustain; and the only way to achieve this effectively where the products are highly valuable is to have exceptionally tight control over marketing. There must be no illicit avenues through which products such as horns or ivory can be moved. This shifts the conservation focus into the international economic realm of customs regulations, embargos and cartels, and other factors impinging on the values and uses of animal products. If control as strict as this cannot be achieved, megaherbivore populations must drift inevitably downwards in the direction of extinction (Martin & Martin 1982; Parker & Amin 1983).

Summary

Megaherbivores pose two contrasting problems for conservationists, (i) overabundance associated with vegetation destruction; (ii) dwindling numbers due to illicit exploitation. Problems of elephant overabundance have been widespread in Africa, resulting in a severe impact on certain woody plant species. Hippos in Uganda, and white rhinos in Umfolozi, have attained biomass levels associated with elimination of tall grass cover and soil erosion. Vegetation disturbance need not be detrimental to ecosystem productivity or species diversity; but where dispersal is restricted or prevented, populations may remain at high biomass levels long enough for critical thresholds for recovery to be surpassed. By creating dispersal sinks within conservation areas, such risks could be reduced, with minimal interference with natural population processes in the remainder of the conservation area.

With regard to dwindling populations, the problem is that human hunting can easily exceed the maximum replacement rate. Populations may be reduced to low numbers within a short period, as has happened with elephants and rhinos through most of Africa. Without effective control of illicit markets, populations of these great beasts will be driven inexorably towards extinction.

17

Epilogue: the megaherbivore syndrome

In the preceding chapters of this book I have documented a variety of aspects of the ecology of those large mammalian herbivores that exceed 1000 kg in adult body mass. I have analyzed how these ecological features are related to the allometric trends evident among smaller species of large herbivore. I have pointed out a number of phenomena that appear to be characteristic of these so-called megaherbivores. I now want to draw together these threads to assess the degree to which megaherbivores share in common a distinct set of coadapted features, which can be referred to as the megaherbivore syndrome.

Faunal patterns

Only in parts of Africa and tropical Asia are the faunal communities of today representative of those that prevailed during the Pleistocene and earlier times in the geological record. In Africa the five extant species of megaherbivore make up only a small fraction of the total species diversity of large (> 5 kg) herbivores present continent-wide. The distribution of some 79 herbivore species in different ranges of body size suggests the existence of three modes in species richness: (1) at a body size of about 100–200 kg, made up largely of ruminant artiodactyls occupying savanna habitats; (2) at a body size of about 20 kg, consisting predominantly of forest duikers; (3) a small outlying blip in the megaherbivore size range, with most of these species being non-ruminants (Fig. 17.1).

In terms of animal biomass distribution, a contrasting pattern emerges. The major peak lies in the megaherbivore size range, with the secondary peak formed by medium-sized ruminants (Fig. 17.2). However, in terms of relative energy turnover, as indexed by $M^{0.75}$, the two peaks are closely equivalent. Thus in terms of ecological efficiency, as measured by the fraction of primary production metabolized through the herbivore trophic level, one adaptive mode is formed by the diversity of medium-sized

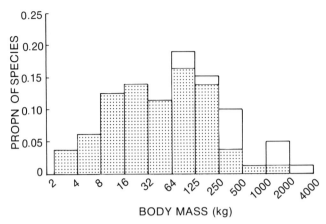

Fig. 17.1 Distribution of sub-Saharan African species of large herbivore among different body size ranges (N = 79 species). Shaded = ruminants, unshaded = non-ruminants. Taken from species listed in Macdonald (1984).

ruminants with individually specialized but complementary diets. The other adaptive peak is made up by a few megaherbivore species, exhibiting a fairly generalized tolerance for a wide range of vegetation components and habitat types.

These adaptive modes are based on alternative digestive strategies for utilizing the energy content of plant cell walls. The ruminant strategy achieves a high efficiency of microbial degradation of cell wall through compartmentalization of the foregut region, connecting orifices being designed to delay passage until a certain degree of breakdown of food residues has been achieved. Remastication of the ingested herbage is an important aid to this end. The megaherbivore strategy is to obtain a sufficiently long retention time for cell wall fermentation simply as a consequence of the allometry of large size, with relatively minor structural modifications of the gut. This route is especially readily followed by cecalids. In fact non-ruminants appear to fall within three quite distinct size ranges: (i) a size of about 100 kg, made up by largely omnivorous suids (plus, on other continents, tapirs and peccaries); (ii) a size range of 250–500 kg, within which zebras and other equids, having a fast passage rate of moderately well digested herbage, are concentrated; (iii) megaherbivores, which approach the digestive efficiency of ruminants. It is at a body mass of about 1000 kg that the mean retention time of the digesta in cecalids becomes similar to that achieved by many medium-sized ruminants.

Distinctions between grazers, subsisting mostly on grasses and similar fine-leaved, slowly fermenting plants, and browsers, eating the foliage and

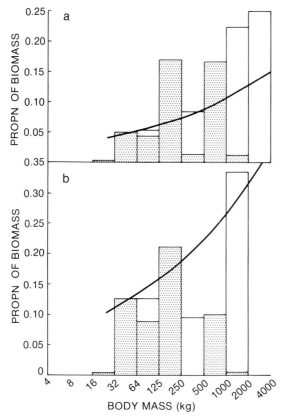

Fig. 17.2 Distribution of the combined population biomass of large herbivore species among different body size ranges. Curves indicate allometric trend in relation to $M^{0.25}$. (a) Amalgamated data from 18 conservation areas in African savanna regions (as listed in Fig. 14.3). (b) Local data for the Umfolozi–Hluhluwe Reserve in Natal, based on the helicopter census totals for 1970 corrected on the basis of intensive ground counts of sample areas for smaller species (see Table 12.1).

supporting stems of woody and herbaceous dicotyledons and non-graminaceous monocotyledons, exist among megaherbivores as they do among smaller ungulates. Despite including a high proportion of grass parts in their year-round diets, both surviving elephant species are basically browsers in their dietary adaptations. When feeding on grasses they select either young leaves or underground parts where soluble carbohydrates are stored, rather than leaves of all ages as grazers like white rhinos and hippos do. Among extinct megaherbivores there were likewise species adapted for either grass or woody browse diets. Specialist grazers were represented among most orders, including in particular mammoths among the elephants. Browsers were prevalent among rhinoceroses and ground sloths. Thus rumen microbial fermentation is not a necessary mechanism for detoxification of the secondary metabolites that are particularly a feature of dicotyledonous plants.

Large size confers other nutritional advantages. Because of the decrease in mass-specific metabolic rate as body size increases, very large animals can subsist on somewhat lower quality forage than smaller animals. As a result, a higher fraction of vegetation becomes nutritionally acceptable food, thereby expanding the food resource base. For the same reason, very large mammals can survive for somewhat longer on a starvation diet than could smaller animals. This provides a mechanism for bridging periods in the annual cycle when food availability is inadequate to meet maintenance requirements. Specific adaptations of food-gathering structures, such as wide mouths or nasal appendages in the form of trunks, may be necessary to enable very large animals to maintain an adequate rate of food intake where standing plant biomass is low. As an outcome of these adaptive features, the geographic ranges of many species of megaherbivore were exceptionally vast. Species such as the African elephant represent supreme examples of habitat generalists.

Lastly, very large size is advantageous for avoiding predation after maturity. Notably, female giraffe, which generally attain a body mass of about 800 kg, are susceptible to lion predation; while black rhinos, which generally weigh over 1000 kg, incur very little predation as adults.

All of these factors provide a suite of advantages to large body size, which come into operation above a body mass threshold of 1000 kg. This explains the trends towards attainment of such large size that are commonly evident among lineages of cecalid herbivores in the fossil record.

Social and life history patterns

With freedom from predation as adults, selective pressures towards group formation are reduced. Thus while medium-sized ungulates commonly form large herds, megaherbivores are typically solitary as adults, although temporary associations may be formed among immature animals. However, where species move over vast and changing home ranges, there are additional benefits for young animals in remaining associated with experienced matriarchs. This leads to the formation of persistent family units by modern elephants and perhaps also by the extinct mammoths.

Once body mass exceeds 1000 kg, gestation time generally exceeds one year. This feature, plus the fact that critical stages of offspring nutrition tend to be stretched through different times of the year, results in reproduction being largely decoupled from the seasonal cycle. Nevertheless, proximate effects of maternal nutrition on ovarian activity remain; in particular ovulation is suppressed during times when food intake is at submaintenance levels. Hence births may be concentrated in certain seasons of the year, but these seasons differ for different species of megaherbivore depending on gestation time. Birth intervals and time to sexual maturity extend over several years. Both maternal conceptions and growth of the young towards maturity may be deferred when conditions are unfavorable, leading to some plasticity in the timing of these critical life history features.

Because of the extended lifespan, males in particular may defer attainment of sexual maturity. Males furthermore exhibit various social strategies to decrease the incidence of damaging combats during adulthood, which may take the form of either spatial or temporal restrictions on dominance. A corrolary of large size is that fighting success becomes strongly dependent on power rather than on agility, promoting a marked size dimorphism between males and females in many species of megaherbivore. Females commonly exhibit behavioral patterns during courtship which allow them to exert some choice over mates, which may serve to eliminate male phenotypes that are not synchronously adapted to prevailing conditions.

If maternal investments in reproduction are related to resting metabolic rates, and allowance is made for neonatal survival, megaherbivores appear no more K-selected than other medium-sized ungulates in the tropics. Among megaherbivores, hippos appear relatively more r-selected for their size than either rhinoceroses or elephants, which may be a consequence of the restricted habitats that hippos are forced to occupy on account of their semi-aquatic habits. The overall time scaling of reproductive events appears related directly to potential lifespan, so that the maximum number of offspring that a female megaherbivore can produce is similar to the potential maximum for smaller ungulates giving birth to single young.

Demographic patterns

As a consequence of 2–4 year birth intervals, the maximum rates of growth in numbers of megaherbivore populations lie in the range 6–10% per annum. Hence populations may take a long time to recover from depression, in particular that caused by past human hunting. In effect the time scaling of population dynamics is shifted to a supra-annual scale. Even where reproduction is suppressed in one year, the effects only become apparent in reduced recruitment some years later. Thus populations exhibit considerable inertia in responding to events occurring on a one or two year time scale, although they may respond to longer term cycles in climate or habitat conditions.

While predation may commonly depress populations of medium-sized herbivores somewhat below the ultimate limits set by environmental resources, megaherbivore populations tend towards saturation densities at which further increase becomes restricted directly by nutritional limitations. These nutritional restrictions manifest themselves in terms of greatly extended birth intervals and deferred attainment of maturity. Population stasis may thus be achieved with relatively minor increases in mortality rates, at least among adults.

Dispersal movements by medium-sized ungulates are largely sex-specific and thus related to inbreeding avoidance. However, among megaherbivores dispersal is important as a fast-acting response to redistribute populations in relation to changing resource distributions, and is thus commonly shown by both sexes. In particular, dispersal movements enable animals to avoid situations where local resource depression may become severe enough to precipitate starvation mortality during short term droughts. However, if such movements are prevented by fences or human disturbance, population crashes can result, although such episodes would have been much less frequent for megaherbivores than they are for smaller species in the absence of such spatial restrictions.

Having evolved in the absence of significant predation on adults, megaherbivore populations do not have the capability to sustain the levels of harvest that can potentially be inflicted by organized bands of skilled human hunters. Thus a wave of extinctions of megaherbivores spread through Eurasia as humans expanded their range northwards during the course of the late Pleistocene, and then lept southwards through the Americas in a brief blitz following the entry of these skilled hunters. Megaherbivores survived only in Africa and in tropical Asia, where they were not hunted to any significant extent during the Pleistocene, perhaps because species had coevolved as human hunters gradually increased their competency as predators on large mammals. This situation has now changed as humans with modern weapons exploit remaining elephant and rhino populations for their ivory and horns.

Community and ecosystem patterns

Megaherbivores typically form at least half of the total biomass of large herbivores, and account for about half of the energy metabolized through the large herbivore community. Their impact on vegetation structure and composition can far outweigh that of smaller herbivores.

In the absence of predation, sustained grazing pressure by medium-sized grazers or browsers may depress plant populations, in particular by suppressing reproduction, either in terms of seed production or seedling survival. Concentrated trampling may amplify such effects, in particular around water points or other localities where animals concentrate. Because of their dietary fiber tolerance, megaherbivores can exert a somewhat greater impact on mature plants, and besides feeding they cause additional breakage to such trees or grasses. Thereby they can induce major transformations of vegetation structure. Savanna woodlands may be transformed into coppice regrowth, while medium-height savanna grasslands are converted into a mosaic of short grass areas. Through such effects megaherbivores can accelerate the recycling of the nutrients locked up in plant structural tissues. The result may be the transformation of high biomass, low productivity and low nutrient content vegetation communities into regenerating communities with a much greater production of better quality plant types and tissues. A mosaic interspersion of habitats at different stages of regeneration may be created.

While such structural changes are largely beneficial to the feeding of megaherbivores, other effects may be detrimental. Reserve food sources used during droughts may be severely damaged or eliminated, making them unavailable for such emergencies in the future. Fire may interact with elephant damage to eliminate woody plant regeneration, particularly on clayey soil substrates. However, in the past when the impacts of both browsing and grazing megaherbivores were coupled in the same areas, fire could have played a much reduced role than it does now, although there would always have been regions far from water where fire would have been the main consumer of accumulated plant biomass.

By promoting lowered and more nutrient-rich vegetation, and a higher mosaic diversity of habitats, the impacts of megaherbivores on vegetation are by and large favorable for smaller species of herbivore. Nevertheless, some species may be affected adversely by structural changes in vegetation cover induced by megaherbivores. Elimination of megaherbivore impact on vegetation was probably an important precipitating factor in the late Pleistocene extinctions of large mammalian species in the Americas and northern Eurasia. The kinds of forest and grassland formations that occur in North and South America today are probably very different from those that prevailed during the Pleistocene and earlier times in the presence of abundant megaherbivores, including both grazers and browsers.

Conservation

At the time of writing those megaherbivore species that survived the late Pleistocene extinctions in Africa and tropical Asia are declining rapidly due to resurging exploitation, as well as human competition for space. Within conservation areas where dispersal is inhibited by fencing or surrounding human activities, megaherbivores could induce progressive habitat changes that would eventually become inimical to the persistence of other animal and plant species. The pre-emptive culling programs carried out by some conservation authorities generate much anxious debate among the public from both ecological and animal welfare considerations. At the same time, many pragmatic conservationists contend that the only workable basis for conserving such large and destructive animals in Third World countries is through economic exploitation for products such as ivory.

The message from this book is that such economic exploitation can be sustained only if it is exceptionally tightly controlled, because of the low rates of harvest that megaherbivore populations can support. Over much of the Africa of today the prospects for achieving such control are somewhat dismal, given the conditions of rural impoverishment and political instability that are widely prevalent. The situation in tropical Asian countries is not that much better. Populations of rhinoceroses in particular are plummetting rapidly to levels where genetic inbreeding becomes a major threat to the long term viability of the remaining remnants. Elephant numbers are undergoing progressive attrition. Only hippos and giraffes are not currently a source of concern: hippos because of their aquatic daytime refuges, from which they are difficult to harvest; and giraffe – which fall questionably into the megaherbivore category – because they currently offer no valuable products.

I hope that this book arouses a full appreciation of the rich biology of these giants among our fellow inhabitants of this planet, and of the important messages for us from their biology. For today *Homo sapiens* has usurped the megaherbivore role over most of the planet. While much of the habitat transformation induced by human activities has been beneficial to our food resource base, opportunities for dispersal across oceans have been instrumental in alleviating short term situations of resource depletion. The stage is now being reached where all productive habitats are becoming saturated; and under such conditions the resource reserves that might be needed for future crises are being rapidly eliminated. Unless we learn from the messages of the past, our future viability as a species might be no more hopeful in the long term than it is for elephants and rhinos.

Appendix I

Scientific names, abbreviations and body mass data for mammal species referred to in the text, tables and figures

Mean and maximum live mass reported for sexually mature animals of each sex are given (where sufficient data are available). A range of values is indicated where body mass varies significantly between different populations of the species. Figures in brackets represent guestimates. The asymptotic body mass for each sex is taken to be the average of the mean and maximum masses. The population mean mass per individual for all age classes is taken to be three-quarters of the mean adult female mass.

Table I.1

Species	Abbreviation	Body mass (kg)		Source
		Ad♀ mean/max	Ad♂ mean/max	
Ass, African wild	WA	(200)	(200)	—
Equus asinus				
Ass, Asian wild	AA	(145)	(145)	8
Equus hemionus				
Barasingha	Bar/Ba	145	212/260	36
Cervus duvauceli				
Bison, American	Bi	495	808/907	26
Bison bison				
Blackbuck	Bla	35/39	38/43	36
Antilope cervicapra				
Blesbok	Bl	60/86	70/100	7,28,40
Damaliscus dorcas phillipsi				
Bontebok	Bo	55	61/64	40
Damaliscus d. dorcas				
Buffalo, African	AB	520/636	650/860	13,33,35,38,43
Syncerus caffer				
Buffalo, water	WB	—	(450)	—
Bubalus bubalis				
Bushbuck	Bb	30–44/34–45	45–55/54–59	15,30,40
Tragelaphus scriptus				
Bushpig				
Potamochoerus porcus				
Camel, Bactrian	Ca	545	—	8
Camelus bactrianus				
Camel, dromedary	Ca	545	545	8
Camelus dromadarius				

Common name	Scientific name				
Caribou	*Rangifer tarandus groenlandicus*	Cu	90	180	46
Cattle	*Bos taurus/indicus*	C	(450)	(450)	—
Cheetah	*Acinonyx jubatus*				
Deer, mule	*Odocoileus hemionus*	MD	55	84/120	45,46
Deer, red	*Cervus elaphus*	ReD/Red	90	120–160/299	29,46
Deer, roe	*Capreolus capreolus*		22	23	46
Deer, spotted	*Axis axis*	SD/SpD	50	86	36,46
Deer, white-tailed	*Odocoileus virginianus*	WD	52	86/136	45,46
Dikdik	*Madoqua kirkii*	Di	4.0–5.3/6.5	4.0–5.1/6.0	15,35,40
Dog, African hunting	*Lycaon pictus*				
Duiker, blue	*Cephalophus monticola*	BD	4.7	4.2	40
Duiker, grey	*Sylvicapra grimmii*	GD	12–21/17–25	12–19/17–21	15,40,43
Duiker, red	*Cephalophus natalensis*	RD	14/18	14/17	17,40
Eland	*Taurotragus oryx*	El	320–445/370–470	500–605/570–945	12,28,40,44
Elephant, African	*Loxodonta africana*	AE	2800/4000	5000/8000	11,21,22,40,43
Elephant, Asian	*Elephas maximus*	IE	2500/4160	4000/5400	37

Table I.1 (cont.)

Species	Abbreviation	Body mass (kg)		Source
		Ad♀ mean/max	Ad♂ mean/max	
Elk, American *Cervus canadensis*	Ek	250	290–350	9,46
Gaur *Bos gaurus*	Gau	590/700	880/940	36
Gazelle, Grant's *Gazella granti*	GG	45/67	65/81	19,35
Gazelle, Thomson's *Gazella thomsoni*	TG	17–20/23	21–23/29	19,23,35
Gerenuk *Litocranius walleri*	Ge	30/45	40/52	15,19,24
Giraffe *Giraffa camelopardalis*	Gi	825/1125	1200/1400	40,43
Goat *Capra hircus* domesticated Goat, Rocky mountain *Oreamnos americanus*	G	(40)	—	—
Guanaco *Lama guanicoe*	Gu	90	90	8
Hartebeest, Coke's *Alcelaphus buselaphus cokei*	CH/Ha	125/135	140/196	15,35,42
Hartebeest, Lichtenstein's *Alcelaphus lichtensteini*	LH	166/181	177/204	40
Hartebeest, red *Alcelaphus buselaphus caama*	RH/Ha	120/136	152/156	40

		1350/2311–2352	1500/2143–2669	3,27,28,40,43
Hippopotamus	Hi			
Hippopotamus amphibius				
Hippopotamus, pigmy	PH	(160)	(200)	8
Choeropsis liberiensis				
Horse, domestic	H	320	350	—
Equus caballus				
Hyena, (spotted)				
Crocuta crocuta				
Impala	Im	40–48/54	55–63/76	13,15,35,40
Aepyceros melampus				
Klipspringer	Kl/Ks	13/17	11/17	28,40,43
Oreotragus oreotragus				
Kob	Ko	63/77	94/121	15,23
Kobus kob				
Kudu, lesser	LK/Lk	56/83	92/108	15,24,25
Tragelaphus imberbis				
Kudu, (greater)	GK	170/204	257/286	18,43
Tragelaphus strepsiceros				
Lechwe	Le	80/97	105/129	40
Kobus leche				
Moose	Mo	343–448	430–550	2,46
Alces alces				
Muskox				
Ovibos moschatus				
Nilgai	Ni	120	240	45
Boselaphus tragocamelus				
Nyala	Ny	63/80	107/143	13,40
Tragelaphus angasi				
Oribi	Or/Oi	10–16/17–20	10–16/17–21	15,28,31,40
Ourebia ourebi				

Table I.1 (*cont.*)

Species	Abbreviation	Body mass (kg)		Source
		Ad♀ mean/max	Ad♂ mean/max	
Oryx	Oy	162–210/188–227	176–235/209–240	24,39,40
Oryx gazella				
Peccary, collared				
Dicotyles tajacu				
Pronghorn				
Antilocapra americana				
Puku	Pu	65/78	75/84	40,43
Kobus vardoni				
Reedbuck, Bohor	BRb	40/55	49/65	15,19
Redunca redunca				
Reedbuck, (southern)	Rb	38–54/51–64	52–80/68–104	19,39,40,43
Redunca arundinum				
Reedbuck, mountain	MR	22–29/35	25–30/38	15,19,40
Redunca fulvorufula				
Reindeer	Rd	100	—	45
Rangifer tarandus				
Rhebuck, grey	GR	25	25	31
Pelea capreolus				
Rhinoceros, black	BR	932–1080/1134–1316	931–1124/1022–1316	14,28
Diceros bicornis				
Rhinoceros, Indian	IR	1600	2100	20
Rhinoceros unicornis				
Rhinoceros, Javan	JR	(1300)	(1300)	—
Rhinoceros sondaicus				
Rhinoceros, Sumatran	SR	(800)	(800)	10
Dicerorhinus sumatrensis				

		1600/1800	2200/2400	this book
Rhinoceros, white	WR			
Ceratotherium simum				
Roan, antelope	RA	260/280	280/300	43
Hippotragus equinus				
Sable, antelope	SA	220/232	235/263	43
Hippotragus niger				
Sambar	Sam	165	285/350	36
Cervus unicolor				
Sheep, Dall	DS	—	100	8
Ovis canadensis dalli				
Sheep, domestic	S	45	—	—
Ovis aries				
Sheep, Soay	SS	25	35	5
Ovis aries feral				
Springbok	Sp/Sb	28–35/32–43	32–42/35–48	32,34,40
Antidorcas marsupialis				
Steenbok	St	11/14	11/15	15,28,40
Raphicerus campestris				
Suni	Su	5.4/8.2	5.0	16,40
Nesotragus moschatus				
Tapir, American	AT	135	160	8
Tapirus terrestris				
Tapir, Asian	IT	(160)	275	8
Tapir indicus				
Thar, Himalayan				
Hemitagus jemlahicus				
Topi	To	100/131	120/152	6,28,35
Damaliscus lunatus korrigum				
Tsessebe	Ts	127/144	140/155	40
Damaliscus l. lunatus				

Table I.1 (*cont.*)

Species	Abbreviation	Body mass (kg)		Source
		Ad♀ mean/max	Ad♂ mean/max	
Vicuna *Lama vicugna*				
Warthog *Phacochoerus aethiopicus*	Wh	58/76	80/107	13,24,35,40,43
Waterbuck *Kobus elipsiprimnus*	Wb/Wa	180/214	240/287	13,24,35,41,43
Wildebeest *Connochaetes taurinus*	Wi	163–223/209–253	210–252/271–295	1,4,13,24,35
Wildebeest, black *Connochaetes gnou*	BW	140	180	40
Zebra, Grevy's *Equus grevyi*	GZ	380/450	400/450	19,23
Zebra, mountain *Equus zebra*	MZ	235–276/257–300	300/350	40
Zebra, plains *Equus burchelli*	BZ	220–310/242–330	250–320/284–340	35,40,43

Reference sources: 1 – Attwell 1982; 2 – Belovsky & Jordan 1978; 3 – Bere 1959; 4 – Berry, 1980; 5 – Doney *et al.* 1974; 6 – Duncan 1975; 7 – du Plessis 1972; 8 – Foose 1982; 9 – Greer & Howe 1964; 10 – Groves & Kurt 1972; 11 – Hanks 1972b; 12 – Hillman 1979; 13 – Hitchins 1968; 14 – P. M. Hitchins personal communication 1983; 15 – Hofman 1973; 16 – Hoppe 1977b; 17 – Hoppe *et al.* 1981; 18 – Huntley 1971; 19 – Kingdon 1982; 20 – Laurie 1982; 21 – Laws 1966; 22 – Laws *et al.* 1975; 23 – Ledger 1963; 24 – Ledger 1964; 25 – Leuthold 1979; 26 – Lott 1979; 27 – Mackie 1973; 28 – Meinertzhagen 1938; 29 – Mitchell *et al.* 1977; 30 – Morris & Hanks 1974; 31 – Oliver *et al.* 1978; 32 – Penzhorn 1974; 33 – Pienaar 1969; 34 – Robinson 1979; 35 – Sachs 1967; 36 – Schaller 1967; 37 – Shoshani & Eisenberg 1982; 38 – Sinclair 1977; 39 – Smithers 1971; 40 – Smithers 1983; 41 – Spinage 1982; 42 – Stanley-Price 1974; 43 – Wilson 1968; 44 – Wilson 1969a; 45 – Macdonald 1984; 46 – Geist & Bayer 1988.

Appendix II

Simulation model of the white rhino population

The model was formulated in PASCAL for implementation on an Apple II microcomputer. The basis of the model is a population made up of 46 year groups, which were grouped into functional age classes differing in their mortality, natality and dispersal rates. The age classes were as follows: old – 36–45 y; adult – 11–35 y; subadult – 6–10 y; immature – 3–5 y; juvenile – 1–2 y; infant – 0 y.

The sex ratio was considered to be 50:50 throughout all age classes. Demographic parameters operated on the year groups in the following order: first emigration, then mortality, then natality. Thus the number of animals entering age group 0 was calculated by multiplying the number of females surviving within the age classes, OLD, ADULT and SUBADULT by the age-class specific natality rates, by a factor of 0.5 to adjust for the sex ratio, and finally by the infant survival rate.

(a) Simple model of expanding and stable populations
In the initial use of the model in Chapters 11 and 13, fixed values were assigned to demographic parameters as in Table II.1 (all rates expressed per annum):

Table II.1

Variable	Expanding population	Stable population
EMIGRATION	0	0
OLDMORT	0.05	0.08
ADMORT	0.015	0.03
SAMORT	$0.5 \times (0.015 + 0.03)$	$0.5 \times (0.03 + 0.06)$
IMMORT	0.03	0.06
JUVMORT	0.035	0.08
INFMORT	0.08	0.24
OLDNATAL	0.30	0.15
ADNATAL	0.40	0.25
SANATAL	0.45	0.07

The class of subadults is assumed to operate as a mix of males, still exhibiting immature mortality rates, and females exhibiting adult mortality rates. Subadult females have a varying natality rate depending on the age at first parturition; a natality of 0.07 means that females first give birth in year class 10.

Arbitrary starting numbers were chosen for each year group, and iteration was carried out over a 100 year period to allow the age structure to stabilize.

(b)   Density dependent demographic variables

In the logistic formulation the rate of population increase is a function of the factor $(1 - N/K)$, where K is a constant representing the asymptotic density of the population, and N is a variable representing the population density. Caughley (1976c) suggests that, for plant–herbivore systems, K reflects the amount of edible and accessible food available, and N reflects the rate at which it is used up by the population. Hence the ratio K/N represents the amount of food available per unit of herbivores. At the asymptotic density, the rate of food gained by the herbivore population is just adequate for rates of recruitment to balance rates of loss.

In the absence of dispersal the net rate of population increase is simply the difference between the overall birth rate (B) and death rate (D). In the logistic formulation these component rates are linearly related to N, i.e. $B = B_{max} - aN$ and $D = D_{min} + cN + P$, where a and c are constants, and P represents the density-independent component of mortality. The density K is the point where these lines intersect, i.e. $B = D$. When N is zero, $B = B_{max}$ and $D = D_{min}$, and the maximum or intrinsic rate of population increase, R_{max}, is given by the difference $B_{max} - D_{min}$ (see Pianka 1974, p. 86).

However, for a realistic population model the simple linear dependence of mortality and natality rates on population density needs to be modified. Firstly, following Caughley's interpretation, mortality and natality rates will be related to the ratio K/N rather than to N. Secondly, because of the wide difference in individual weights, K and N will be expressed in terms of biomass rather than density. Thirdly, lower and upper bounds for mortality and natality respectively, reached at some biomass below K, will be introduced. Fourthly, where mortality rates are as low as they are in white rhinos their values increase too slowly, over the range of densities in excess of K, with a linear formulation, to generate the starvation induced mortality which must eventually set in at high biomass levels. Potential alternative formulations include (i) a breakpoint, with a steeper rate of increase in mortality once populations exceed this threshold density; (ii) a power function relating mortality to biomass. In the white rhino model the latter function will be assumed.

In order to formulate specific functions for demographic variables, values for at least two points are needed. Mortality and natality rates were known for the real white rhino population in 1970, when it was increasing at near its maximal rate. A realistic combination of changed values which would achieve population stasis was found by trial and error. However, the biomass level at which the latter values would prevail was unknown. Arbitrarily it will be assumed that, with no depression of grass production and with no emigration, the white rhino population would tend to stabilize at a biomass about 50% higher than that prevailing in 1970; i.e. a mean biomass (M) of 6500 kg km^{-2} over the 450 km^2 area of UGR. Thus at $M = 0.67\,K$, the 1970 values for mortality and natality rates prevail. At $M = K$, the values producing population stasis prevail, where $K = 6500$ kg km^{-2}. Based on these two points, functions relating demographic variables to the ratio K/M were derived, where M represents the prevailing population biomass. Because of physiological lags due to accumulated body reserves, the demographic responses of adults were taken to be functions of biomass levels over the preceding three year period, those of juveniles to depend on biomass levels over the preceding two years, while infant mortality responds directly to the currently prevailing conditions. Because of its dependence on growth rates, subadult natality will be related to mean biomass over a five year period.

Emigration rates are difficult to estimate. During the phase of white rhino population increase, emigration rates of 3% per annum for adult males and 7.5% per annum for subadults were indicated (Chapter 11). It will be assumed that these represent individuals dispersing out of saturated zones, and that 75% of the population occurred in such high density zones. Thus at saturation densities dispersal rates are estimated to be 5% per adult male per year, or 2.5% per adult per year (assuming a 50:50 sex ratio), and 10% per subadult or immature per year. At two-thirds of the saturation biomass, dispersal rates are assumed to be about half of the above values. At biomasses in excess of the saturation biomass, some adult females with calves start dispersing. Emigration rates will be assumed to depend on currently prevailing conditions with no lag effects. In the model, emigration may be either allowed or prevented as an option at the beginning of the run. The functions used for demographic variables in the white rhino model are given in Table II.2, where M represents the prevailing population biomass the preceding year, $M2$ represents the mean population biomass over the preceding two years, $M3$ represents the mean population biomass over the preceding three years, and $M5$ represents the mean population biomass over the preceding five years:

Table II.2

Variable	Function	Upper or lower bound	
OLDMORT	exp $(-2.526 - 1.159 \times \ln (K/M3))$	0.05	
ADMORT	exp $(-3.507 - 1.709 \times \ln (K/M3))$	0.01	
IMMORT	exp $(-2.813 - 1.709 \times \ln (K/M3))$	0.02	
SAMORT	$(0.5 \times \text{ADMORT} + \text{IMMORT})$		
JUVMORT	exp $(-2.526 - 2.039 \times \ln (K/M2))$	0.02	
INFMORT	exp $(-1.427 - 2.709 \times \ln (K/M))$	0.05	
OLDNATAL	$-0.15 + 0.30 \times (K/M3)$		0.35
ADNATAL	$-0.05 + 0.30 \times (K/M3)$		0.45
SANATAL	$-0.75 + 0.80 \times (K/M5)$		0.50
ADEMIGR	$-0.01 + 0.03/(K/M)$		0
IMEMIGR	$-0.05 + 0.15/(K/M)$		0
SAEMIGR	$-0.05 + 0.15/(K/M)$		0
JUVEMIGR	$-0.02 + 0.02/(K/M)$		0

The bounds for mortality and natality rates have been set slightly lower and higher respectively than the rates prevailing in the 1970 population. Thus the model population has an intrinsic rate of increase of 10.5% per annum.

(c) Vegetation interaction

Caughley (1982) showed that a many-plant system could be modelled as if it consisted of a single plant species whose rate of increase was the harmonic mean of the specific rates of increase of each component species. Hence a simple logistic equation will be used to represent all grasses.

No grass production measurements are available for Umfolozi. Based on measurements carried out at the Matopos research station in Zimbabwe (Dye 1983), and at Nylsvley in the northern Transvaal (Grunow *et al.* 1980), the saturation biomass of grass was taken to be 200 g m^{-2}. Not all of this would be available for consumption by white rhinos. The consumptive demands of other grazing ungulates combined are about equal to those of white rhinos, and to this has to be added the grass removed by insects such as grasshoppers and harvester termites. On a regional basis some grassland areas are relatively inaccessible to white rhinos, for example on steep hillsides or in thickets. It will be assumed that, of the total grass production, 65% is consumed by other herbivores, while 30% of the remainder is inaccessible to white rhinos, so that the amount available for white rhino consumption is 50 g m^{-2}.

The food demands of white rhinos amount to 1.5% of body mass (drymass/livemass) per day, or five times the standing biomass of white rhinos over the course of a year. Some reduction in eating rates at low grass biomass will be allowed according to a type 2 curve of Holling (1965). Thus

EATRATE: $= 5 \times (2 \times \text{VHFACTOR}/1 + \text{VHFACTOR})$

if VHFACTOR < 1

where VHFACTOR represents the quantity of grass available per white rhino, as determined by the factor (grass biomass)/(white rhino biomass), normalized to vary between 1 and 0.

Based on standard agricultural recommendations, the threshold for overutilization of grass will be taken as 50% of available grass production. If the white rhino population consumes less than this proportion, they have no influence on future grass production (i.e. they simply remove part of the annual growth, which is replaced completely the next season). If the consumptive demands of the white rhino population exceed this threshold utilization, the grass population is depressed to the extent that consumption exceeds this threshold; thus

GRASS: $= \text{GRASS} \times (1-\text{XSUTIL})$

where GRASS represents the grass population, and XSUTIL the fraction by which consumption exceeded the 50% threshold. However, the depressed grass population then grows back at the rate given by the logistic equation. To prevent the grass population going to extinction, an ungrazable reserve amounting to 20% of the saturation biomass of grass is incorporated.

(d)   Variable rainfall regime

In the preceding formulation the annual grass biomass was assumed to be constant, apart from the influence of white rhino consumption. However, grass production is influenced by rainfall, and for rainfall regimes below 700 mm per year annual grass production is almost linearly related to the current season's rainfall (Dye 1983). Thus to obtain the amount of grass produced in a particular year, the potential saturation biomass of grass is multiplied by the ratio (annual rainfall/mean rainfall).

(e)   Medium-sized ungulate

To simulate a population of medium-sized ungulates, for example wildebeest, the main changes made to the model were as follows: (i) the number of year groups was reduced; (ii) different upper or lower bounds were chosen for demographic variables; (iii) new values of demographic variables that would yield a static population in the absence of dispersal were found; (iv) the period over which biomass was averaged in the functions for demographic variables was reduced. The differences in parameter values between the two versions of the model are listed in Table II.3. For demographic variables the value for a static population is given first, followed by the upper or lower bound in brackets, expressed per annum:

Table II.3

Parameter or variable	White rhino	Wildebeest
Unit adult body mass (kg)	1600	180
Saturation biomass	6500	4643
Eating rate (% per day)	1.5	2.0
Age ranges (y):		
Old	36–45	11–15
Adult	11–35	3–10
Subadult	6–10	2
Immature	3–5	1
Juvenile	0–2	0
OLDMORT	0.08 (0.05)	0.20 (0.05)
ADMORT	0.03 (0.01)	0.08 (0.03)
IMMORT	0.06 (0.02)	0.16 (0.04)
JUVMORT	0.08 (0.02)	0.50 (0.06)
INFMORT	0.24 (0.05)	
OLDNATAL	0.15 (0.30)	1.00 (1.00)
ADNATAL	0.25 (0.45)	1.00 (1.00)
SANATAL	0.05 (0.50)	0 (1.00)
ADEMIGR	0.02 (0.05)	0.02 (0.04)
IMEMIGR	0.10 (0.25)	0.10 (0.20)
JUVEMIGR	0 (0.02)	0.01 (0.02)
Biomass averaging period (y):		
Adults	3	2
Immatures	2	1
Juveniles	1	1
Subadult natality	5	3
R_{max} (% per year)	10.5	30.0

References

Agenbroad, L. D. (1984). Hot Springs, South Dakota: entrapment and taphonomy of Columbian mammoth. In *Quaternary Extinctions*, ed. P. S. Martin & R. G. Klein, pp. 113–27. Tucson, Arizona: University of Arizona Press.

Alexandre, D.-Y. (1978). Le role disseminateur des elephants en foret de Tai, Cote D'Ivoire. *La Terre et la Vie*, **32**, 47–72.

Allen-Rowlandson, T. S. (1980). The social and spatial organisation of the greater kudu in the Andries Vosloo Game Reserve, Eastern Cape, M.Sc. thesis, Rhodes University, Grahamstown.

Anderson, A. (1984). The extinction of moa in southern New Zealand. In *Quaternary Extinctions*, ed. P. S. Martin & R. G. Klein, pp. 691–707. Tucson, Arizona: University of Arizona Press.

Anderson, E. (1984). Who's who in the Pleistocene: a mammalian bestiary. In *Quaternary Extinctions*, ed. P. S. Martin & R. G. Klein, pp. 40–89. Tucson, Arizona: University of Arizona Press.

Anderson, G. D. & Walker, B. H. (1974). Vegetation composition and elephant damage in the Sengwa Research Area, Rhodesia. *J. Sth. Afr. Wildl. Mgmt. Ass.*, **4**, 1–14.

Anderson, J. L. (1978). Aspects of the ecology of nyala in Zululand. D. Phil. thesis, London University, London.

Anderson, J. L. (1980). The social organization and aspects of behaviour of the nyala. *Z. Saugetierk*, **45**, 90–123.

Anderson, J. L. (1983). Sport hunting in national parks: sacrilege or salvation? In *Management of Large Mammals in African Conservation Areas*, ed. R. N. Owen-Smith, pp. 271–280. Pretoria: Haum.

Ansell, W. F. H. (1963). Additional breeding data on Northern Rhodesian mammals. *Puku*, **1**, 9–28.

Arman, P. & Field, C. R. (1973). Digestion in the hippopotamus. *E. Afr. Wildl. J.*, **11**, 9–18.

Arman, P. & Hopcraft, D. (1975). Nutritional studies in East African herbivores. Digestibilities of dry matter, crude fibre and crude protein in antelope, cattle and sheep. *Brit. J. Nutr.*, **33**, 255–64.

Attwell, C. A. M. (1977). Reproduction and population ecology of the blue wildebeest in Zululand. Ph.D. thesis, University of Natal, Pietermaritzburg.

Attwell, C. A. M. (1982). Growth and condition of blue wildebeest in Zululand. *S. Afr. J. Wildl. Res.*, **12**, 63–70.

Attwell, C. A. M. & Hanks, J. (1980). Reproduction of the blue wildebeest in Zululand, South Africa. *Saugetierk. Mitt.*, **28**, 264–81.

Backhaus, D. (1959). Beobachtungen uber das Freileben von Lelwel-Kuhantelopen und Gelegenheit-Beobachtungen an Sennar-Pferdeantilopen. *Z. Saugetierk.*, **24**, 1–34.

Bainbridge, W. R. (1965). Distribution of seed in elephant dung. *Puku*, **3**, 173–5.

Barnes, R. F. W. (1979). Elephant ecology in Ruaha National Park, Tanzania. Ph.D. thesis, Cambridge University.

Barnes, R. F. W. (1980). The decline of the baobab tree in Ruaha National Park, Tanzania. *Afr. J. Ecol.*, **18**, 243–52.

Barnes, R. F. W. (1982a). Elephant feeding behaviour in Ruaha National Park, Tanzania. *Afr. J. Ecol.*, **20**, 123–36.

Barnes, R. F. W. (1982b). Mate-searching behaviour of elephant bulls in a semi-arid environment. *Anim. Behav.*, **30**, 1217–23.

Barnes, R. F. W. (1982c). A note on elephant mortality in Ruaha National Park, Tanzania. *Afr. J. Ecol.*, **20**, 137–41.

Barnes, R. F. W. (1983a). The elephant problem in Ruaha National Park, Tanzania. *Biol. Cons.*, **26**, 127–48.

Barnes, R. F. W. (1983b). Effects of elephant browsing on woodlands in a Tanzanian national park: measurements, models and management. *J. Appl. Ecol.*, **20**, 521–40.

Barnes, R. F. W. (1985). Woodland changes in Ruaha National Park (Tanzania) between 1976 and 1982. *Afr. J. Ecol.*, **23**, 215–22.

Barnes, R. F. W. & Douglas-Hamilton, I. (1982). The numbers and distribution patterns of large mammals in the Ruaha-Rungwa area of southern Tanzania. *J. Appl. Ecol.*, **19**, 411–25.

Barrow, J. (1801). *An account of travels in the interior of Southern Africa, in the years 1797 and 1798*. London.

Bell, R. H. V. (1969). The use of herbaceous layer by grazing ungulates in the Serengeti National Park, Tanzania. Ph.D. thesis, University of Manchester.

Bell, R. H. V. (1970). The use of the herb layer by grazing ungulates in the Serengeti. In *Animal Populations in Relation to their Food Resources.* ed. A. Watson, Symp. Brit. Ecol. Soc., pp. 111–24. Oxford: Blackwell.

Bell. R. H. V. (1971). A grazing ecosystem in the Serengeti. *Sci. Amer.*, **225**, 86–93.

Bell, R. H. V. (1981a). An outline of a management plan for Kasungu National Park, Malawi. In *Problems in Management of Locally Abundant Wild Mammals*, ed. P. A. Jewell, S. Holt & D. Hart, pp. 69–89. New York: Academic Press.

Bell, R. H. V. (1981b). An estimate of nyala numbers in Lengwe National Park by estimating the drinking rates of recognisable animals. *Mimeogr. Rep.*, Dept. of National Parks & Wildlife, Malawi.

Bell, R. H. V. (1982). The effect of soil nutrient availability on community structure in African ecosystems. In *The Ecology of Tropical Savannas*, ed. B. J. Huntley & B. H. Walker, pp. 193–216. Berlin: Springer-Verlag.

Bell, R. H. V. (1984). Notes of elephant-woodland interactions. In *The Status and Conservation of Africa's Elephants and Rhinos*, ed. D. H. M. Cumming & P. Jackson, pp. 98–103. Gland, Switzerland: IUCN.

Belovsky, G. E. & Jordan, P. A., (1978). The time-energy budget of a moose. *Theor. Pop. Biol.*, **14**, 76–104.

Benedict, F. G. (1936). *Physiology of the elephant.* Carnegie Institute, Washington D.C., Publ. No. 474, 302 pp.

Bere, R. M. (1959). Queen Elizabeth National Park, Uganda. The hippopotamus problem and experiment. *Oryx*, **5**, 116–24.

Berger, J. (1981). The role of risks in animal combat: zebra and onager fights. *Z. Tierpsychol.*, **56**, 297–304.

Berry, H. H. (1980). Behavioural and eco-physiological studies on blue wildebeest at the Etosha National Park. Ph.D. thesis, University of Cape Town.

Berry, H. H., Siegfried, W. R. & Crowe, T. M. (1982). Activity patterns in a population of free-ranging wildebeest at Etosha National Park. *Z. Tierpsychol.*, **59**, 229–46.

Berry, P. S. M. (1978). Range movements of giraffe in the Luangwa Valley, Zambia. *E. Afr. Wildl. J.*, **16**, 77–84.

Bigalke, R. C. (1970). Observations on springbok populations. *Zoologica Africana*, **5**, 59–70.

Bigalke, R. C. (1972). Observations on the behaviour and feeding habits of the springbok. *Zoologica Africana*, **7**, 333–59.

Blankenship, L. H. & Qvortrup, S. A. (1974). Resource management on a Kenya ranch. *J. Sth. Afr. Wildl. Mgmt. Ass.*, **4**, 185–90.

Blueweiss, L., Fox, H,. Kudzama, V., Nakashima, D., Peters, R. & Sams, S. (1978). Relationships between body size and some life history parameters. *Oecologia*, **37**, 257–72.

Borner, M. (1979). A field study of the Sumatran rhinoceros. Ph.D. thesis, University of Basel.

Bosman, P. & Hall-Martin, A. (1986). *Elephants of Africa*. Cape Town: Struik.

Botkin, D. B., Mellilo, J. M. & Wu, L. S.-Y. (1981). How ecosystem processes are linked to large mammal population dynamics. In *Dynamics of Large Mammal Populations*, ed. C. W. Fowler & T. D. Smith, pp.373–88. New York: John Wiley.

Bourliere, F. (1961). Le sex-ratio de la Giraffe. *Mammalia*, **25**, 467–71.

Bourliere, F. (1965). Densities and biomasses of some ungulate populations in Eastern Congo and Ruanda, with notes on population structure and lion/ungulate ratios. *Zoologica Africana*, **1**, 199–207.

Bourliere, F. & Verschuren, J. (1960). *Introduction a l'ecologie des Ongules du Parc National Albert*. Brussels.

Boyce, M. S. (1979). Seasonality and patterns of natural selection for life histories. *Am. Nat.*, **114**, 569–83.

Bradbury, J. W. & Vehrencamp, S. L. (1977). Social organization and foraging in emballonurid bats. III Mating systems. *Behav. Ecol. Sociobiol.*, **2**, 1–17.

Brain, C. K. (1985). Temperature induced environmental changes in Africa as evolutionary stimuli. In *Species and Speciation*, ed. E. S. Vrba, pp. 45–52. Transvaal Museum Monograph No. 4, Pretoria: Transvaal Museum.

Brooks, P. M. (1974). The effect of extensive removals on the population composition of the square-lipped rhinoceros in the H.G.R.–Corridor–U.G.R. Complex in Zululand, and recommendations on the age and sex ratios of future removals. Unpublished report, Natal Parks Board.

Brooks, P. M. & Macdonald, I. A. W. (1983). An ecological case history of the Hluhluwe–Corridor–Umfolozi Game Reserve Complex, Natal, South Africa. In *Management of Large Mammals in African Conservation Areas*, ed. R. N. Owen-Smith, pp. 51–78. Pretoria: Haum.

Brown, J. H., Davidson, D. W., Munger, J. C. & Inouye, R. S. (1986). Experimental community ecology: the desert granivore system. In *Community Ecology*, ed. J. Diamond & T. J. Case, pp. 41–62. New York: Harper & Row.

Brown, J. H. & Maurer, B. A. (1986). Body size, ecological dominance and Cope's rule. *Nature*, **324**, 248–50.

Brown, J. L. & Orians, G. H. (1970). Spacing patterns in mobile animals. *Ann. Rev. Ecol. Syst.*, **1**, 239–62.

Bryant, J. P., Chapin, F. S. III, & Klein, D. R. (1983). Carbon/nutrient balance of boreal plants in relation to vertebrate herbivory. *Oikos*, **40**, 357–68.

Bryant, J. P., Wieland, G. D., Clausen, T. & Kuropat, P. (1985). Interactions of snowshoe hares and feltleaf willow in Alaska. *Ecology* **66**:1564–1573.

Buechner, H. K. & Dawkins, H. C. (1961). Vegetation change induced by elephants and fire in Murchison Falls National Park, Uganda. *Ecology*, **42**, 752–66.

Buechner, H. K., Morrison, J. A. & Leuthold, W. (1966). Reproduction in Uganda kob with special reference to behaviour. In *Comparative Biology of Reproduction in Mammals*, ed. I. W. Rowlands, Symp. Zool. Soc. London 15:69–88.

Buechner, H. K. & Roth, H. D. (1974). The lek system in Uganda kob antelope. *Am. Zool.*, **14**, 145–62.

Bullock, W. R. (1962). The weight of the African elephant. *Proc. Zool. Soc. Lond.*, **138**, 133–5.

Buss, I. O. (1961). Some observations on food habits and behavior of the African elephant. *J. Wildl. Manage.*, **25**, 131–48.

Buss, I. O. & Estes, J. A. (1971). The functional significance of movements and positions of the pinnae of the African elephant. *J. Mammal.*, **52**, 21–7.

Buss, I. O. & Savidge, J. M. (1966). Change in population number and reproductive rate of elephants in Uganda. *J. Wildl. Manage.*, **30**, 791–809.

Calder, W. A. III (1984). *Size, Function and Life History*. Cambridge, Mass.: Harvard University Press. 431pp.

Caughley, G. (1970). Eruption of ungulate populations, with emphasis on Himalayan thar in New Zealand. *Ecology*, **51**, 53–72.

Caughley, G. (1976a). Plant-herbivore systems. In *Theoretical Ecology. Principles and Applications*, ed. R. M. May, pp. 94–113. Oxford: Blackwell.

Caughley, G. (1976b). Wildlife management and the dynamics of ungulate populations. In *Applied Biology*, vol. 1, ed. T. H. Coaker, pp. 183–246. London: Academic Press.

Caughley, G. (1976c). The elephant problem – an alternative hypothesis. *E. Afr. Wildl. J.*, **14**, 265–83.

Caughley, G. (1977). *Analysis of Vertebrate Populations*. New York: Wiley. 234pp.

Caughley, G. (1981). Overpopulation. In *Problems in Management of Locally Abundant Mammals*, ed. P. A. Jewell, S. Holt & D. Hart, pp. 7–20. New York: Academic Press.

Caughley, G. & Goddard, J. (1975). Abundance and distribution of elephants in the Luangwa Valley, Zambia. *E. Afr. Wildl. J.*, **13**, 39–48.

Caughley, G. & Krebs, C. J. (1983). Are big mammals simply small mammals writ large? *Oecologia*, **59**, 7–17.

Cave, A. J. E. (1962). Burchell's original specimens of *Rhinoceros simus*. *Proc. Zool. Soc. London.*, **139**, 691–700.

Cave, A. J. E., & Allbrook, D. B. (1958). Epidermal structures in a rhinoceros *Cerototherium simum*. *Nature*, **182**, 196–7.

Chacon, E. & Stobbs, T. H. (1976). Influence of progressive defoliation of a grass sward on the eating behaviour of cattle. *Austr. J. Agric. Res.*, **27**, 709–27.

Chalmers, G. (1963). Breeding data: steinbok. *E. Afr. Wildl. J.*, **1**, 121–2.

Child, G. (1972). Observations on a wildebeest die-off in Botswana. *Arnoldia* (*Rhodesia*), **5**, 31, 1–13.

Child, G., Roth, H. H. & Kerr, M. (1968). Reproduction and recruitment patterns in warthog populations. *Mammalia*, **32**, 6–29.

Churcher, C. S. (1978). Giraffidae. In *Evolution of African Mammals*, ed. V. J. Maglio & H. B. S. Cooke, pp. 509–32. Cambridge, Mass: Harvard University Press.

Cifelli, R. L. (1981). Patterns of evolution among the Artiodactyla and Perissodactyla (Mammalia). *Evol.*, **35**, 433–40.

Clemens, E. T. & Maloiy, G. M. (1982). The digestive physiology of three East African herbivores: the elephant, rhinoceros and hippopotamus. *J. Zool. Lond.*, **198**, 141–56.

Clough, G. (1969). Some preliminary observations on reproduction in the warthog. *J. Reprod. Fert. Suppl.*, **6**, 323–37.

Clough, G. & Hassam, A. G. (1970). A quantitative study of the daily activity of the warthog in the Queen Elizabeth National Park, Uganda. *E. Afr. Wildl. J.*, **8**, 19–24.

Clutton-Brock, T. H. (1984). Reproductive effort and terminal investment in iteroparous animals. *Am. Nat.*, **123**, 212–29.

Clutton-Brock, T. H. & Albon, S. D. (1982). Parental investment in male and female offspring in mammals. In *Current Problems in Sociobiology*, ed. Kings College Sociobiology Group, pp. 223–48. Cambridge: Cambridge University Press.

Clutton-Brock, T. H., Albon, S. D. & Guinness, F. E. (1981). Parental investment in male and female offspring in polygynous mammals. *Nature*, **289**, 487–9.

Clutton-Brock, T. H., Albon, S. D. & Guinness, F. E. (1984). Maternal dominance, breeding success and birth sex ratios in red deer. *Nature*, **308**, 358–60.

Clutton-Brock, T. H., Albon, S. D. & Guinness, F. E. (1985). Parental investment and sex differences in juvenile mortality in birds and mammals. *Nature*, **313**, 131–3.

Clutton-Brock, T. H., Guinness, F. E. & Albon, S. D. (1982). *Red Deer. Behaviour and Ecology of the Two Sexes*. University of Chicago Press, Chicago. 378pp.

Clutton-Brock, T. H. & Harvey, P. H. (1977a). Primate ecology and social organization, *J. Zool. Lond.*, **183**, 1–39.

Clutton-Brock, T. H. & Harvey, P. H. (1977b). Species differences in feeding and ranging behaviour in primates. In *Primate Ecology*, ed. T. H. Clutton-Brock, pp. 557–84. London: Academic Press.

Clutton-Brock, T. H., Harvey, P. H. & Rudder, B. (1977). Sexual dimorphism, socionomic sex ratio and body weight in primates. *Nature*, **269**, 797–800.

Coe, M. J. (1972). Defaecation by African elephants. *E. Afr. Wildl. J.*, **10**, 165–74.

Coe, M. J., Cumming, D. H. M. & Phillipson, J. (1976). Biomass and production of large African herbivores in relation to rainfall and primary production. *Oecologia*, **22**, 341–54.

Coetzee, B. J., Engelbrecht, A. H., Joubert, S. P. J. & Retief, P. F. (1979). Elephant impact on *Sclerocarya* trees in *Acacia nigrescens* tropical plains thornveld of the Kruger National Park. *Koedoe*, **22**, 39–60.

Colbert, E. H. (1969). *Evolution of the Vertebrates*. 2nd edn. New York: John Wiley, 479pp.

Condy, P. R. (1973). The population status, social behaviour and daily activity pattern of the white rhinoceros in Kyle National Park, Rhodesia. M.Sc. thesis, University of Rhodesia.

Connell, J. H. (1978). Diversity in tropical rain forests and coral reefs. *Science*, **199**, 1302–10.

Conybeare, A. (1980). Buffalo numbers, home range and movement in the Sengwa Wildlife Research Area, Zimbabwe. *S. Afr. J. Wildl. Res.*, **10**, 89–93.

Cooper, S. M., & Owen-Smith, N. (1986). Effects of plant spinescence on large mammalian herbivores. *Oecologia*, **68**, 446–55.

Coppens, Y., Maglio, V. J., Madden, C.T. & Beden, M. (1978). Proboscidea. In *Evolution of African Mammals*, ed. V. J. Maglio & H. B. S. Cooke, pp. 336–66. Cambridge, Mass: Harvard University Press.

Corfield, T. F. (1973). Elephant mortality in Tsavo National Park, Kenya. *E. Afr. Wildl. J.*, **11**, 339–68.

Coryndon, S. C. (1978). Hippopotamidae. In *Evolution of African Mammals*, ed. V. J. Maglio & H. B. S. Cooke, pp. 483–96. Cambridge, Mass: Harvard University Press.

Craig, G. C. (1984). Foetal mass and date of conception in African elephants: a revised formula. *S. Afr. J. Sci.*, **80**, 512–16.

Craighead, J. J., Craighead, F. C., Ruff, R. L. & O'Gara, B. W. (1973). Home ranges and activity patterns of non-migratory elk of the Madison Drainage herd as determined by biotelemetry. *Wildl. Monogr.*, **33**, 1–50.

Crook, J. H. & Goss-Custard, J. (1972). Social ethology. *Ann. Rev. Psychol.*, **23**, 277–312.

Croze, H. (1972). A modified photogrammetric technique for assessing age structure of elephant populations and its use in Kidepo National Park. *E. Afr. Wildl. J.*, **10**, 91–115.

Croze, H. (1974a). The Seronera bull problem. I The elephants. *E. Afr. Wildl. J.*, **12**, 1–28.

Croze, H. (1974b). The Seronera bull problem. II The trees. *E. Afr. Wildl. J.*, **12**, 29–47.

Croze, H., Hillman, A. K. K. & Lang, E. (1981). Elephants and their habitats: how do they tolerate each other. In *Dynamics of Large Mammal Populations*, ed. C. W. Fowler & T. D. Smith, pp. 297–316. New York: Wiley.

Cumming, D. H. M. (1975). *A Field Study of the Ecology and Behaviour of Warthog.* Museum Memoir No. 7, pp. 1–179. Salisbury; Trustees of the National Museums and Monuments of Rhodesia.

Cumming, D. H. M. (1981a). The management of elephant and other large mammals in Zimbabwe. In *Management of Locally Abundant Wild Mammals*, ed. P. A. Jewell, S. Holt & D. Hart, pp. 91–118. New York: Academic Press.

Cumming, D. H. M. (1981b). Elephant and woodlands in Chizarira National Park, Zimbabwe. In *Problems in Management of Locally Abundant Wild Mammals*, ed. P. A. Jewell, S. Holt & D. Hart, pp. 347–9. New York: Academic Press.

Cumming, D. H. M. (1982). The influence of large herbivores on savanna structure in Africa. In *The Ecology of Tropical Savannas*, ed. B. J. Huntley & B. H. Walker, pp. 217–45. Berlin: Springer–Verlag.

Cumming, D. H. M. (1983). The decision-making framework with regard to the culling of large mammals in Zimbabwe. In *Management of Large Mammals in African Conservation Areas*, ed. R. N. Owen-Smith, pp. 173–86. Pretoria: Haum.

Cumming, D. H. M., & Jackson, P. (eds.). (1984). *The Status and Conservation of Africa's Elephants and Rhinos.* IUCN, Gland, Switzerland, 191pp.

Dagg, A. I. & Foster, J. B. (1976). *The Giraffe. Its Biology, Behavior and Ecology.* New York: Van Nostrand Reinhold.

Damuth, J. (1981a). Population density and body size in mammals. *Nature*, **290**, 699–700.

Damuth, J. (1981b). Home range, home range overlap and energy use among animals. *Biol. J. Linn. Soc.*, **15**, 185–93.

Darling, F. F. (1937). *A Herd of Red Deer. A Study in Animal Behaviour.* London: Oxford University Press.

Darwin, C. (1871). *The Descent of Man, and Selection in Relation to Sex.* New York: Appleton. 374pp.

David, J. H. M. (1973). The behaviour of bontebok with special reference to territorial behaviour. *Z. Tierpsychol.*, **33**, 38–107.

David, J. H. M. (1978). Observations on territorial behaviour of springbok in the Bontebok National Park, Swellendam. *Zoologica Africana*, **13**, 123–41.

Davis, S. (1985). Tiny elephants and giant mice. *New Scientist*, 3 Jan., 25–7.

Demment, M. W. (1982). The scaling of ruminoreticulum size with body weight in East African ungulates. *Afr. J. Ecol.*, **20**, 43–8.

Demment, M. W. & Van Soest, P. J. (1985). A nutritional explanation for body-size patterns of ruminant and nonruminant herbivores. *Am. Nat.*, **125**, 641–72.

de Vos, A. & Dowsett, R. J. (1966). The behaviour and population structure of three species of the genus *Kobus. Mammalia*, **30**, 30–55.

de Vos, V., Bengis, R. C. & Coetzee, H. J. (1983). In *Management of Large Mammals in African Conservation Areas*, ed. R. N. Owen-Smith, pp. 213–32. Pretoria: Haum.

Dittrich, L. (1972). Gestation periods and age of sexual maturity of some African antelopes. *Int. Zoo Yrb.*, **12**, 184–7.

Dittrich, L. (1976). Age of sexual maturity in the hippopotamus. *Int. Zoo Yrb.*, **16**, 171–3.

Doney, J. M., Ryder, M. L., Gunn, R. G. & Grubb, P. (1974). Colour, conformation, affinities, fleece and patterns of inheritance in Soay sheep. In *Island Survivors*, ed. P. A. Jewell, C. Milner & J. Morton Boyd, pp. 88–125. London: Athlone Press.

Dougall, H. W. & Sheldrick, D. L. W. (1964). The chemical composition of a day's diet of an elephant. *E. Afr. Wildl. J.*, **2**, 51–9.

Douglas-Hamilton, I. (1971). Radio-tracking of elephants. In *Proceedings of a Symposium of Biotelemetry*, pp. 335–342. Pretoria: CSIR.

Douglas-Hamilton, I. (1972). On the ecology and behaviour of the African elephant. PhD thesis, Oxford University.

Douglas-Hamilton, I. (1987). African elephant population trends and their causes. *Oryx*, **21**, 11–14.

Downing, B. H. (1972). A plant ecological survey of the Umfolozi Game Reserve, Zululand. PhD thesis, University of Natal.

Downing, B. H. (1979). Grass protein content and soils as factors affecting area selective grazing by wild herbivores in the Umfolozi Game Reserve, Zululand. *Proc. Grassl. Soc. Sth. Afr.*, **14**, 85–8.

Dowsett, R. J. (1966). Behaviour and population structure of hartebeest in the Kafue National Park. *Puku*, **4**, 147–54.

Dublin, H. (1984). The Serengeti-Mara ecosystem. *Swara*, **7**, 8–13.

Duckworth, B. J. (1972). The distribution and movement of buffalo herds in the Kalahari sand area of Wankie National Park. Certif. in Field Ecol. thesis, University of Rhodesia.

Dunbar, R. I. M. (1979). Energetics, thermoregulation and the behavioural ecology of the klipspringer. *Afr. J. Ecol.*, **17**, 217–30.

Dunbar, R. I. M. & Dunbar, E. P. (1974). Social organization and ecology of the klipspringer in Ethiopia. *Z. Tierpsychol.*, **35**, 481–93.

Duncan, P. (1975). Topi and their food supply. PhD thesis, University of Nairobi.

Duncan, P. (1980). Time budgets of Camargue horses. II Time-budgets of adult horses and weaned subadults. *Behaviour*, **72**, 26–49.

Dunham, K. M. (1982). The foraging behaviour of impala. *S. Afr. J. Wildl. Res.*, **12**, 36–40.

Dunham, K. M. (1986). Movements of elephant cows in the unflooded Middle Zambezi Valley, Zimbabwe, *Afr. J. Ecol.*, **24**, 287–91.

du Plessis, S. S. (1969). The past and present geographical distribution of the Perissodactyla and Artiodactyla in Southern Africa. MSc. thesis, University of Pretoria.

du Plessis, S. S. (1972). Ecology of blesbok with special reference to productivity. *Wildl. Monogr.*, **30**, 1–70.

du Preez. J. S. & Grobler, J. D. (1977). Drinking times and behaviour at waterholes of some game species in Etosha National Park. *Madoqua*, **10**, 61–70.

Du Toit, P. J., Louw, J. G. & Malan, A. J. (1940). A study of the mineral content and feeding value of the natural pastures in the Union of South Africa. *Onderstepoort J. Vet. Sci.*, **14**, 123–327.

Dye, P. J. (1983). Prediction of variation in grass growth in a semi-arid induced grassland. PhD thesis, University of the Witwatersrand.

East, R. (1981). Species area curves and populations of large mammals in African savanna reserves. *Biol. Cons.*, **21**, 111–26.

East, R. (1984). Rainfall, soil nutrient status and biomass of large African savanna animals. *Afr. J. Ecol.*, **22**, 245–70.

Eisenberg, J. F. & Lockhart, M. (1972). *An Ecological Reconnaissance of Wilpattu National Park, Ceylon*. Smithsonian Contributions to Zoology No. 101, Washington D.C.: Smithsonian Institution Press.

Eisenberg, J. F., McKay, G. M. & Jainudeen, M. R. (1971). Reproductive behaviour of the Asiatic elephant. *Behaviour*, **38**, 193–225.

Eisenberg, J. F. & Seidensticker, J. (1976). Ungulates in southern Asia. A consideration of biomass estimates for selected habitats. *Biol. Cons.*, **10**, 293–308.

Ellerman, J. R., Morrison-Scott, T. S. C. & Hayman A. W. (1953). *South African Mammals 1758–1951*. London: British Museum (Nat. Hist.) Trustees.

Eltringham, S. K. (1974). Changes in the large mammal community of Mweya Peninsula, Rwenzori National Park, following removal of hippopotamus. *J. Appl. Ecol.*, **11**, 855–65.

Eltringham, S. K. (1977). The numbers and distribution of elephant in the Rwenzori National Park and Chambura Game Reserve, Uganda. *E. Afr. Wildl. J.*, **15**, 19–39.

338   *References*

Eltringham, S. K. (1980). A quantitative assessment of range usage by large African mammals with particular reference to the effect of elephants on trees. *Afr. J. Ecol.*, **18**, 53–71.

Eltringham, S. K. (1982). *Elephants*. Poole, Dorset: Blandford Books.

Eltringham, S. K. & Din, N. A. (1977). Estimates of the population size of some ungulate species in the Rwenzori National Park, Uganda. *E. Afr. Wildl. J.*, **15**, 305–16.

Eltringham, S. K. & Woodford, M. H. (1973). The numbers and distribution of buffalo in the Rwenzori National Park, Uganda. *E. Afr. Wildl. J.*, **11**, 151–64.

Emlen, S. T. & Oring, L. W. (1977). Ecology, sexual selection and the evolution of mating systems. *Science*, **197**, 215–23.

Estes, R. D. (1966). Behaviour and life history of the wildebeest. *Nature*, **212**, 999–1000.

Estes, R. D. (1967). The comparative behavior of Grant's and Thomson's gazelles. *J. Mammal.*, **48**, 189–209.

Estes, R. D. (1974). Social organization of the African bovids. In *The Behaviour of Ungulates and its Relation to Management*, ed. V. Geist & F. Walther, pp. 166–205. IUCN public new series no. 24, Morges: IUCN.

Estes, R. D. (1976). The significance of breeding synchrony in the wildebeest. *E. Afr. Wildl. J.*, **14**, 135–52.

Estes, R. D. & Estes, R. K. (1974). The biology and conservation of the giant sable antelope, *Proc. Acad. Nat. Sci. Philadelphia*, **126**, 73–104.

Farentinos, R. C. (1972). Social dominance and mating activity in the tassle-eared squirrel. *Anim. Behav.*, **20**, 316–26.

Fenchel, T. (1974). Intrinsic rate of natural increase: the relationship with body size. *Oecologia*, **14**, 317–26.

Ferrar, A. A. (1983). *Handbook for the Management of African Large Mammals*. Pretoria: CSIR.

Field, C. R. (1970). A study of the feeding habits of hippopotamus in the Queen Elizabeth National Park, Uganda, with management implications. *Zoologica Africana*, **5**, 71–86.

Field, C. R. (1971). Elephant ecology in the Queen Elizabeth National Park, Uganda. *E. Afr. Wildl. J.*, **9**, 99–123.

Field, C. R. & Blankenship, L. H. (1973). Nutrition and reproduction of Grant's and Thomson's gazelles, Coke's hartebeest and giraffe in Kenya. *J. Reprod. Fert. Suppl*, **19**, 287–301.

Field, C. R. & Laws, R. M. (1970). The distribution of the larger herbivores in the Queen Elizabeth National Park, Uganda. *J. Appl. Ecol.*, **7**, 273–94.

Field, C. R. & Ross, I.C. (1976). The savanna ecology of Kidepo Valley National Park. II Feeding ecology of elephant and giraffe. *E. Afr. Wildl. J.*, **14**, 1–16.

Finch, V. A. (1972). Thermoregulation and heat balance of the East African eland and hartebeest. *Am. J. Phys.*, **222**, 1374–9.

Fisher, D. C. (1984). Mastodon butchery by North American Paleo-Indians. *Nature*, **308**, 271–2.

Fisher, R. A. (1930). *The Genetical Theory of Natural Selection*. Oxford: Oxford University Press. 291pp.

Floody, O. R. & Arnold, A. P. (1975). Uganda kob: territoriality and the spatial distributions of sexual and agonistic behaviors at a territorial ground. *Z. Tierpsychol.*, **37**, 192–212.

Foose, T. J. (1982). Trophic strategies of ruminant vs nonruminant ungulates. PhD thesis, University of Chicago.

Fortelius, M. (1982). Ecological aspects of the dental functional morphology of the Plio-Pleistocene rhinoceroses of Europe. In *Teeth: Form, Function and Evolution*. ed. B. Kurten, New York: Columbia University Press.

Foster, J. B. (1967). The square-lipped rhino. *E. Afr. Wildl. J.*, **5**, 167–70.

Foster, J. B. & Coe, M. J. (1968). The biomass of game animals of Nairobi National Park 1960–1966. *J. Zool. Lond.*, **155**, 413–28.

Foster, J. B. & Dagg, A. I. (1972). Notes on the biology of the giraffe. *E. Afr. Wildl. J.*, **10**, 1–16.

Fowler, C. W. (1981). Density dependence as related to life history strategy. *Ecology*, **62**, 602–10.

Frame, G. W. (1971). The black rhinoceros. *Animals*, **13**, 693–9.

Frame, G. W. (1980). Black rhinoceros sub-population on the Serengeti Plains, Tanzania. *Afr. J. Ecol.*, **18**, 155–66.

Frame, G. W. & Goddard, J. (1970). Black rhinoceros vocalizations. *E. Afr. Wildl. J.*, **8**, 207.

Gadgil, M. (1971). Dispersal: population consequences and evolution. *Ecology*, **52**, 253–61.

Gaines, M. S. & McLenaghan, L. R. Jr. (1980). Dispersal in small mammals. *Ann. Rev. Ecol. Syst.*, **11**, 163–96.

Garland, T., Jr. (1983). Scaling the ecological cost of transport to body mass in terrestrial mammals. *Am. Nat.*, **121**, 571–87.

Geist, V. (1966). The evolution of horn-like organs. *Behaviour*, **27**, 175–214.

Geist, V. (1971). *Mountain Sheep: A Study in Behavior and Evolution*. Chicago: University of Chicago Press. 383pp.

Geist, V. (1974a). On the relationship of social evolution and ecology in ungulates. *Am. Zool.*, **14**, 205–220.

Geist, V. (1974b). On the relationship of ecology and behaviour in the evolution of ungulates: theoretical considerations. In *The Behaviour of Ungulates and its Relation to Management*, ed. V. Geist & F. R. Walther, pp. 235–46. IUCN public, new series no. 24, Morges: IUCN.

Geist, V. (1981). On the reproductive strategies in ungulates and some problems of adaptation. In *Evolution Today*, ed. G. G. E. Scudder & J. L. Reveal, Proc. 2nd Int. Cong. Syst. Biol., pp. 111–32.

Geist, V. & Bayer, M. (1988). Sexual dimorphism in the Cervidae and its relation to habitat, *J. Zool., Lond.*, **214**, 45–53.

Georgiadis, N. (1985). Growth patterns, sexual dimorphism and reproduction in African ruminants. *Afr. J. Ecol.*, **23**, 75–88.

Getz, L. L. (1972). Social structure and aggressive behavior in a population of *Microtus pennsylvanicus*. *J. Mammal.*, **53**, 310–17.

Gilbert, B. M. & Martin, L. D. (1984). Late Pleistocene fossils of Natural Trap Cave, Wyoming, and the climatic model of extinction. In *Quaternary Extinctions*, ed. P. S. Martin & R. G. Klein, pp. 138–147. Tucson: University of Arizona Press.

Gingerich, P. G. (1984). Pleistocene extinctions in the context of origination-extinction equilibria in Cenozoic mammals. In *Quaternary Extinctions*, ed. P. S. Martin & R. G. Klein, pp. 211–22. Tucson: University of Arizona Press.

Glover, J. (1963). The elephant problem at Tsavo. *E. Afr. Wildl. J.*, **1**, 30–9.

Goddard, J. (1966). Mating and courtship of the black rhinoceros. *E. Afr. Wildl. J.*, **4**, 69–76.

Goddard, J. (1967). Home range behaviour and recruitment rates of two black rhinoceros populations. *E. Afr. Wildl. J.*, **5**, 133–50.

Goddard, J. (1968). Food preferences of two black rhinoceros populations. *E. Afr. Wildl. J.*, **6**, 1–18.

Goddard, J. (1969). Aerial census of black rhinoceros using stratified random sampling. *E. Afr. Wildl. J.*, **7**, 105–14.

Goddard, J. (1970a). Age criteria and vital statistics of a black rhinoceros population. *E. Afr. Wildl. J.*, **8**, 105–21.

Goddard, J. (1970b). A note on age at sexual maturity in wild black rhinoceros. *E. Afr. Wildl. J.*, **8**, 208.

Gogan, P. J. P. (1973). Some aspects of nutrient utilization by Burchell's zebra in the Serengeti-Mara region, East Africa. MSc thesis, Texas A&M University.

Gold, A. (1973). Energy expenditure in animal locomotion. *Science*, **181**, 275–6.

Goodman, D. (1981). Life history analysis of large mammals. In *Dynamics of Large Mammal Populations*, ed. C. W. Fowler & T. D. Smith, pp. 415–436. New York: Wiley.

Gosling, L. M. (1974). The social organization of Coke's hartebeest. In *The Behaviour of Ungulates and Its Relation to Management*, ed. V. Geist & F. R. Walther, pp. 488–511. IUCN public, new series no. 24, Morges: IUCN.

Gould, S. J. & Vrba, E. S. (1982). Exaptation – a missing term in the science of form. *Paleobiology*, **8**, 4–15.

Graham, R. W. & Lundelius, E. L. (1984). Coevolutionary disequilibrium and Pleistocene extinctions. In *Quaternary Extinctions*, ed. P. S. Martin & R. G. Klein, pp. 223–49. Tucson: University of Arizona Press.

Grayson, D. K. (1977). Pleistocene avifaunas and the overkill hypothesis. *Science*, **195**, 691–3.

Greer, K. R. & Howe, R. E. (1964). Winter weights of northern Yellowstone elk. *Trans. N. Am. Wildl. Conf.*, **29**, 237–48..

Grimsdell, J. J. R. & Field, C. R. (1976). Grazing patterns of buffalos in the Rwenzori National Park, Uganda. *E. Afr. Wildl. J.*, **14**, 339–44.

Grobler, J. H. (1973). Biological data on tsessebe in Rhodesia. *Arnoldia (Rhodesia)*, **6**, 12, 1–16.

Grobler, J. H. (1974). Aspects of the biology, population ecology and behaviour of the sable in the Rhodes Matopos National Park, Rhodesia. *Arnoldia (Rhodesia)*, **7**, 6, 1–36.

Grobler, J. H. (1980). Breeding behaviour and aspects of social behaviour of sable in the Rhodes Matopos National Park, Zimbabwe. *S. Afr. J Wildl. Res.*, **10**, 150–2.

Groves, C. P. (1967). Geographic variation in the black rhinoceros. *Z. Saugetierk*, **32**, 267–76.

Groves, C. P. & Kurt, F. (1972). *Dicerorhinus sumatrensis. Mammalian species No. 21.* American Soc. of Mammalogists. 6pp.

Grubb, P. (1974). Mating activity and the social significance of rams in a feral sheep community. In *The Behaviour of Ungulates and its Relation to Management*, ed. V. Geist & F. Walther, pp. 457–76. IUCN public, new series no. 24, Morges: IUCN.

Grubb, P. and Jewell, P. A. (1974). Movement, daily activity and home range of Soay sheep. In *Island Survivors*, ed. P. A. Jewell, C. Milner & J. Morton Boyd, pp. 160–194. London: Athlone Press.

Gruhn, R. & Bryan, A. L. (1984). The record of Pleistocene megafaunal extinctions at Taima-taima, northern Venezuela. In *Quaternary Extinctions*, ed. P. S. Martin & R. G. Klein, pp. 128–37. Tucson, Arizona: University of Arizona Press.

Grunow, J. O., Groeneveld, H. T. & du Toit, H. C. (1980). Above ground dry matter dynamics of the grass layer in the Nylsvley tree savanna. *J. Ecol.*, **68**, 877–84.

Guggisberg, C. A. W. (1961). *Simba. The Life of the Lion*. London: Andre Deutch.

Guggisberg, C. A. W. (1966). *S.O.S. Rhino*. London: Andre Deutsch. 174pp.

Guidon, N. & Delibrias, G. (1986). Carbon-14 dates point to man in the Americas 32,000 years ago. *Nature*, **321**, 769–72.

Guilday, J. E. (1984). Pleistocene extinction and environmental change: case study of the Appalachians. In *Quaternary Extinctions*, ed. P. S. Martin & R. G. Klein, pp. 250–8. Tucson: University of Arizona Press.

Guthrie, R. D. (1982). Mammals of the mammoth steppe as paleoenvironmental indicators. In *Paleoecology of Beringia*, ed. D. M. Hopkins, J. V. Matthews, C. E. Schweger & S. B. Young, pp. 307–26. New York: Academic Press.

Guthrie, R. D. (1984). Mosaics, allelochemics and nutrients. An ecological theory of late Pleistocene megafaunal extinctions. In *Quaternary Extinctions*, ed. P. S. Martin & R. G. Klein, pp. 259–98. Tucson: University of Arizona Press.

Guy, P. (1974). Feeding behaviour of the African elephant in the Sengwa Research Area of Rhodesia. MSc thesis, University of Rhodesia.

Guy, P. R. (1975). The daily food intake of the African elephant. *Arnoldia (Rhodesia)*, 7, 26, 1–6.

Guy, P. R. (1976a). The feeding behaviour of elephant in the Sengwa Area, Rhodesia. *S. Afr. J. Wildl. Res.*, 6, 55–64.

Guy, P. R. (1976b). Diurnal activity patterns of elephant in the Sengwa area, Rhodesia, *E. Afr. Wildl. J.*, 14. 285–95.

Guy, P. R. (1981). Changes in the biomass and productivity of woodlands in the Sengwa Research Area, Zimbabwe, *J. Appl. Ecol.*, 18, 507–19.

Hall-Martin, A. J. (1974). Food selection by Transvaal lowveld giraffe as determined by analysis of stomach contents. *J. Sth. Afr. Wildl. Mgmt. Ass.*, 4, 191–202.

Hall-Martin, A. J. (1975). Aspects of the ecology and sociality of the giraffe. *Die Soogdiernavorsingsinstituut 1966–1975*, pp. 48–56. Pretoria: University of Pretoria.

Hall-Martin, A. J. (1980). Elephant survivors. *Oryx*, 15, 355–62.

Hall-Martin, A. J. (1984). Conservation and management of elephants in the Kruger National Park, South Africa. In *The Status and Conservation of Africa's Elephants and Rhinos*, ed. D. H. M. Cumming & P. Jackson, pp. 104–18. Gland: IUCN.

Hall-Martin, A. J. (1985). The nabab of Aukoerebis. *African Wildlife*, 39, 244–7.

Hall-Martin, A. J. (1986). Recruitment in a small black rhino population. *Pachyderm*, 7, 6–8.

Hall-Martin, A. J. (1987). Kruger's magnificent seven still capture the imagination. *Custos*, 16, 26–8.

Hall-Martin, A. J. & Basson, W. D. (1975). Seasonal chemical composition of the diet of Transvaal lowveld giraffe. *J. Sth. Afr. Wildl. Mgmt. Ass.*, 5, 19–21.

Hall-Martin, A. J., Erasmus, T. & Botha, B. P. (1982). Seasonal variation of diet and faeces composition of black rhinoceros in the Addo Elephant National Park. *Koedoe*, 25, 63–82.

Hall-Martin, A. J. & Penzhorn, B. L. (1977). Behaviour and recruitment of translocated black rhinoceros. *Koedoe*, 20, 147–62.

Hall-Martin, A. J. & Skinner, J. D. (1978). Observations on puberty and pregnancy in female giraffe. *J. Sth. Afr. Wildl. Mgmt. Ass.*, 8, 91–4.

Hall-Martin, A. J., Skinner, J. D. & Smith, A. (1977). Observations on lactation and milk composition of the giraffe. *S. Afr. J. Wildl. Res.*, 7, 67–71.

Hall-Martin, A. J., Skinner, J. D. & Van Dyk, J. M. (1975). Reproduction in the giraffe in relation to some environmental factors. *E. Afr. Wildl. J.*, 13, 237–45.

Hamilton, W. D. & May, R. M. (1977). Dispersal in stable habitats. *Nature*, 269, 578–81.

Hanks, J. (1969a). Growth in weight of the female African elephant in Zambia. *E. Afr. Wildl. J*, 7, 7–10.

Hanks, J. (1969b). Seasonal breeding of the African elephant in Zambia. *E. Afr. Wildl. J.*, 7, 167.

Hanks, J. (1972a). Growth of the African elephant. *E. Afr. Wildl. J.*, 10, 251–272.

Hanks, J. (1972b). Reproduction of elephant in the Luangwa Valley, Zambia. *J. Reprod. Fert.*, 30, 13–26.

Hanks, J. (1973). Reproduction in the male African elephant in the Luangwa Valley, Zambia, *J. Sth. Afr. Wildl. Mgmt. Ass.*, 3, 31–9.

Hanks, J. (1979). *A Struggle for Survival*. Cape Town: Struik.

Hanks, J. & McIntosh, E. A. (1973). Population dynamics of the African elephant. *J. Zool., Lond.*, 169, 29–38.

Harestad, A. S. & Bunnell, F. L. (1979). Home range and body weight – a reevaluation. *Ecology*, **60**, 389–402.

Harrington, G. N. & Ross, I. C. (1974). The savanna ecology of Kidepo Valley National Park. I The effects of burning and browsing on the vegetation. *E. Afr. Wildl. J.*, **12**, 93–106.

Harris, W. C. (1838). *Narrative of an Expedition into Southern Africa During the Years 1836 and 1837*. London: John Murray.

Harvey, P. H. & Clutton-Brock, T. H. (1981). Primate home range size and metabolic needs. *Behav. Ecol. Sociobiol.*, **8**, 151–5.

Hatton, J. C. & Smart, N. O. E. (1984). The effect of long term exclusion of large herbivores on soil nutrient status in Murchison Falls National Park, Uganda. *Afr. J. Ecol.*, **22** 23–30.

Hawkins, R. E., Klimstra, W. D. & Autry, D. C. (1971). Dispersal of deer from Crab Orchard National Wildlife Refuge. *J. Wildl. Manage.*, **35**, 216–20.

Haynes, C. V. (1982). Were Clovis pregenitors in Beringia? In *Paleoecology of Beringia*, ed. D. M. Hopkins, J. V. Matthews, C. E. Schweger & S. B. Young, pp. 383–98. New York: Academic Press.

Heller, E. (1913). The white rhinoceros. *Smithsonian Miscellaneous Collections* **61**, 1–77.

Hendrichs, H. (1971). Freilandsbeobachtungen zum Sozialsystem der Afrikanischen Elefanten. In *Dikdik und Elefanten*, ed. H. Hendrichs, pp. 77–173. Munich: R. Piper Verlag.

Hendrichs, H. (1975). Changes in a population of dikdik. *Z. Tierpsychol.*, **38**, 55–69.

Hendrichs, H. & Hendrichs, U. (1971). Freilandsuntersuchunen zur Okologie und Ethologie der Zwerg-Antilope *Madoqua (Rhynchotragus) kirki* Gunther 1880. In *Dikdik und Elefanten*, ed. H. Hendrichs, pp. 9–73. Munich: R. Piper Verlag.

Hennemann, W. W. III. (1983). Relationship among body mass, metabolic rate and the intrinsic rate of natural increase in mammals. *Oecologia*, **56**, 104–8.

Hillman, J. C. (1979). The biology of eland in the wild. PhD thesis, University of Nairobi.

Hillman-Smith, A. K. K., Owen-Smith, N., Anderson, J. L., Hall-Martin, A. J. & Selaladi, J. P. (1986). Age estimation of the white rhinoceros. *J. Zool. Lond.*, **210**, 355–79.

Hillman-Smith, A. K. K., Oyisenzoo, K. M. & Smith, F. (1986). A last chance to save the northern white rhino? *Oryx*, **20**, 20–6.

Hintz, H. F., Schryver, H. F. & Stevens, C. E. (1978). Digestion and absorption in the hindgut of nonruminant herbivores. *J. Anim. Sci.*, **46**, 1803–7.

Hirst, S. M. (1975). Ungulate–habitat relationships in a South African savanna. *Wildl. Monogr.* No. 44, 60pp.

Hitchins, P. M. (1968). Liveweights of some mammals from Hluhluwe Game Reserve, Zululand. *Lammergeyer (Natal)*, **9**, 42.

Hitchins, P. M. (1971). Preliminary findings in a radiotelemetric study on the black rhinoceros in Hluhluwe Game Reserve, Zululand. *Symposium in Biotelemetry*, Pretoria, 1971, pp. 79–100. Pretoria: CSIR.

Hitchins, P. M. (1979). The effects of the black rhinoceros on the vegetation of the NE area of Hluhluwe Game Reserve. Mimeog. paper presented at the Symposium/Workshop on the Vegetation Dynamics of the Hluhluwe-Corridor-Umfolozi Complex, 10–12 August 1979, 7 pp. Natal Parks Board.

Hitchins, P. M., & Anderson, J. L. (1983). Reproductive characteristics and management of the black rhinoceros in the Hluhluwe/Corridor/Umfolozi Game Reserve Complex, *S. Afr. J. Wildl. Res.*, **13**, 78–85.

Hitchins, P. M. & Vincent, J. (1972). Observations on range extension and dispersal of impala in Zululand. *J. Sth. Afr. Wildl. Mgmt Ass.*, **2**, 3–8.

Hofmann, R. R. (1973). *The Ruminant Stomach*. East Afr. Monogr. Biol. Vol. 2. Nairobi: East African Literature Bureau.

Hofmann, R. R. & Stewart, D. R. M. (1972). Grazer or browser: a classification based on the stomach structure and feeding habits of East African ruminants. *Mammalia*, **36**, 226–40.

Holling, C. S. (1965). The functional response of predators to prey density and its role in mimicry and population regulation. *Mem. Entomol. Soc. Can.*, **45**, 1–60.

Hoogerwerf, A. (1970). *Udjong Kulon. Land of the Last Javan Rhinoceros.* Leiden: E. J. Brill.

Hooijer, D. A. (1969). Pleistocene East African rhinoceroses. In *Fossil Vertebrates of Africa*, ed. L. B. Leakey, vol. 1, pp. 71–98. London: Academic Press.

Hooijer, D. A. (1978). Rhinocerotidae. In *Evolution of African mammals*, ed. V. J. Maglio & H. B. S. Cooke, pp. 371–8. Cambridge, Mass: Harvard University Press.

Hooijer, D. A. & Patterson, B. (1972). Rhinoceroses from the Pliocene of north-western Kenya. *Bull. Mus. Compar. Zool. (Harvard Univ.)*, **144**, 1–26.

Hope, G. (1984). Australian environmental change: timing, direction, magnitude, and rates. In *Quaternary Extinctions*, ed. P. S. Martin & R. G. Klein, pp. 681–90. Tucson: University of Arizona Press.

Hopkins, D. M. (1982). Aspects of the paleogeography of Beringia during the late Pleistocene. In *Paleoecology of Beringia*, ed. D. M. Hopkins, J. V. Matthews, C. E. Schweger & S. B. Young, pp. 3–28. New York: Academic Press.

Hopkins, D. M., Matthews, J. V., Schweger, C. E. & Young S. B. (eds.) (1982). *Paleoecology of Beringia.* New York: Academic Press.

Hoppe, P. P. (1977a). Comparison of voluntary food and water consumption and digestion in Kirk's dikdik and suni. *E. Afr. Wildl. J.*, **15**, 41–8.

Hoppe, P. P. (1977b). Rumen fermentation and body weight in African ruminants. *Proc. XIIIth Congress of Game Biologists*, pp. 141–50.

Hoppe, P. P., Gwynne, M. D. & van Hoven, W. (1981). Nutrients, protozoa and volatile fatty acids in the rumen of Harvey's red duiker. *S. Afr. J. Wildl. Res.*, **11**, 110–1.

Hoppe, P. P., Qvortrup, S. A. & Woodford, M. H. (1977a). Rumen fermentation and food selection in East African sheep, goats, Thomson's gazelle, Grant's gazelle and impala. *J. Agric. Sci. Camb.*, **89**, 129–35.

Hoppe, P. P., Qvortrup, S. A. & Woodford, M. H. (1977b). Rumen fermentation and food selection in East African zebu cattle, wildebeest, Coke's hartebeest, and topi. *J. Zool. Lond.*, **181**, 1–9.

Horn, H. S. (1978). Optimal tactics of reproduction and life history. In *Behavioural Ecology. An Evolutionary Approach*, ed. J. R. Krebs & N. B. Davies, pp. 411–30. Oxford: Blackwell.

Horton, D. R. (1984). Red kangaroos: last of the Australian megafauna. In *Quaternary Extinctions*, ed. P. S. Martin & R. G. Klein, pp. 639–80. Tucson: University of Arizona Press.

Houston, D. B. (1971). Ecosystems of National Parks. *Science*, **172**, 648–51.

Houston, D. B. (1982). *The Northern Yellowstone Elk. Ecology and Management.* New York: Macmillan. 474pp.

Hubback, T. (1939). The Asiatic two-horned rhinoceros. *J. Mammal.*, **20**, 1–20.

Huntley, B. J. (1967). *Ceratotherium simum* (Burchell). A literature survey. B.Sc. Hons. report, University of Pretoria.

Huntley, B. J. (1971). Seasonal variation in the physical condition of mature male blesbok and kudu. *J. Sth. Afr. Wildl. Mgmt. Ass.*, **1**, 17–19.

Hutchinson, G. E. (1975). Variations on a theme by Robert MacArthur. In *Ecology and Evolution of Communities*, ed. M. L. Cody & J. M. Diamond, pp. 492–521. Cambridge, Mass: Harvard University Press.

Hvidberg-Hansen, H. (1970). Contribution to the knowledge of the reproductive physiology of the Thomson's gazelle. *Mammalia*, **34**, 551–63.

Irby, L. R. (1977). Studies on mountain reedbuck populations with special reference to the Loskop Dam Nature Reserve. *S. Afr. J. Wildl. Res.*, **7**, 73–86.

Irby, L. R. (1981). Mountain reedbuck activity patterns in the Loskop Dam Nature Reserve. *S. Afr. J. Wildl. Res.*, **11**, 115–20.

Ishwaran, R. (1981). Comparative study on elephant populations in Gal Oya, Sri Lanka. *Biol. Cons.*, **21**, 303–13.

Ishwaran, N. (1983). Elephant and woody plant relationships in Gal Oya, Sri Lanka. *Biol. Cons.*, **26**, 255–70.

IUCN (1980). *World Conservation Strategy*. Gland, Switzerland: IUCN.

Jachmann, H. (1980). Population dynamics of the elephants in the Kasungu National Park, Malawi. *Netherlands J. Zool.*, **30**, 622–34.

Jachmann, H. (1985). Estimating age in African elephants. *Afr. J. Ecol.*, **23**, 199–202.

Jachmann, H. (1986). Notes on the population dynamics of the Kasungu elephants. *Afr. J. Ecol.*, **24**, 215–26.

Jachmann, H. & Bell, R. H. V. (1984). Why do elephants destroy woodland? *African Elephant and Rhino Group Newsletter*, **3**, 9–10.

Jachmann, H. & Bell, R. H. V. (1985). Utilization by elephants of the *Brachystegia* woodlands of the Kasungu National Park, Malawi. *Afr. J. Ecol.*, **23**, 245–58.

Jacobsen, N. H. G. (1973). Distribution, home range and behaviour patterns of bushbuck in the Lutope and Sengwa valleys, Rhodesia, *J. Sth. Afr. Wildl. Mgmt. Ass.*, **4**, 75–93.

Jainudeen, M. R., McKay, G. M. & Eisenberg, J. F. (1972). Observations on musth in the domesticated Asian elephant. *Mammalia*, **36**, 247–61.

Janis, C. (1976). The evolutionary strategy of the Equidae and the origins of rumen and caecal digestion. *Evolution*, **30**, 757–76.

Jarman, P. J. (1968). The effect of the creation of Lake Kariba upon the terrestrial ecology of the middle Zambezi valley. PhD thesis, University of Manchester.

Jarman, P. J. (1974). The social organization of antelope in relation to their ecology. *Behaviour*, **48**, 215–67.

Jarman, P. J. & Jarman, M. V. (1973a). Daily activity of impala. *E. Afr. Wildl. J.*, **11**, 75–92.

Jarman, P. J. & Jarman, M. V. (1973b). Social behaviour, population structure and reproductive potential in impala. *E. Afr. Wildl. J.*, **11**, 329–38.

Jarman, P. J. & Jarman, M. V. (1974). A review of impala behaviour and its relevance to management. In *The Behaviour of Ungulates and its Relation to Management*, ed. V. Geist & F. Walther, pp. 871–81. IUCN public, new series no. 24, Morges: IUCN.

Jarman, P. J. & Sinclair, A. R. E. (1979). Feeding strategy and the patterning of resource partitioning in ungulates. In *Serengeti. Dynamics of an Ecosystem*, ed. A. R. E. Sinclair & M. Norton-Griffiths, pp. 130–63. Chicago: University of Chicago Press.

Jewell, P. A. (1966). The concept of home range in mammals. *Symp. Zool. Soc. London*, **18**, 85–109.

Jewell, P. A. (1972). Social organization and movements of topi during the rut at Ishasha, Queen Elizabeth Park, Uganda. *Zoologica Africana*, **7**, 233–55.

Joubert, E. (1970). The taxonomic status of the rhinoceros in South West Africa. *Madoqua (South West Africa)*, **2**, 27–38.

Joubert, E. (1972). The social organization and associated behaviour in the Hartmann zebra. *Madoqua (South West Africa)*, **6**, 17–56.

Joubert, E. (1974). Notes on the reproduction in Hartmann zebra in South West Africa. *Madoqua*, **1**, 31–6.

Joubert, E. & Eloff, F. C. (1971). Notes on the ecology and behaviour of the black rhinoceros in South West Africa. *Madoqua (South West Africa)*, **3**, 5–54.

Joubert, S. C. J. (1972). Territorial behaviour of the tsessebe in the Kruger National Park. *Zoologica Africana*, **7**, 141–56.

Joubert, S. C. J. (1974). The social organization of the roan antelope and its influence on the spatial distribution of herds in the Kruger National Park. In *The Behaviour of Ungulates and its Relation to Management*, ed. V. Geist & F. Walther, pp. 661–675. IUCN public., new series no. 24, Morges: IUCN.

Joubert, S. C. J. (1975). The mating behaviour of the tsessebe. *Z. Tierpsychol.*, **37**, 182–91.

Joubert, S. C. J. (1983). A monitoring programme for an extensive national park. In *Management of Large Mammals in African Conservation Areas*, ed. R.N. Owen-Smith, pp. 201–212. Pretoria: Haum.

Joubert, S. C. J. & Bronkhorst, P. J. L. (1977). Some aspects of the history and population ecology of the tsessebe in the Kruger National Park. *Koedoe*, **20**, 125–45.

Jungius, H. (1971a). Studies on the food and feeding behaviour of the reedbuck in the Kruger National Park. *Koedoe*, **14**, 65–97.

Jungius, H. (1971b). The biology and behaviour of the reedbuck in the Kruger National Park. *Mammalia Depicta*, Berlin: Paul Parey. 106pp.

Keith, L. B. & Wyndberg, L. A. (1978). A demographic analysis of the snowshoe hare cycle. *Wildl. Monogr.* No. 59, 70pp.

Kingdon, J. (1979). *East African Mammals. An Atlas of Evolution in Africa.* vol. III B (Large Mammals). London: Academic Press. 435pp.

Kingdon, J. (1982). *East African Mammals. An Atlas of Evolution in Africa.* vol. III C. Bovids. London: Academic Press.

Kirby, F. V. (1920). The white rhinoceros in Zululand. *Ann. Durban Mus.*, **2**, 223–42.

Kleiber, M. (1961). *The Fire of Life.* New York: John Wiley.

Klein, R. G. (1977). The ecology of early man in southern Africa. *Science*, **197**, 115–26.

Klein, R. G. (1984a). Mammalian extinctions and Stone Age people in Africa. In *Quaternary Extinctions*, ed. P. S. Martin & R. G. Klein, pp. 553–73. New York: Academic Press.

Klein, R. G. (1984b). The large mammals of southern Africa: late Pleistocene to Recent. In *Southern African Prehistory and Paleoenvironments*, ed. R. G. Klein, pp. 107–46. Rotterdam: Balkema.

Klingel, H. (1965). Notes on the biology of the plains zebra. *E. Afr. Wildl. J.*, **3**, 86–8.

Klingel, H. (1967). Soziale Organisation und Verhalten freilebender Steppenzebras. *Z. Tierpsychol.*, **24**, 580–624.

Klingel, H. (1969). Reproduction in the plains zebra, behaviour and ecological factors. *J. Reprod. Fert. Suppl.* **6**, 339–45.

Klingel, H. (1974). Soziale Organization und Verhalten des Grevy-Zebras. *Z. Tierpsychol.*, **36**, 37–70.

Klingel, H. (1975). Das Verhalten der Pferde (Equidae). *Handbuch der Zoologie*, **8**, 10, 65pp.

Klingel, H. & Klingel, U. (1966). Rhinoceroses of Ngorongoro Crater. *Oryx*, **8**, 302–6.

Klös, H.-G. & Frese, R. (1978). Population trends in African rhinoceroses living in zoos and safari parks. *Int. Zoo Yrb.*, **18**, 231–4.

Klös, H.-G. & Frese, R. (1981). *International Studbook of the Black Rhinoceros*, 45pp., and *International Studbook of the White Rhinoceros*, 52pp. Berlin Zoological Garden.

Kok, O. B. (1975). *Behaviour and Ecology of the Red Hartebeest.* Orange Free State Prov. Admin., Nature Cons. Misc. Publ. No. 4.

Kok, O. B. & Opperman, D. P. J. (1980). Voedingsgedrag van kameelperde in die Willem Pretorius Wildtuin, Oranje-Vrystaat. *S. Afr. J. Wildl. Res.*, **10**, 45–55.

Konigkramer, A. J. (1983). Influencing public attitudes: the socio-politics of culling in Natal. In *Management of Large Mammals in African Conservation Areas*, ed. R. N. Owen-Smith, pp. 233–40. Pretoria; Haum.

Kortlandt, A. (1976). Tree destruction by elephants in Tsavo National Park and the role of man in African ecosystems. *Netherlands J. Zool.*, **26**, 319–44.

Krebs, C. R., Gaines, M. S., Keller, B. L., Myers, J. H. & Tamarin, R. H. (1973). Population cycles in small rodents. *Science*, **179**, 35–41.

Kuhme, W. (1963). Erganzende Beobachtungen an afrikanischen Elefanten im Freigehege. *Z. Tierpsychol.*, **20**, 66–79.

Kurt, F. (1974). Remarks on the social structure and ecology of the Ceylon elephant in the Yala National Park. In *The Behaviour of Ungulates and its Relation to Management*, ed. V. Geist & F. Walther, pp. 618–34. IUCN public., new series no. 24, Morges: IUCN.

Kurtén, B. (1968). *Pleistocene Mammals of Europe*. Chicago: Aldine.

Kurtén, B. & Anderson, E. (1980). *Pleistocene Mammals of North America*. New York: Columbia University Press.

Lamprey, H. F. (1983). Pastoralism yesterday and today: the overgrazing problem. In *Ecosystems of the World. 13 Tropical Savannas*, ed. F. Bourliere, pp. 643–66. Amsterdam: Elsevier.

Lamprey, H. F., Bell, R. H. V., Glover, P. E. & Turner, M. J. (1967). Invasion of the Serengeti National Park by elephants. *E. Afr. Wildl. J.*, **5**, 151–66.

Lang, E. M. (1961). Beobachtungen am Indischen Panzernashorn. *Zool. Garten.*, **25**, 369–409.

Lang, E. M. (1967). Einige biologische Daten vom Panzernashorn. *Rev. Suisse Zool.*, **74**, 603–7.

Lang, E. M. (1980). Observations on growth and molar change in the African elephant. *Afr. J. Ecol.*, **18**, 217–34.

Lang, H. (1920), The white rhinoceros of the Belgian Congo (General Account of habits, physical features, distribution, habitat). *Zool. Soc. Bull.*, **23**, 67–92.

Lang, H. (1923). Recent and historical notes on the square-lipped rhinoceros. *J. Mammal.*, **4**, 155–63.

Langer, P. (1976). Functional anatomy of the stomach of hippopotamus. *S. Afr. J. Sci.*, **72**, 12–6.

Langer, P. (1984). Anatomical and nutritional adaptations in wild herbivores. In *Herbivore Nutrition in the Tropics and Subtropics*, ed. F. M. C. Gilchrist & R. I. Mackie, pp. 185–203. Craighall, South Africa: The Science Press.

Langman, V. A. (1973). Radio-tracking giraffe for ecological studies. *J. Sth. Afr. Wildl. Mgmt. Ass.*, **3**, 75–8.

Langman, V. A. (1977). Cow-calf relationships in giraffe. *Z. Tierpsychol.*, **43**, 264–86.

Lark. R. M. (1984). A comparison between techniques for estimating the ages of African elephants. *Afr. J. Ecol.*, **22**, 69–72.

Laurie, W. A. (1978). The ecology and behaviour of the greater one-horned rhinoceros. PhD thesis, Cambridge University.

Laurie, W. A. (1982). Behavioural ecology of the greater one-horned rhinoceros. *J. Zool., Lond.*, **196**, 307–42.

Laws, R. M. (1966). Age criteria for the African elephant. *E. Afr. Wildl. J.*, **4**, 1–37.

Laws, R. M. (1968a). Dentition and aging of the hippopotamus. *E. Afr. Wildl. J.*, **6**, 19–52.

Laws, R. M. (1968b). Interactions between elephant and hippopotamus populations and their environments. *E. Afr. Agric. For. J.*, **33**, 140–47.

Laws, R. M. (1969a). Aspects of reproduction in the African elephant. *J. Reprod. Fert., Suppl.*, **6**, 193–217.

Laws, R. M. (1969b). The Tsavo Research Project. *J. Reprod. Fert., Suppl.*, **6**, 495–531.

Laws, R. M. (1970). Elephants as agents of habitat and landscape change in East Africa. *Oikos*, **21**, 1–15.

Laws, R. M. (1974). Behaviour, dynamics, and management of elephant populations. In

The Behaviour of Ungulates and its Relation to Management, ed. V. Geist & F. Walther, pp. 513–29. IUCN public., new series no. 24, Morges: IUCN.

Laws, R. M. (1981a). Experiences in the study of large mammals. In *Dynamics of Large Mammal Populations*, ed. C. W. Fowler & T. D. Smith, pp. 19–45. New York: Wiley.

Laws, R. M. (1981b). Large mammal feeding strategies and related overabundance problems. In *Management of Locally Abundant Wild Mammals*, ed. P. A. Jewell, S. Holt & D. Hart, pp. 217–232. New York: Academic Press.

Laws, R. M. (1984). Hippopotamuses. In *The Encyclopaedia of Mammals*, ed. D. Macdonald, vol. 2, pp. 506–11. London: George Allen & Unwin.

Laws, R. M. & Clough, G. (1966). Observations on reproduction in the hippopotamus. In *Comparative Biology of Reproduction in Mammals*, ed. I. W. Rowlands, pp. 117–40. Symp. Zool. Soc. London No. 15.

Laws, R. M. & Parker, I. S. C. (1968). Recent studies on elephant populations in East Africa. In *Comparative Nutrition of Wild Animals*, ed. M. A. Crawford, pp. 319–59. Symp. Zool. Soc. London No. 21.

Laws, R. M., Parker, I. S. C. & Johnstone, R. C. B. (1970). Elephants and habitats in North Bunyoro, Uganda. *E. Afr. Wildl. J.*, **8**, 163–80.

Laws, R. M., Parker, I. S. C. & Johnstone, R. C. B. (1975) *Elephants and their Habitats. The Ecology of Elephants in North Bunyoro, Uganda*. Oxford: Clarendon Press.

Leader-Williams, N. (1980). Population dynamics and regulation of reindeer introduced into South Georgia. *J. Wildl. Manage.*, **44**, 640–57.

Ledger, H. P. (1963). Weights of some East African mammals. *E. Afr. Wildl. J.*, **1**, 123–4.

Ledger, H. M. (1964). Weights of some East African mammals:2. *E. Afr. Wildl. J.*, **2**, 159.

Ledger, H. P. (1968). Body composition as a basis for a comparative study of some East African mammals. In *Comparative Nutrition of Wild Animals*, ed. M. A. Crawford, pp. 289–310. Symp. Zool. Soc. London No. 21.

Lee, P. C. (1987). Allomothering among African elephants. *Anim. Behav.*, **35**, 278–91.

Lee, P. C. & Moss, C. J. (1986). Early maternal investment in male and female elephant calves. *Behav. Ecol. Sociobiol.*, **18**, 353–61.

Lent, P. C. (1974). Mother-infant relationships in ungulates. In *The Behaviour of Ungulates and its Relation to Management*, ed. V. Geist & F. Walther, pp. 14–55. IUCN public., new series no. 24, Morges: IUCN.

Leopold, A. S. (1968). Ecological objectives in park management. *E. Afr. Agric. For. J.*, **33**, 168–72.

Leopold, A. S., Cain, S. A., Cottam, C. M., Gabrielson, I. N. & Kimball, T. L. (1963). Wildlife management in the national parks. *Trans. N. Amer. Wildl. & Nat. Res. Conf.*, **28**, 28–45.

Leuthold, B. M. (1979). Social organization and behaviour of giraffe in Tsavo East National Park. *Afr. J. Ecol.*, **17**, 19–34.

Leuthold, B. M. & Leuthold, W. (1972). Food habits of giraffe in Tsavo National Park, Kenya. *E. Afr. Wildl. J.*, **10**, 129–41.

Leuthold, B. M. & Leuthold, W. (1978). Ecology of giraffe in Tsavo East National Park, Kenya. *E. Afr. Wildl. J.*, **16**, 1–20.

Leuthold, W. (1966). Variations in territorial behaviour of Uganda kob. *Behaviour*, **27**, 215–58.

Leuthold, W. (1970). Observations on the social organisation of impala. *Z. Tierpsychol.*, **27**, 693–721.

Leuthold, W. (1971). Freilandsbeobachtungen an Giraffengazellen in Tsavo National Park, Kenya. *Z. Saugetierk.*, **36**, 19–37.

Leuthold, W. (1972). Home range, movements and food of a buffalo herd in Tsavo National Park. *E. Afr. Wildl. J.* **10**, 237–243.

Leuthold, W. (1974). Observations on home range and social organisation of lesser kudu. In *The Behaviour of Ungulates and its Relation to Management*, ed. V. Geist & F. Walther, pp. 206–34. IUCN public., new series no. 24, Morges: IUCN.

Leuthold, W. (1976a). Age structure of elephants in Tsavo National Park, Kenya. *J. Appl. Ecol.*, **13**, 435–44.

Leuthold, W. (1976b). Group size in elephants of Tsavo National Park and possible factors influencing it. *J. Anim. Ecol.*, **45**, 425–39.

Leuthold, W. (1977a). *African Ungulates. A Comparative Review of their Ethology and Behavioural Ecology.* Zoophysiology and Ecology No. 8, New York: Springer Verlag. 307pp.

Leuthold, W. (1977b). A note on group size and composition in the oribi. *Z. Saugetierk.*, **42**, 233–5.

Leuthold, W. (1977c). Spatial organization and strategy of habitat utilization of elephants in Tsavo National Park, Kenya. *Z. Saugetierk.*, **42**, 358–79.

Leuthold, W. (1977d). Changes in tree populations of Tsavo East National Park, Kenya. *E. Afr. Wildl. J.*, **15**, 61–9.

Leuthold, W. (1979). The lesser kudu *Tragelephus imberbis* (Blyth, 1869). Ecology and behaviour of an African antelope. *Saugetierk. Mitt.*, **27**, 1–75.

Leuthold, W. & Leuthold, B. M. (1975). Patterns of social grouping in ungulates of Tsavo National Park, Kenya. *J. Zool. Lond.*, **175**, 405–20.

Leuthold, W. & Leuthold, B. M. (1976). Density and biomass of ungulates in Tsavo East National Park, Kenya. *E. Afr. Wildl. J*, **14**, 49–58.

Leuthold, W. & Sale, J. B. (1973). Movements and patterns of habitat utilization of elephants in Tsavo National Park, Kenya. *E. Afr. Wildl. J.*, **11**, 369–84.

Lewis, D. M. (1984). Demographic changes in Luangwa Valley elephants. *Biol. Cons.*, **29**, 7–14.

Lewis, D. M. (1986). Disturbance effects on elephant feeding: evidence for compression in Luangwa Valley, Zambia. *Afr. J. Ecol.*, **24**, 227–41.

Lidicker, W. Z. (1975). The role of dispersal in the population ecology of small mammals. In *Small Mammals: Their Productivity and Population Dynamics*, ed. F. B. Golley, K. Petrusewicz & L. Ryszkowski, pp. 103–28. Cambridge: Cambridge University Press.

Lightfoot, C. J. (1978). The feeding ecology of giraffe in the Rhodesian middleveld as a basis for the determination of carrying capacity. MSc thesis, University of Rhodesia.

Lindemann, H. (1982). African rhinoceroses in captivity. Thesis, University of Copenhagen.

Lindstedt, S. L. & Boyce, M. S. (1985). Seasonality, fasting endurance, and body size in mammals. *Am. Nat.*, **125**, 873–8.

Lindstedt, S. L. & Calder, W. A. III. (1981). Body size, physiological time and longevity of homeothermic animals. *Q. Rev. Biol.*, **56**, 1–16.

Liu, T.-S. & Li, X.-G. (1984). Mammoths in China. In *Quarternary Extinctions*, ed. P. S. Martin & R. G. Klein, pp. 517–27. New York: Academic Press.

Livingstone, D. A. (1975). Late quarternary climatic change in Africa. *Ann. Rev. Ecol. Syst.*, **6**, 249–80.

Lock, J. M. (1972). The effect of hippopotamus grazing in grasslands. *J. Ecol.*, **60**, 445–67.

Lock, J. M. (1985). Recent changes in the vegetation of Queen Elizabeth National Park, Uganda. *Afr. J. Ecol.*, **23**, 63–5.

Lomnicki, A. (1978). Individual differences between animals and the natural regulation of their numbers. *J. Anim. Ecol.*, **47**, 461–75.

Lomnicki, A. (1982). Individual heterogeneity and population regulation. In *Current Problems in Sociobiology*, ed. King's College Sociobiology Group, Cambridge, pp. 153–67. Cambridge: Cambridge University Press.

Lomolino, M. V. (1985). Body size of mammals on islands: the island rule reexamined. *Am. Nat.*, **125**, 310–6.

Lorius, C., Jouzel, J., Ritz, C., Merlivat, L., Barkov, N. I., Korotkevich, Y. S. & Kotlyakov, V. M. (1985). A 150,000 year climatic record from Antarctic ice. *Nature*, **316**, 591–6.

Lott, D. F. (1974). Sexual and aggressive behaviour of American bison. In *The Behaviour of Ungulates and its Relation to Management*. ed. V. Geist & F. Walther, pp. 382–94. IUCN public., new series no. 24, Morges: IUCN.

Lott, D. F. (1979). Dominance relations and breeding rate in mature male American bison. *Z. Tierpsychol.*, **49**, 418–32.

Loutit, B. D., Louw, G. N. & Seely, M. K. (1987). Preliminary observations of food preferences and chemical composition of the diet of the desert-dwelling black rhinoceros. *Madoqua*, **15**, 35–54.

Low, W. A., Tweedie, R. L., Edwards, C. B. H., Hodder, R. M., Malafant, K. W. J. & Cunningham, R. B., (1981). The influence of environment on daily maintenance behaviour of free-ranging shorthorn cows in central Australia. II Multivariate analysis of duration and incidence of activities. *Appl. Anim. Ethol.*, **7**, 27–38.

MacArthur, R. H. & E. O. Wilson, (1967). *The Theory of Island Biogeography*. Princeton, New Jersey: Princeton University Press. 203pp.

Macdonald, D. (ed.) (1984). *The Encyclopaedia of Mammals*. vol. 2. London: George Allen & Unwin. 895pp.

Mace, G. M. & Harvey, P. H. (1983). Energy constraints on home range size. *Am. Nat.*, **121**, 120–32.

Mackie, C. S. (1973). Interactions between the hippopotamus and its environment on the Lundi River. Certificate in Field Ecology thesis, University of Rhodesia.

Mackie, C. S. (1976). Feeding habits of the hippopotamus on the Lundi River, Rhodesia. *Arnoldia (Rhodesia)*, **7**, 34, 1–6.

Maglio, V. J. (1973). Origin and evolution of the Elephantidae. *Trans. Am. Phil Soc.*, **n.s. 63**, 3, 1–149.

Maglio, V. J. (1978). Patterns of faunal evolution. In *Evolution of African Mammals*, ed. V. J. Maglio & H. B. S. Cooke, pp. 603–19. Cambridge, Mass: Harvard University Press.

Maglio, V. J. & Cooke, H. B. S. (eds.). (1978). *Evolution of African Mammals*. Cambridge, Mass: Harvard University Press.

Malpas, R. C. (1977). The condition and growth of elephants in Uganda. *J. Appl. Ecol.*, **14**, 489–504.

Malpas, R. C. (1978). The ecology of the African elephant in Uganda. PhD thesis, University of Cambridge.

Marcus, L. F. & Berger, R. (1984). The significance of radiocarbon dates for Rancho La Brea. In *Quaternary Extinctions*, ed. P. S. Martin & R. G. Klein, pp. 159–83. Tucson: University of Arizona Press.

Markgraf, V. (1985). Late Pleistocene faunal extinctions in southern Patagonia. *Science*, **228**, 1110–2.

Marshall, P. J. & Sayer, J. A. (1976). Population ecology and response to cropping of a hippopotamus population in eastern Zambia. *J. Appl. Ecol.*, **13**, 391–404.

Martin, E. B. & Martin, C. (1982). *Run Rhino Run*. London: Chatto & Windus.

Martin, P. S. (1967). Prehistoric overkill. In *Pleistocene Extinctions: The Search for a Cause*, ed. P. S. Martin & H. E. Wright, pp. 75–120. New Haven, Conn.: Yale University Press.

Martin, P. S. (1973). The discovery of America. *Science*, **179**, 969–74.

Martin, P. S. (1982). The pattern and meaning of holarctic mammoth extinction. In *The Paleoecology of Beringia*, ed. D. M. Hopkins, J. V. Matthews, C. S. Schweger & S. B. Young, pp. 398–408. New York: Academic Press.

Martin, P. S. (1984a). Prehistoric overkill: the global model: In *Quaternary Extinctions*, ed. P. S. Martin & R. G. Klein, pp. 354–403. Tucson: University of Arizona Press.

Martin, P. S. (1984b). Catastrophic extinctions and late Pleistocene blitzkrieg: two radiocarbon tests. In *Extinctions*, ed. M. Nitecki, pp. 153–89. Chicago: University of Chicago Press.

Martin, P. S. & Guilday, J. E. (1967). A bestiary for Pleistocene biologists. In *Pleistocene Extinctions: The Search for a Cause*, ed. P. S. Martin & H. E. Wright, pp. 1–62. New Haven, Conn.: Yale University Press.

Martin, P. S. & Klein, R. G. (1984). *Quaternary Extinctions: a Prehistoric Revolution*. Tucson: University of Arizona Press. 892pp.

Martin, P. S. & Wright, H. E. (eds.). (1967). *Pleistocene Extinctions: The Search for a Cause*. New Haven, Conn.: Yale University Press, 453pp.

Martin, R. B. (1978). Aspects of elephant social organisation. *Rhodesia Science News*, **12**, 184–7.

Martin, R. B. & Taylor, R. D. (1983). Wildlife conservation in a regional land-use context: the Sebungwe region of Zimbabwe. In *Management of Large Mammals in African Conservation Areas*, ed. R. N. Owen-Smith, pp. 249–70. Pretoria: Haum.

Martin, R. D. & MacLarnon, A. M. (1985). Gestation period, neonatal size and maternal investment in placental mammals. *Nature*, **313**, 220–3.

Mason, D. R. (1982). Studies on the biology and ecology of the warthog in Zululand. DSc thesis, University of Pretoria.

Mason, D. R. (1985a). Postnatal growth and physical condition of warthogs in Zululand. *S. Afr. J. Wildl. Res.*, **15**, 89–97.

Mason, D. R. (1985b). *Monitoring of Ungulate Population Structure in the Kruger National Park – Report on Survey during August, September and October 1984*. Skukuza: Kruger National Park Dept. of Research and Information Memorandum.

Mason, D. R. & van der Walt, P. F. (1984). Sex and age data from cropping of buffalo in the Kruger National Park. *Koedoe*, **27**, 73–8.

Matthews, J. V. (1982). East Beringia during late Wisconsin time: a review of the biotic evidence. In *Paleoecology of Beringia*, ed. D. M. Hopkins, J. V. Matthews, C. E. Schweger & S. B. Young, pp. 127–50. New York: Academic Press.

Maynard Smith, J. (1980). A new theory of sexual investment. *Behav. Ecol. Sociobiol.*, **7**, 247–51.

McCullagh, K. (1969). The growth and nutrition of the African elephant. II The chemical nature of the diet. *E. Afr. Wildl. J.*, **7**, 91–7.

McCullough, D. R. (1969). The Tule elk. Its history, behavior and ecology. *Univ. Calif. Public. Zool.*, **88**, 1-191.

McKay, G. M. (1973). *Behavior and ecology of the Asiatic elephant in southeastern Ceylon*. Smithsonian Contrib. Zool. no. 125, 113 pp.

McNab, B. K. (1963). Bioenergetics and the determination of home range size. *Am. Nat.*, **97**, 133–40.

McNaughton, S. J. (1976). Serengeti migratory wildebeest: facilitation of energy flow by grazing. *Science*, **191**, 92–3.

McNaughton, S. J. (1979). Grazing as an optimization process: grass-ungulate relationships in the Serengeti. *Am. Nat.*, **113**, 691–703.

McNaughton, S. J. (1984). Grazing lawns: animals in herds, plant form and coevolution. *Am. Nat.*, **124**, 863–86.

McNaughton, S. J. (1985). Ecology of a grazing ecosystem: the Serengeti. *Ecol. Monogr.*, **55**, 259–94.

Mead, J. J. & Meltzer, D. J. (1984). North American late Quarternary extinctions and the radiocarbon record. In *Quaternary Extinctions*, ed. P. S. Martin & R. G. Klein, pp. 440–50. Tucson: University of Arizona Press.

Meinertzhagen, R. (1938). Some weights and measurements of large mammals. *Proc. Zool. Soc. London.*, **A108**, 433–9.

Melton, D. A. (1978). Undercounting bias of helicopter censuses in the Umfolozi Game Reserve. *Lammergeyer (Natal)*, **26**, 1–6.

Melton, D. A. (1985). The status of elephants in northern Botswana. *Biol. Cons.*, **31**, 317–33.

Mentis, M. T. (1972). A review of some life history features of the large herbivores of Africa. *Lammergeyer (Natal)*, **16**, 1–89.

Merz, G. (1981). Recherches sur la biologie de nutrition et les habitats preferes de l'elephant de foret. *Mammalia*, **45**, 299–312.

Merz, G. (1986a). Counting elephants in tropical rain forests with particular reference to Tai National Park, Ivory Coast. *Afr. J. Ecol.*, **24**, 61–8.

Merz, G. (1986b). Movement patterns and group size of the African forest elephant in the Tai National Park, Ivory Coast. *Afr. J. Ecol.*, **24**, 133–6.

Millar, J. S. (1977). Adaptive features of mammalian reproduction. *Evol.*, **31**, 370–86.

Millar, J. S. & Zammuto, R. M. (1983). Life histories of mammals: an analysis of life tables. *Ecology*, **64**, 631–5.

Miller, K. R. (1983). Matching conservation goals to diverse conservation areas: a global perspective. In *Management of Large Mammals in African Conservation Areas*, ed. R. N. Owen-Smith, pp. 1–12. Pretoria: Haum.

Milton, K. & May, M. (1976). Body weight, diet and home range area in primates. *Nature*, **259**, 459–62.

Mitchell, B., Staines, B. W. & Welch, D. (1977). *Ecology of Red Deer*. Banchory: Institute of Terrestrial Ecology.

Moen, A. N. (1973). *Wildlife Ecology*. San Francisco: Freeman. 458 pp.

Morris, N. E. & Hanks, J. (1974). Reproduction in the bushbuck. *Arnoldia (Rhodesia)*, **7**, 1, 1–8.

Moss, C. J. (1983). Oestrous behaviour and female choice in the African elephant. *Behaviour*, **86**, 167–96.

Mossman, A. S. & Mossman, H. W. (1962). Ovulation, implantation and fetal sex ratio in impala. *Science*, **137**, 869.

Mueller-Dombois, D. (1972). Crown distortion and elephant distribution in the woody vegetation of Ruhunu National Park, Ceylon. *Ecology*, **53**, 208–26.

Mukinya, J. G. (1973). Density, distribution, population structure and social organization of the black rhinoceros in Masai Mara Game Reserve. *E. Afr. Wildl. J.*, **11**, 385–400.

Mukinya, J. G. (1977). Feeding and drinking habits of the black rhinoceros in Masai Mara Game Reserve. *E. Afr. Wildl. J.*, **15**, 125–38.

Müller-Beck, H. (1982). Late Pleistocene man in northern Alaska and the mammoth-steppe biome. In *Paleoecology of Beringia*, ed. D. M. Hopkins, J. V. Matthews, C. E. Schweger & S. B. Young, pp. 329–52. New York: Academic Press.

Murray, M. G. (1980). Social structure of an impala population. D. Phil. thesis, University of Rhodesia.

Murray, M. G. (1982a). The rut of impala: aspects of seasonal mating under tropical conditions. *Z. Tierpsychol.*, **59**, 319–37.

Murray, M. G. (1982b). Home range, dispersal and clan system of impala. *Afr. J. Ecol.*, **20**, 253–269.

Murray, P. (1984). Extinctions downunder: a bestiary of extinct Australian late Pleistocene monotremes and marsupials. In *Quarternary Extinctions*, ed. P. S. Martin & R. G. Klein, pp. 600–28. Tucson: University of Arizona Press.

Mwalyosi, R. B. B. (1977). A count of large mammals in Lake Manyara National Park. *E. Afr. Wildl. J.*, **15**, 333–5.

Mwalyosi, R. B. B. (1987). Decline of *Acacia tortilis* in Lake Manyara National Park, Tanzania. *Afr. J. Ecol.*, **25**, 51–4.

Myers, N. (1972). National Parks in savanna Africa. *Science*, **178**, 1255–63.

Myers, N. (1973). Tsavo National Park, Kenya, and its elephants: an interim appraisal. *Biol. Cons.*, **5**, 123–32.

Nagy, K. A. (1987). Field metabolic rate and food requirement scaling in mammals and birds. *Ecol. Monogr.*, **57**, 111–28.

Napier Bax, P. & Sheldrick, D. L. W. (1963). Some preliminary observations on the food of elephant in the Tsavo Royal National Park (East) of Kenya. *E. Afr. Wildl. J.*, **1**, 40–53.

Naylor, J. N., Caughley, G. J., Abel, N. O. & Liberg, O. N. (1973). *Luangwa Valley Conservation and Development Project, Zambia. Game Management and Habitat manipulation.* Rome: FAO Report.

Nge'the, J. C. (1976). Preference and daily intake of five East African grasses by zebras. *J. Range Manage*, **29**, 5–10.

Nge'the, J. C. & Box, T. W. (1976). Botanical composition of eland and goat diets in an *Acacia* grassland community in Kenya. *J. Range Manage.*, **29**, 290–93.

Norton, P. M. (1981). Activity patterns of the klipspringer in two areas of the Cape Province. *S. Afr. J. Wildl. Res.*, **11**, 126–34.

Norton-Griffiths, M. (1975). The numbers and distribution of large mammals in Ruaha National Park, Tanzania, *E. Afr. Wildl. J.*, **13**, 121–40.

Novellie, P. A., Manson, J. & Bigalke R. C. (1984). Behavioural ecology and communiction in the Cape grysbok. *S. Afr. J. Zool.*, **19**, 22–30.

O'Connor, T. G. & Campbell, B. M. (1986). Hippopotamus habitat relationships on the Lundi River, Gonarezhou National Park, Zimbabwe. *Afr. J. Ecol*, **24**, 7–26.

Oftedal, O. T. (1984). Milk composition, milk yield and energy output at peak lactation. In *Lactation Strategies*, Symp. Brit. Ecol. Soc., No. 51, pp. 33–85.

Oliver, M. D. N., Short, N. R. M. & Hanks, J. (1978). Population ecology of oribi, grey rhebuck and mountain reedbuck in Highmoor State Forest Land, Natal. *S. Afr. J. Wildl. Res.*, **8**, 95–105.

Olivier, R. C. D. (1978). On the ecology of the Asian elephant with particular reference to Malaya and Sri Lanka. PhD thesis, University of Cambridge.

Olivier, R. C. D. (1982). Ecology and behavior of living elephants: bases for assumptions concerning the extinct woolly mammoths. In *Paleoecology of Beringia*, ed. D. M. Hopkins, J. V. Matthews, C. E. Schweger & S. B. Young, pp. 291–306. New York: Academic Press.

Olivier, R. C. D. & Laurie, W. A. (1974). Habitat utilisation by hippopotamus in the Mara River. *E. Afr. Wildl. J.*, **12**, 249–72.

Ottochilo, W. K. (1986). Population estimates and distribution patterns of elephants in the Tsavo ecosystem, Kenya, in 1980. *Afr. J. Ecol.*, **24**, 53–7.

Owaga, M. L. (1975). The feeding ecology of wildebeest and zebra in the Athi-Kaputei plains. *E. Afr. Wildl. J.*, **13**, 375–83.

Owen-Smith, G. (1986). The Kaokoveld, South West Africa/Namibia's threatened wilderness. *Afr. Wildl.*, **40**, 104–13.

Owen-Smith, N. (1971a). Territoriality in the white rhinoceros. *Nature*, **231**, 294–6.

Owen-Smith, N. (1971b). The contribution of radio telemetry to a study of the white rhinoceros. *Symposium on Biotelemetry, Pretoria, 1971*. Pretoria: Council for Scientific and Industrial Research.

Owen-Smtih, N. (1972). Territoriality: the example of the white rhinoceros. *Zoologica Africana*, **7**, 273–80.

Owen-Smtih, N. (1973). The behavioural ecology of the white rhinoceros. PhD thesis, University of Wisconsin.

Owen-Smith, N. (1974a). The social system of the white rhinoceros. In *The Behaviour of Ungulates and its Relation to Management*, ed. V. Geist & F. Walther, pp. 341–51. IUCN public., new series no. 24, Morges: IUCN.

Owen-Smith, N. (1974b). Minisender decken Verhalten von Nashornen auf. *Umschau*, **74**, 119–20.

Owen-Smith, N. (1975). The social ethology of the white rhinoceros. *Z. Tierpsychol.*, **38**, 337–84.

Owen-Smith, N. (1977). On territoriality in ungulates and an evolutionary model. *Q. Rev. Biol.*, **52**, 1–38.

Owen-Smith, N. (1979). Assessing the foraging efficiency of a large herbivore, the kudu. *S. Afr. J. Wildl. Res.*, **9**, 102–10.

Owen-Smith, N. (1981). The white rhino overpopulation problem, and a proposed solution. In *Problems in Management of Locally Abundant Wild Mammals*, ed. P. A. Jewell, S. Holt & D. Hart, pp. 129–50. New York: Academic Press.

Owen-Smith, N. (1982). Factors influencing the consumption of plant products by large herbivores. In *The Ecology of Tropical Savannas*, ed. B. J. Huntley & B. H. Walker, pp. 359–404. Berlin: Springer-Verlag.

Owen-Smith, N. (1983). Dispersal and the dynamics of large herbivore populations in enclosed areas. In *Management of Large Mammals in African Conservation Areas*, ed. R. N. Owen-Smith, pp. 127–43. Pretoria: Haum.

Owen-Smith, N. (1984). Spatial and temporal components of the mating systems of kudu bulls and red deer stags. *Anim. Behav.*, **32**, 321–32.

Owen-Smith, N. (1985). Niche separation among African ungulates. In *Species and Speciation*, ed. E. S. Vrba, pp. 167–71. Transvaal Mus. Monogr. No. 4, Pretoria: Transvaal Museum.

Owen-Smith, N. (1988 in press). Morphological factors and their consequences for resource partitioning among African savanna ungulates: a simulation modelling approach. In *Patterns in the Structure of Mammalian Communities*. ed. D. W. Morris, Z. Abramsky, B. J. Fox & M. L. Willig, Lubbock, Texas: Texas Tech Museum Special Publication Series.

Owen-Smith, N. & Novellie, P. (1982). What should a clever ungulate eat? *Am. Nat.*, **119**, 151–78.

Packer, C. (1983). Sexual dimorphism: the horns of African antelopes. *Science*, **221**, 1191–3.

Paraa, R. (1978). Comparison of foregut and hindgut fermentation in herbivores. In *The Ecology of Arboreal Folivores*, ed. G. G. Montgomery, pp. 205–29. Washington, D.C.: Smithsonian Institution Press.

Parker, G. A. (1974). Assessment strategy and the evolution of animal conflicts. *J. Theor. Biol.*, **47**, 223–43.

Parker, I. S. C. (1983). The Tsavo story: an ecological case history. In *Management of Large Mammals in African Conservation Areas*, ed. R. N. Owen-Smith, pp. 37–50. Pretoria: Haum.

Parker, I. S. C. & Amin, M. (1983). *Ivory Crisis*. London: Chatto & Windus.

Patterson, H. E. H. (1985). The Recognition Concept of species. In *Species and Speciation*, ed. E. S. Vrba, pp. 21–30. Transvaal Mus. Monogr. No. 4, Pretoria: Transvaal Museum.

Payne, K. B., Langbauer, W. R. Jr. & Thomas, E. M. (1986). Infrasonic calls of the Asian elephant. *Behav. Ecol. Sociobiol.*, **18**, 297–302.

Pellew, R. A. (1981). The giraffe and its *Acacia* food resource in the Serengeti National Park. PhD thesis, University of Cambridge.

Pellew, R. A. (1983a). The giraffe and its food resource in the Serengeti. II. Response of the population to changes in the food supply. *Afr. J. Ecol.*, **21**, 269–83.

Pellew, R. A. (1983b). Modelling and the systems approach to management problems: the *Acacia*/elephant problem in the Serengeti. In *Management of Large Mammals in African Conservation Areas*, ed. R. N. Owen-Smith, pp. 93–114. Pretoria: Haum.

Pellew, R. A. (1983c). The impacts of elephant, giraffe and fire upon *Acacia tortilis* woodlands of the Serengeti. *Afr. J. Ecol.*, **21**, 41–74.

Pellew, R. A. (1984a). Giraffe and okapi. In *The Encyclopaedia of Mammals*, vol. 2, ed. D. Macdonald, pp. 534–41. London: George Allen & Unwin.

Pellew, R. A. (1984b). The feeding ecology of a selective browser, the giraffe. *J. Zool. Lond.*, **202**, 57–81.

Pellew, R. A. (1984c). Food consumption and energy budgets of the giraffe. *J. Appl. Ecol.*, **21**, 141–59.

Pennycuick, L. (1975). Movements of the migratory wildebeest population in the Serengeti. *E. Afr. Wildl. J.*, **13**, 65–87.

Penzhorn, B. L. (1974). Sex and age composition and dimensions of the springbok population in the Mountain Zebra National Park. *J. Sth. Afr. Wildl. Mgmt. Ass.*, **4**, 63–6.

Penzhorn, B. L. (1975). Behaviour and population ecology of the Cape mountain zebra in the Mountain Zebra National Park. DSc thesis, University of Pretoria.

Penzhorn, B. L., Robbertse, P. J. & Olivier, M. C. (1974). The influence of the African elephant on the vegetation of the Addo Elephant National Park. *Koedoe*, **17**, 137–58.

Perry, J. S. (1953). The reproduction of the African elephant. *Philos. Trans. Roy. Soc. Lond.*, **B237**, 93–149.

Peters, R. H. (1983). *The Ecological Implications of Body Size*. Cambridge: Cambridge University Press. 329pp.

Peters, R. H. & Raelson, J. V. (1984). Relations between individual size and mammalian population density. *Am. Nat.*, **124**, 498–517.

Peters, R. H. & Wassenberg, K. (1983). The effect of body size on animal abundance. *Oecologia*, **60**, 89–96.

Petrides, G. A. (1974). The overgrazing cycle as a characterisitc of tropical savanna and grasslands in Africa. *Proc. 1st Int. Congr. Ecol.*, pp. 86–91. Wageningen.

Petrides, G. A. & Swank, W. G. (1965). Population densities and range-carrying capacity for large mammals in Queen Elizabeth National Park, Uganda. *Zoologica Africana*, **1**, 209–25.

Phillips, A. M. III. (1984). Shasta ground sloth extinction: fossil packrat midden evidence from the western Grand Canyon. In *Quarternary Extinctions*, ed. P. S. Martin & R. G. Klein, pp. 148–58. Tucson: University of Arizona Press.

Phillipson, J. (1975). Rainfall, primary production and carrying capacity of Tsavo National Park (East), Kenya. *E. Afr. Wildl. J.*, **13**, 171–201.

Pianka, E. R. (1970). On r- and K-selection. *Am. Nat.*, **104**, 592–7.

Pianka, E. R. (1974). *Evolutionary Ecology*. New York: Harper & Row. 356pp.

Pickett, S. T. A. & White, P. S. (eds.) (1985). *The Ecology of Natural Disturbance and Patch Dynamics*. New York: Academic Press. 496pp.

Pienaar, U. de V. (1963). The large mammals of the Kruger National Park – their distribution and present-day status. *Koedoe*, **6**, 1–37.

Pienaar, U. de V. (1969). Observations on development biology, growth and some aspects of the population ecology of African buffalo in the Kruger National Park. *Koedoe*, **12**, 29–52.

Pienaar, U. de V. (1970). The recolonisation history of the square-lipped (white) rhinoceros in the Kruger National Park (October 1961-November 1969). *Koedoe*, **13**, 157–69.

Pienaar, U. de V. (1982). The Kruger Park saga (1898–1981). Mimeographed paper presented at the workshop on Management of Large Mammals in African

Conservation Areas, Kruger National Park, September 1982, 38pp.

Pienaar, U. de V. (1983). Management by intervention: the pragmatic/economic option. In *Management of Large Mammals in African Conservation Areas*, ed. R. N. Owen-Smith, pp. 23–36. Pretoria: Haum.

Pienaar, U. de V., van Wyk, P. & Fairall, N. (1966a). An experimental cropping scheme of hippopotami in the Letaba River of the Kruger National Park. *Koedoe*, **9**, 1–33.

Pienaar, U. de V., van Wyk, P. & Fairall, N. (1966b). An aerial census of elephant and buffalo in the Kruger National Park, and the implications thereof on intended management schemes. *Koedoe*, **9**, 40–107.

Pilgram, T. & Western, D. (1986). Managing African elephant for ivory production through ivory trade restrictions. *J. Appl. Biol.*, **23**, 515–29.

Player, I. C. & Feely, J. M. (1960). A preliminary report on the square-lipped rhinoceros. *Lammergeyer (Natal)*, **1**, 3–23.

Poole, J. H. & Moss, C. J. (1981). Musth in the African elephant. *Nature*, **292**, 830–1.

Pratt, D. M. & Anderson, V. H. (1979). Giraffe cow-calf relationships and social development of the calf in the Serengeti. *Z. Tierpsychol.*, **51**, 233–51.

Ralls, K. (1977). Sexual dimorphism in mammals: avian models and unanswered questions. *Am. Nat.*, **111**, 917–38.

Rawlins, C. G. C. (1979). The breeding of white rhinos in captivity – a comparative survey. *Zool. Gart.*, *NF*, **49**, 1–7.

Reed, C. A. (1970). The extinctions of mammalian megafauna in the Old World late Quarternary. *BioScience*, **20**, 284–8.

Reiss, M. J. (1985). The allometry of reproduction: why larger species invest relatively less in their offspring. *J. Theor. Biol.*, **113**, 529–44.

Reiss, M. J. (1986). Sexual dimorphism in body size – are larger species more dimorphic? *J. Theor. Biol.*, **121**, 163–72.

Remmert, H. (1982). The evolution of man and the extinction of animals. *Naturwissenschaften*, **69**, 524–7.

Reuterwall, C. (1981). Temporal and spatial variability of the calf sex ratio in Scandinavian moose. *Oikos*, **37**, 39–45.

Reynolds, R. J. (1960). White rhinos in captivity. *Int. Zoo Ybk.*, **2**, 42–43.

Rhoades, D. F. (1985). Offensive-defensive interactions between herbivores and plants: their relevance in herbivore population dynamics and ecological theory. *Am. Nat.*, **125**, 205–38.

Riney, T. (1964). The impact of introductions of large herbivores on the tropical environment. IUCN public., new series no. 4, pp. 261–273.

Ripley, S. D. (1952). Territorial and sexual behaviour in the Great Indian rhinoceros: a speculation. *Ecology*, **33**, 570–3.

Ritchie, A. T. A. (1963). The black rhinoceros. *E. Afr. Wildl. J.*, **1**, 54–62.

Ritchie, J. C. & Cwynar, L. C. (1982). The late Quarternary vegetation of the north Yukon. In *Paleoecology of Beringia*, ed. D. M. Hopkins, J. V. Matthews Jr., C. E. Schweger & S. B. Young, pp. 113–26. New York: Academic Press.

Robbins, C. T. & Robbins, B. L. (1979). Fetal and neonatal growth patterns and maternal reproductive effort in ungulates and subungulates. *Am. Nat.*, **114**, 101–16.

Robinette, W. L. & Archer, A. L. (1971). Notes on ageing criteria and reproduction of Thomson's gazelle. *E. Afr. Wildl. J.*, **9**, 83–98.

Robinette, W. L. & Child, G. F. T. (1964). Notes on biology of the lechwe. *Puku*, **2**, 84–117.

Robinette, W. L., Jones, D. A., Geshwiler, J. S. & Low, J. B. (1957). Differential mortality by sex and age among mule deer. *J. Wildl. Manage.*, **21**, 1–16.

Robinson, T. J. (1979). Influence of a nutritional factor on the size differences of the three springbok subspecies. *S. Afr. J. Wildl. Res.*, **14**, 13–15.

Rodgers, D. A. & Elder, W. H. (1977). Movements of elephants in Luangwa Valley, Zambia, *J. Wildl. Manage.*, **41**, 56–62.

Rodgers, W. A. (1977). Seasonal changes in group size amongst five wild herbivore species. *E. Afr. Wildl. J.*, **15**, 175–90.

Roff, D. A. (1974). The analysis of a population model demonstrating the importance of dispersal in a heterogeneous environment. *Oecologia*, **15**, 259–75.

Roff, D. A. (1975). Population stability and the evolution of dispersal in a heterogeneous environment. *Oecologia*, **19**, 217–37.

Rosenzweig, M. L. (1968). Net primary production of terrestrial communities: prediction from climatological data. *Am. Nat.*, **102**, 74–6.

Roth, H. H., Kerr, M. A. & Posselt, J. (1972). Studies on the utilization of semi-domesticated eland in Rhodesia. V Reproduction and herd increase. *Z. Tierzuchtung und Zuchtungsbiologie*, **89**, 69–83.

Rubenstein, D. I. (1980). On the evolution of alternative mating strategies. In *Limits to Action*, ed. J. E. Staddon, pp. 65–100. New York: Academic Press.

Rutherford, M. C. (1980). Annual plant production-precipitation relations in arid and semi-arid regions. *S. Afr. J. Sci.*, **76**, 53–6.

Sachs, R. (1967). Liveweights and body measurements of Serengeti game animals. *E. Afr. Wildl. J.*, **5**, 24–36.

Sadleir, R. M. F. S. (1969). *The Ecology of Reproduction in Wild and Domestic Animals.* London: Methuen. 321pp.

Santiapillai, C., Chambers, M. R. & Ishwaran, N. (1984). Aspects of the ecology of the Asian elephant in the Ruhunu National Park, Sri Lanka. *Biol. Cons.*, **29**, 47–61.

Sauer, J. J. C., Theron, G. K. & Skinner, J. D. (1977). Food preferences of giraffe in the arid bushveld of the western Transvaal. *S. Afr. J. Wildl. Res.*, **7**, 53–9.

Sayer, J. A. & Rahka, A. M. (1974). The age of puberty in the hippopotamus in the Luangwa River in eastern Zambia. *E. Afr. Wildl. J.*, **12**, 227–37.

Sayer, J. A. & van Lavieren, L. P. (1975). The ecology of the Kafue lechwe population of Zambia before the operation of hydroelectric dams on the Kafue River. *E. Afr. Wildl. J.*, **13**, 9–37.

Schaller, G. B. (1967). *The Deer and the Tiger. A Study of Wildlife in India.* Chicago: University of Chicago Press. 370pp.

Schaller, G. B. (1972). *The Serengeti Lion. A Study of Predator-Prey Relationships.* Chicago: University of Chicago Press. 480pp.

Schaurte, W. T. (1969). Uber das geburt eines Breitmaulnashornes in Wildschutzgebiet Krugersdorp in Transvaal. *Saugetierk. Mitt.*, **17**, 158–60.

Schenkel, R. (1966). Zum Problem der Territorialitat und des Markierens bei Saugern – am Beispiel des Schwartzen Nashorns und des Lowens. *Z. Tierpsychol.*, **23**, 593–626.

Schenkel, R. & Schenkel-Hulliger, L. (1969a). *Ecology and Behaviour of the Black Rhinoceros. A Field Study.* Mammalia Depicta, Berlin-Hamburg: Paul Parey. 101pp.

Schenkel, R. & Schenkel-Hulliger, L. (1969b). The Javan rhinoceros in Udjong Kulon Nature Reserve, its ecology and behaviour. *Acta Tropica*, **26**, 98–135.

Schmidt-Nielsen, K. (1984). *Scaling. Why is Animal Size So Important?* Cambridge: Cambridge University Press. 241pp.

Schomber, H. W. (1966). Die Verbreitung und der Bestand des zentralafrikanischen Breitmaulnashorn. *Saugetierk. Mitt.*, **14**, 214–27.

Schweger, C. E. (1982). Late Pleistocene vegetation of eastern Beringia: pollen analysis of dated alluvium. In *Paleoecology of Beringia*, ed. D. M. Hopkins, J. V. Matthews Jr., C. E. Schweger & S. B. Young, pp. 95–112. New York: Academic Press.

Scotcher, J. S. B. (1982). Interrelations of vegetation and eland in Giant's Castle Game Reserve, Natal. PhD thesis, University of the Witwatersrand.

Scotcher, J. S. B., Stewart, D. R. M. & Breen, C. M. (1978). The diet of hippopotamus in Ndumu Game Reserve, Natal, as determined by faecal analysis. *S. Afr. J. Wildl. Res.*, **8**, 1–11.

Scott, L. (1984). Palynological evidence for Quarternary paleoenvironments in southern Africa. In *Southern African Prehistory and Paleoenvironments*, ed. R. G. Klein, pp. 65–80. Rotterdam: Balkema.

Selous, F. C. (1899). The white or square-lipped rhinoceros. In *Great and Small Game of Africa*, ed. H. A. Bryden, pp. 52–67. London: Rowland Ward.

Shackleton, N. J., Hall, M. A., Line, J. & Shuxi, Cang (1983). Carbon isotope data in core V19–30 confirm reduced carbon dioxide concentration in the ice age atmosphere. *Nature*, **306**, 319–22.

Sherry, B. J. (1975). Reproduction of elephant in Gonarezhou, South Eastern Rhodesia. *Arnoldia (Rhodesia)*, **7**, 29, 1–13.

Sherry, B. Y. (1978). Growth of elephants in the Gonarezhou National Park, south-eastern Rhodesia. *S. Afr. J. Wildl. Res.*, **8**, 49–58.

Short, J. (1981). Diet and feeding behaviour of the forest elephant. *Mammalia*, **45**, 177–86.

Short, R. V. (1966). Oestrous behaviour, ovulation and the formation of the corpus luteum in the African elephant. *E. Afr. Wildl. J.*, **4**, 56–69.

Shoshani, J. & Eisenberg, J. F. (1982). *Elephas maximus.* Mammalian species No. 182, 8pp. The American Society of Mammalogists.

Sidney, J. (1966). The past and present distribution of some African ungulates. *Trans. Zool. Soc. Lond.*, **30**, 5–397.

Siegfried, W. R. & Davies, B. R. (1982). *Conservation of Ecosystems: Theory and Practice.* S. Afr. Nat. Sci. Progr. Rep. No. 61, Pretoria: CSIR. 97pp.

Sikes, S. K. (1971). *The Natural History of the African Elephant.* London: Weidenfeld & Nicolson, 397pp.

Simpson, C. D. (1968). Reproduction and population structure in the greater kudu in Rhodesia. *J. Wildl. Manage.*, **32**, 149–61.

Simpson, C. D. (1974). Food studies of the Chobe bushbuck. *Arnoldia (Rhodesia)*, **6**, 32, 1–9.

Simpson, G. L. (1980). *Splendid Isolation.* New Haven: Yale University Press.

Sinclair, A. R. E. (1974). The social organisation of the East African buffalo. In *The Behaviour of Ungulates and its Relation to Management*, ed. V. Geist & F. Walther, pp. 676–89. IUCN public., new series no. 24, Morges: IUCN.

Sinclair, A. R. E. (1975). The resource limitation of trophic levels in tropical grassland ecosystems. *J. Anim. Ecol.*, **44**, 497–520.

Sinclair, A. R. E. (1977). *The African Buffalo. A Study of Resource Limitation of Populations.* Chicago: University of Chicago Press. 355pp.

Sinclair, A. R. E. (1983). Management of conservation areas as ecological baseline controls. In *Management of Large Mammals in African Conservation Areas*, ed. R. N. Owen-Smith, pp. 13–22. Pretoria: Haum.

Sinclair, A. R. E. (1985). Does interspecific competition or predation shape the African ungulate community? *J. Anim. Ecol.*, **54**, 899–918.

Sinclair, A. R. E. & Norton-Griffiths, M. (1979). *Serengeti. Dynamics of an Ecosystem.* Chicago: University of Chicago Press. 389pp.

Skinner, J. D., Scorer, J. A. & Millar, R. P. (1975). Observations on the reproductive physiological status of mature herd bulls, bachelor bulls, and young bulls in the hippopotamus. *Gen. Compar. Endocr.*, **26**, 92–5.

Skinner, J. D. & van Zyl, J. H. M. (1969). Reproductive performance of the common eland in two environments. *J. Reprod. Fert. Suppl.*, **6**, 319–22.

Smart, N. O. E., Hatton, J. C. & Spence, D. H. N. (1985). The effect of long-term exclusion of large herbivores on vegetation in Murchison Falls National Park, Uganda. *Biol. Cons.*, **33**, 229–45.

Smith, N. S. & Buss, I. O. (1973). Reproductive ecology of the female African elephant. *J. Wildl. Manage.*, **37**, 524–33.

Smithers, R. H. N. (1971). *The Mammals of Botswana*. Salisbury: Trustees of the National Museums of Rhodesia.

Smithers, R. H. N. (1983). *The Mammals of the Southern African Subregion*. Pretoria: University of Pretoria. 367pp.

Smuts, G. L. (1974). Growth, reproduction and population characteristics of Burchell's zebra in the Kruger National Park. DSc thesis, University of Pretoria.

Smuts, G. L. (1975a). Home range sizes for Burchell's zebra from the Kruger National Park. *Koedoe*, **18**, 139–146.

Smuts, G. L. (1975b). Pre- and post-natal growth phenomena of Burchell's zebra from the Kruger National Park. *Koedoe*, **18**, 69–102.

Smuts, G. L. (1975c). Reproduction and population characteristics of elephants in the Kruger National Park. *J. Sth. Afr. Wildl. Manage. Assoc.*, **5**, 1–10.

Smuts, G. L. (1976). Reproduction in the zebra mare from the Kruger National Park. *Koedoe*, **19**, 89–132.

Smuts, G. L. (1978). Interrelations between predators, prey and their environment. *BioScience*, **28**, 316–20.

Smuts, G. L. (1979). Diet of lions and spotted hyenas assessed from stomach contents. *S. Afr. J. Wildl. Res.*, **9**, 19–25.

Smuts, G. L. & Whyte, I. C. (1981). Relationships between reproduction and environment in the hippopotamus in the Kruger National Park. *Koedoe*, **24**, 169–85.

Sokal, R. R. & Rohlf, F. J. (1969). *Biometry. The Principles and Practice of Statistics in Biological Research*. San Francisco: Freeman. 776pp.

Soule, M. E. (1980). Thresholds for survival; maintaining fitness and evolutionary potential. In *Conservation Biology*, ed. M. E. Soule & B. A. Wilcox, pp. 151–70. Sunderland, Mass: Sinauer.

Sowls, L. K. (1974). Social behaviour of the collared peccary. In *The Behaviour of Ungulates and its Relation to Management*, ed. V. Geist & F. Walther, pp. 144–65. IUCN public., new series no. 24, Morges: IUCN.

Spinage, C. A. (1968). A quantitative study of the daily activity of the Uganda defassa waterbuck. *J. Zool. Lond.*, **159**, 329–61.

Spinage, C. A. (1969). Naturalistic observations on the reproductive and material behaviour of the Uganda defassa waterbuck. *Z. Tierpsychol.*, **26**, 39–47.

Spinage, C. A. (1970). Population dynamics of the Uganda defassa waterbuck in the Queen Elizabeth Park, Uganda. *J. Anim. Ecol.*, **39**, 51–78.

Spinage, C. A. (1973). A review of ivory exploitation and elephant population trends in Africa. *E. Afr. Wildl. J.*, **11**, 281–9.

Spinage, C. A. (1974). Territoriality and population regulation in the Uganda defassa waterbuck. In *The Behaviour of Ungulates and its Relation to Management*, ed. V. Geist & F. Walther, pp. 635–43. IUCN public., new series no. 24, Morges: IUCN.

Spinage, C. A. (1982). *A Territorial Antelope. The Uganda Waterbuck*. London: Academic Press. 334pp.

Spinage, C. A., Guinness, F., Eltringham, S. K. & Woodford, M. H. (1972). Estimation of large mammal numbers in the Akagera National Park and Mutara Hunting Area, Rwanda. *Terre et Vie*, **4**, 561–70.

Stanley-Price, M. R. (1974). The feeding ecology of Coke's hartebeest in Kenya. PhD thesis, University of Oxford.

Stanley-Price, M. R. (1977). The estimation of food intake, and its seasonal variation, in the hartebeest. *E. Afr. Wild. J.*, **15**, 107–24.

Steadman, D. W. & Martin, P. S. (1984). Extinctions of birds in the late Pleistocene of North America. In *Quaternary Extinctions*, ed. P. S. Martin & R. G. Klein, p. 466–77. Tucson: University of Arizona Press.

Steele, N. (1968). *Game Ranger on Horseback*. Cape Town: Books of Africa.

Stewart, D. R. M. & Zaphiro, D. R. P. (1963). Biomass and density of wild herbivores in different East African habitats. *Mammalia*, **27**, 483–96.

Struhsaker, T. T. (1967). Behaviour of elk during the rut. *Z. Tierpsychol.*, **24**, 80–114.

Swanepoel, C. M. & Swanepoel, S. M. (1986). Baobab damage by elephant in the middle Zambezi Valley, Zimbabwe, *Afr. J. Ecol.*, **24**, 129–32.

Talbot, L. M. & Talbot, M. H. (1963a). The wildebeest in western Masailand, East Africa. *Wild Monogr.* No. 12, 88pp.

Talbot, L. M. & Talbot, M. H. (1963b). The high biomass of wild ungulates on East African savanna. *Trans. 28th N. Amer. Wildl. Conf.*, pp. 465–76.

Taylor, C. R. (1973). Energy cost of animal locomotion. In *Comparative Physiology*, ed. L. Bolis, K. Schmidt-Nielsen & S. H. P. Maddrell, pp. 23–42. New York: New Holland.

Taylor, C. R. & Maloiy, G. M. O. (1967). The effect of dehydration and heat stress on intake and digestion of food in some East African bovids. *Trans. 8th Int. Congr. Game Biol.*

Taylor, L. R. & Taylor, R. A. J. (1977). Aggregation, migration and population mechanics. *Nature*, **265**, 415–21.

Taylor, R. A. J. (1981a). The behavioural basis of redistribution. I The delta-model concept. *J. Anim. Ecol.*, **50**, 573–86.

Taylor, R. A. J. (1981b). The behavioural basis of redistribution. II Simulations of the delta-model. *J. Anim. Ecol.*, **50**, 587–604.

Thenius, E. (1968). Nashorner. In *Grzimek's Tierleben*, ed. B. Grzimek, XIII Saugetiere 4.

Thenius, E. (1969). Stammegesichte der Saugetiere. Ceratomorpha II (Rhinocerotoidea). *Handbuch der Zoologie*, **8**, 2, 543–54.

Thompson, P. J. (1975). The role of elephants, fire and other agents in the decline of a *Brachystegia boehmii* woodland. *J. Sth. Afr. Wildl. Mgmt. Ass.*, **5**, 11–18.

Thomson, B. R. (1973). Wild reindeer activity. Hardangervidda 1971. Mimeogr. Rep., Statens Vildundersohelse, Direktoratet for Jakt, Vilstell og Ferskvannsfiske, and International Biological Programme (Norway IBP, PT/UM), Trondheim.

Thonney, M. L., Touchberry, R. W., Goodrich, R. D., & Meicke, J. C. (1976). *J. Anim. Sci.*, **43**, 692–704.

Thorbahn, P. F. (1984). Br'er elephant and the briar patch. *Natural History (N. Y.)*, **93**, 70–81.

Thornton, D. D. (1971). The effect of complete removal of hippopotamus on grassland in the Queen Elizabeth National Park, Uganda. *E. Afr. Wildl. J.*, **9**, 47–55.

Trivers, R. L. & Willard, D. E. (1973). Natural selection of parental ability to vary the sex ratio of offspring. *Science*, **179**, 90–2.

Trotter, M. M. & McCulloch, B. (1984). Moas, men and middens. In *Quaternary Extinctions*, ed. P. S. Martin & R. G. Klein, pp. 708–27. Tucson: University of Arizona Press.

Turner, M., & Watson, M. (1964). A census of game in Ngorongoro Crater. *E. Afr. Wildl. J.*, **2** 165–8.

Tyrrell, J. G. (1985). Distribution of elephants in the Tsavo region during the late 19th century. *E. Afr. Wildl. J.*, **23**, 29–33.

Ullrich. W. (1964). Zur biologie der Panzerhashorner in Assam. *D. Zool. Garten.*, *NF*, **21**, 129–51.

Vancuylenberg, B. W. B. (1977). Feeding behaviour of the Asiatic elephant in south-east Sri Lanka in relation to conservation. *Biol. Cons.*, **12**, 33–54.

van Gyseghem, R. (1984). Observations on the ecology and behaviour of the northern white rhinoceros. *Z. Saugetierk*, **49**, 348–58.

van Hoven, W. (1977). Stomach fermentation rate in the hippopotamus. *S. Afr. J. Sci.*, **73**, 216.

van Hoven, W. (1978). Digestive physiology in the stomach complex and hindgut of the hippopotamus. *S. Afr. J. Wildl. Res.*, **8**, 59–64.

van Hoven, W., Prins, R. A. & Lankhorst, A. (1981). Fermentative digestion of the African elephant. *S. Afr. J. Wild. Res.*, **11** 78–86.

van Jaarsveld, A. S. & Knight-Eloff, A. K. (1984). Digestion in the porcupine. *S. Afr. J. Zool*, **19**, 109–112.

Van Lavieren, L. P. & Esser, J. D. (1980). Numbers, distribution and habitat preference of large mammals in Bouba Njida National Park, Cameroon. *Afr. J. Ecol.*, **18**, 141–53.

van Wyk, P. & Fairall, N. (1969). The influence of the African elephant on the vegetation of the Kruger National Park. *Koedoe*, **12**, 57–89.

van Zyl, J. H. M. & Skinner, J. D. (1970). Growth and development of the springbok foetus. *Afr. Wild Life*, **24**, 309–16.

Venter, J. (1979). The ecology of the southern reedbuck on the eastern shores of Lake St. Lucia, Zululand, MSc thesis, University of Natal.

Vereshchagin, N.K. (1967). Primitive hunters and Pleistocene extinctions in the Soviet Union. In *Pleistocene Extinctions*, ed. P. S. Martin & H. E. Wright, pp. 365–98. New Haven: Yale University Press.

Vereshchagin, N. K. & Baryshnikov, G. T. (1984). Quarternary mammalian extinctions in northern Eurasia. In *Quaternary Extinctions*, ed. P. S. Martin & R. G. Klein, pp. 483–516. Tucson: University of Arizona Press.

Verheyen, R. (1954). *Monographic Ethologique de la Hippopotame*. Bruxelles: Inst. des Parcs Nat. du Congo Belge. 91pp.

Verme, L. J. (1983). Sex ratio variation in *Odocoileus*: a critical review. *J. Wildl. Manage.*, **47**, 573–82.

Verme, L. J. & Ozaga, J. J. (1981). Sex ratio of white-tailed deer and the estrous cycle. *J. Wildl. Manage.*, **45**, 710–5.

Vesey-Fitzgerald, D. F. (1960). Grazing succession among East African game animals. *J. Mammal.*, **41**, 161–72.

Vesey-Fitzgerald, D. F. (1973). Animal impact on vegetation and plant succession in Lake Manyara National Park, Tanzania. *Oikos*, **24**, 314–25.

Vesey-Fitzgerald, D. F. (1974). The changing state of *Acacia xanthophloea* groves in Arusha National Park, Tanzania. *Biol. Cons.*, **6**, 40–7.

Viljoen, P. C. (1980). Distribution and numbers of the hippopotamus in the Olifants and Blyde Rivers. *S. Afr. J. Wildl. Res.*, **10**, 129–32.

Vincent, J. (1969). The status of the square-lipped rhinoceros in Zululand. *Lammergeyer (Natal)*, **10**, 12–21.

Vincent, J. (1970). The history of Umfolozi Game Reserve, Zululand. *Lammergeyer (Natal)*, **11**, 7–48.

Volman, T. P. (1984). Early prehistory of southern Africa. In *Southern African Prehistory and Paleoenvironments*, ed. R. G. Klein, pp. 169–220. Rotterdam: Balkema.

Von Ketelholdt, H. F. (1977). The lambing interval of the blue duiker in captivity, with observations on its breeding and care. *S. Afr. J. Wildl. Res.*, **7**, 41–3.

Von Richter, W. (1971). *The Black Wildebeest*. O F.S. Prov. Admin., Nature Conserv. Misc. Rep. No. 2.

Von Richter, W. (1972). Territorial behaviour of the black wildebeest. *Zoologica Africana*, **7**, 207–31.

Vrba, E. S. (1985). African Bovidae: evolutionary events since the Miocene. *S. Afr. J. Sci.*, **81**, 263–6.

Waldo, D. R., Smith, L. W. & Cox, E. L. (1972). Model of cellulose disappearance from the rumen. *J. Dairy Science*, **55**, 125–8.

Walker, B. H., Emslie, R. H., Owen-Smith, N. & Scholes, R. J. (1987). To cull or not to cull: lessons from a southern African drought. *J. Appl. Ecol.* **24**, 381–401.

Walker, B. H. & Goodman, P. S. (1983). Some implications of ecosystem properties for wildlife management. In *Management of Large Mammals in African Conservation Areas*, ed. R. N. Owen-Smith, pp. 79–92. Pretoria: Haum.

Walther, F. R. (1964). Einige Verhaltensbeobachtungen an Thomson-gazelle im Ngorongoro-Krater. *Z. Tierpsychol.*, **21**, 871–90.

Walther, F. R. (1965). Verhaltensstudien an der Grantsgazelle. *Z. Tierpsychol*, **22**, 167–208.

Walther, F. R. (1972a). Territorial behaviour in certain horned ungulates, with special reference to the examples of Thomson's and Grant's gazelles. *Zoologica Africana*, **7**, 303–8.

Walther, F. R. (1972b). Social grouping in Grants gazelle in the Serengetj National Park. *Z. Tierpsychol.*, **31**, 348–403.

Walther, F. R. (1973). Round the clock activity of Thomsons gazelle in the Serengeti National Park. *Z. Tierpsychol.*, **32**, 75–105.

Walther, F. R. (1978). Behavioural observations on oryx antelope invading Serengeti National Park, Tanzania. *J. Mammal.*, **59**, 243–60.

Waltz, E. C. (1982). Alternative mating tactics and the law of diminishing returns: the satellite threshold model. *Behav. Ecol. Sociobiol.*, **10**, 75–83.

Waser, P. M. (1974). Spatial associations and social interactions in a solitary antelope: the bushbuck. *Z. Tierpsychol.*, **37**, 24–36.

Waser, P. M. (1975). Diurnal and nocturnal strategies of the bushbuck. *E. Afr. Wildl. J.*, **13**, 49–64.

Watson, R. M. (1969). Reproduction of wildebeest in the Serengeti region, and its significance to conservation. *J. Reprod. Fert. Suppl.* **6**, 287–310.

Weatherhead, P. J. (1984). Mate choice in avian polygyny: why do females prefer older males? *Am. Nat.*, **123**, 873–5.

Webb, S. D. (1984). Ten million years of mammalian extinctions in North America. In *Quaternary Extinctions*, ed. P. S. Martin & R. G. Klein, pp. 189–210. Tucson: University of Arizona Press.

Weir, J. S. (1972). Spatial distribution of elephants in an African national park in relation to environmental sodium. *Oikos*, **23**, 1–13.

Weir, J. & Davison, E. (1965). Daily occurrence of African game animals at waterholes during dry weather. *Zoologica Africana*, **1**, 353–68.

West, F. H. (1984). The antiquity of man in America. In *Late-Quaternary Environments of the United States.* ed., H. E. Wright, Jr., Vol. 1. The Late Pleistocene, ed. S. C. Porter, pp. 364–82. Minneapolis: University of Minnesota Press.

Western, D. (1975). Water availability and its influence on the structure and dynamics of a savannah large mammal community. *E. Afr. Wildl. J.*, **13**, 265–86.

Western, D. (1979). Size, life history and ecology in mammals. *Afr. J. Ecol.*, **17**, 185–204.

Western, D. & Lindsay, W. K. (1984). Seasonal herd dynamics of a savanna elephant population. *Afr. J. Ecol.*, **22**, 229–44.

Western, D. & Sindiyo, D. M. (1972). The status of the Amboseli rhino population. *E. Afr. Wildl. J.*, **10**, 43–57.

Western, D. & Van Praet, C. (1973). Cyclical changes in the habitat and climate of an East African ecosystem. *Nature*, **241**, 104–6.

Weyerhaeuser, F. J. (1985). Survey of elephant damage to baobabs in Tanzania's Lake Manyara National Park. *Afr. J. Ecol.*, **23**, 235–43.

Wiley, R. H. (1974). Evolution of social organization and life history patterns among grouse. *Q. Rev. Biol.*, **49**, 201–27.

Williamson, B. R. (1975a). Seasonal distribution of elephant in Wankie National Park. *Arnoldia (Rhodesia)*, **7**, 11, 1–16.

Williamson, B. R. (1975b). The condition and nutrition of elephant in Wankie National Park. *Arnoldia (Rhodesia)*, **7**, 12, 1–20.

Williamson, B. R. (1976). Reproduction in female African elephant in the Wankie National Park, Rhodesia. *S. Afr. J. Wildl. Res.*, **6**, 89–93.

Wilson, E. O. (1975). *Sociobiology. The New Synthesis.* Cambridge, Mass.: Harvard University Press. 697pp.

Wilson, V. J. (1966). Observations on Lichtenstein's hartebeest over a three year period, and their response to various tsetse control measures in E. Zambia. *Arnoldia (Rhodesia)*, **2**, 15, 1–14.

Wilson, V. J. (1968). Weights of some mammals from Eastern Zambia. *Arnoldia (Rhodesia)*, **3**, 1–19.

Wilson, V. J. (1969a). Eland in eastern Zambia. *Arnoldia (Rhodesia)* **4**, 12, 1–9.

Wilson, V. J. (1969b). The large mammals of the Matopos National Park. *Arnoldia (Rhodesia)*, **4**, 13, 1–32.

Wilson, V. J. & Child, G. F. T. (1964). Notes on bushbuck from a tsetse fly control area in Northern Rhodesia. *Puku*, **2**, 118–28.

Wilson, V. J. & Child, G. (1965). Notes on klipsinger from tsetse fly control areas in eastern Zambia. *Arnoldia (Rhodesia)*, **1**, 35, 1–9.

Wilson, V. J. & Clarke, J. E. (1962). Observations on the common duiker based on material collected from a tsetse control game elimination scheme. *Proc. Zool. Soc. Lond.*, **138**, 487–97.

Wilson, V. J. & Kerr, M. A. (1969). Brief notes on reproduction in steenbok. *Arnoldia (Rhodesia)*, **4**, 23, 1–5.

Wing, L. D. & Buss, I. O. (1970). Elephants and forests. *Wildl. Monogr.* No. 19, 92pp.

Wirtz, P. (1981). Territorial defence and territory takeover by satellite males in the waterbuck. *Behav. Ecol. Sociobiol.*, **8**, 161–2.

Wirtz, P. (1982). Territory holders, satellite males and bachelor males in a high density population of waterbuck and their association with conspecifics. *Z. Tierpsychol.*, **58**, 277–300.

Wittenberger, J. F. (1980). Group size and polygamy in social mammals. *Am. Nat.*, **115**, 197–222.

Wolff, J. O. (1980). The role of habitat patchiness in the population dynamics of snowshoe hares. *Ecol. Monogr.*, **50**, 111–30.

Work, D. R. (1986). Summary report of aerial surveys conducted in 1983, 1984 and 1985. Unpublished report, Resource Ecology Group, University of the Witwatersrand.

Wright, H. E. Jr. (1977). Quaternary vegetation history – some comparisons between Europe and America. *Ann. Rev. Earth Planet. Sci.*, **5**, 123–58.

Wright, H. E. Jr. (ed.) (1984). *Late-Quaternary Environments of the United States*. Vol. 1. The Late Pleistocene. Minneapolis: University of Minnesota Press.

Wyatt, J. R. & Eltringham, S. K. (1974). The daily activity of the elephant in the Rwenzori National Park, Uganda. *E. Afr. Wildl. J.*, **12**, 273–90.

Yoaciel, S. M. (1981). Change in the populations of large herbivores and in the vegetation community in Mweya Peninsula, Rwenzori National Park, Uganda. *Afr. J. Ecol.*, **19**, 303–12.

Yurtsev, B. A. (1982). Relicts of the xerophyte vegetation of Beringia in northeastern Asia. In *Paleoecology of Beringia*, ed. D. M. Hopkins, J. V. Matthews, C. E. Schweger & S. B. Young, pp. 157–78. New York: Academic Press.

Zahavi, A. (1975). Mate selection – a selection for a handicap. *J. Theor. Biol.*, **53**, 205–14.

Zimmerman, I. (1978). The feeding ecology of Africander steers on mixed bushveld at Nylsvley Nature Reserve, Transvaal. MSc thesis, University of Pretoria.

Index

Acacia 29, 33–4, 37, 40, 227, 231–2, 234, 238–9, 242, 258–9, 278, 300
  albida 228, 258
  gerrardii 227, 233
  hocki 233
  nigrescens 227, 232
  senegal 227
  sieberiana 232–3
  tortilis 227, 234
  xanthophloea 33, 227, 234
Adansonia digitata (baobab) 33, 228–9, 232, 239, 302
adolescence and puberty 138–9
  elephant 139
  giraffe 140
  hippopotamus 140
  rhinoceros 140–4
Aloe africana 230
antelopes 50–1, 68, 92–3, 131–2, 196, 242, 270

Baikaea/Baphia thickets 34, 230–1
baobab *see Adansoia digitata*
biomass
  community 271–4
  effect of rainfall on 271–4
  population regulation 265–71
birth intervals 186–90
  elephant 188
  rhinoceros 188
  see also reproduction, female
bison, American 14, 51, 289
Brachystegia 34, 229, 232, 257, 286
  boehmi 229, 239
  glaucescens 229, 258
browsers 7, 14, 16, 31, 34, 37, 45, 50, 77, 87, 91, 95, 165, 239–42, 257–8, 277–8, 286, 288, 293, 310–12, 315
buffalo, African 14, 29, 68, 93, 131–2, 159, 225, 241–2, 245, 271–2, 275, 278
bushbuck 29, 50, 244

calves *see* infancy and juvenilehood; reproduction, female
Capparis sepiaria 230
cecalids (hindgut fermenters) 7, 16, 73–7, 79, 98, 275–6, 310
Cenchris ciliaris 233–4
Chloris gayana 233
Chrysochloa orientalis 233
Chrysophyllum 35, 231
Colophospermum mopane (mopane) 34, 40, 229–31, 239, 257
Combretum 34, 40, 241
  apiculatum 232
  obovatum 231
Combretum/Commiphora thickets 231–2
Commiphora 34, 241, 258, 299
  see also Combretum
communities 226–45
community biomass 271–4
community effects on other large herbivores 239–44
  elephant 240–1
  giraffe 242
  hippopotamus 241–2
  rhinoceros 242; white 242–4
community patterns 314–15
conception
  elephant 116
  giraffe 118
  hippopotamus 118
conception, age at first 185–6
  elephant 186
  hippopotamus 186
  rhinoceros 186
conservation 297–308
  dispersal sinks 303–6
  objectives 298–9; elephant 299–300; hippopotomus 300–1; rhinoceros, white 300–1
  overexploitation, problems of 307–8
  problems 299–303

vacuum zones 304–6
copulation
elephant 117
giraffe 119
hippopotamus 118
rhinoceros, black 119; Indian 119; white
121–4
Cordea 35, 231
courtship 116, 172–9
elephant 116–18
giraffe 118–19
hippopotamus 118
rhinoceros 119–24
see also mating
Croton macrostachyus 233
Cynodon dactylon 36, 38, 233
Cynometra 231

Deer 51, 197, 261
dentition 6, 7, 9
elephant 9
giraffe 14
hippopotamus 14
rhinoceros, Asian 13; black 11; Javan
12–13; white 11
diet 92
diet composition
elephant 31–6
giraffe 37–8
hippopotamus 36–7
rhinoceros 38–44
diet quality 82–7
digestion 6
elephant 9
giraffe 14
hippopotamus 14
rhinoceros 13
digestive passage rates 95
digestive efficiency 77–9
distribution (of herbivores) 21–9
elephant 21–3
giraffe 25
hippopotamus 23–5
rhinoceros 25; white 25–8
dominance 109
elephant 109–10
giraffe 111
hippopotamus 110
rhinoceros 111–16
dominance, male systems of 167–8
duiker 51, 92, 132, 271

Ecosystem stability and disturbance 278–9
ecosystem patterns 314–15
eland 51, 272
elephant 139
adolescence and puberty 139

African (*Loxodonta africana*) 7, 8
Asian or Indian (*Elephas maximus*) 8, 9
birth intervals 188
body heat 94–5
bush (*Loxodonta africana africana*) 7
community effects on other herbivores
240–1
conception 116
conservation 299–300, 316
copulation 117
courtship and mating 116–18
dentition 9
diet composition 31–6
digestion 9, 80
distribution 21–3
dominance 109–10
evolutionary origins 16–17
foraging 53–4
forest (*Loxodonta africana cyclotis*) 7
gestation period 117
groups 101–3
infancy and juvenilehood 13–14
males, satellite 170–2
mating gains 174–6
mating, female choice in 178
mortality costs 169
mouth parts 90
mud wallowing 46–7
musth 109–10, 139
population growth 212–14
population regulation, interactions with
vegetation 257–60
population size 7–9
population structure 201–3
predators, response to 124–6
reproduction, female 144–5; male 151;
seasonality of 183
sex ratio, offspring 196
temporal patterning of activities 53–4
utilization of space 61–2
vegetation, impact on 227–33
water 46–7
energy 274–6
Euphorbia candelabrum 233
evolutionary origins
elephant 16–17
giraffe 20–1
hippopotamus 19–21
rhinoceros 17–19, 20–1
extinctions
climatic effects on 284–9
effect of human predation in 289–92, 295
pattern of 281–3
role of megaherbivores 292–5

feeding, diurnal and nocturnal 92–3
antelope 92

feeding, diurnal and nocturnal (*cont.*)
  hippopotamus 92
  warthog 92
fertility, changes with *see* reproduction,
    female
food 30–1
  ingestion rate 87–90
  intake and digestion 72–3
  intake, daily 73–6
  retention time 76–7
foraging 53–4, 59–60, 67, 82, 87, 89, 92, 95
  elephant 53–4
  giraffe 55
  hippopotamus 55
  rhinoceros 55–60
  foregut fermenters 7, 16, 73, 76–7, 79,
    98, 275–6

gazelle 51, 270
  Grant's 50, 240
  Thompson's 51
gestation period 117–20, 189, 198
  elephant 117
  giraffe 118
  hippopotamus 118
  rhinoceros, black 119; Indian 119; white
    120
giraffe (*Giraffa camelopardalis*) 14, 15
  adolescence and puberty 140
  community effects on other large
    herbivores 242
  conception 118
  conservation 316
  copulation 119
  courtship and mating 118–19
  dentition 14
  diet composition 37–8
  digestion 14, 81
  distribution 25
  dominance 111
  evolutionary origins 20
  foraging 55
  gestation period 118
  groups 104
  infancy and juvenilehood 134–5
  males, satellite 170–2
  mortality costs 170
  population growth 214
  population structure 204
  predators response to 126
  reproduction, male 152; female 145–6;
    seasonality of 184
  sex ratio, offspring 198
  size 14
  temporal patterning of activities 55
  utilization of space 62–3
  vegetation, impact on 234
  water 47

grazers 7, 14, 16, 42–3, 45, 50, 67, 91, 95,
    165, 234, 236–9, 240–2, 251, 253,
    277–8, 286, 288, 293, 310–12, 315
Grewia flavescens 227
group structure 101, 160
  elephant 101–3
  giraffe 104
  hippopotamus 103–4
  rhinoceros 104–9
grouping patterns 160–7
gut anatomy 71–2

habitat structure 165
hare 51, 260, 270
Heteropogon contortis 233, 234
hindgut fermenters *see* cecalids
hippopotamus (*Hippopotamus amphibius*)
    13
  adolescence and puberty 140
  community effects on other large
    herbivores 241–2
  conception 118
  conservation 300–1, 316
  copulation 118
  courtship and mating 118
  dentition 14
  diet composition 36–7
  digestion 14
  distribution 23–5
  dominance 110
  evolutionary origins 19–20
  foraging 55
  gestation period 118
  groups 103–4
  infancy and juvenilehood 134
  males, satellite 170–2
  mouth parts 90
  nocturnal feeding 92–4
  population growth 214
  population regulation, interaction with
    vegetation 257, 259
  population structure 203–4
  predators, response to 126
  reproduction, male 145; female 151;
    seasonality of 184
  sex ratio, offspring 196
  size 13–14
  temporal patterns of activities 54–5
  utilization of space 62
  vegetation, impact on 233–4
  water 47
  weight 13
Hodotermes mossambicus 238
home range extents 95–8
  Hyparrhenia filipendula 31, 234

impala 51, 93, 238, 242, 245, 261, 270
infancy and juvenilehood 133

elephant 133–4
giraffe 134–5
hippopotamus 134
rhinoceros 135–8

Khaya (mahogany) 35, 231
Kirkia acuminata 232
kuda 50–1, 240, 242, 261, 271

lifespan *see* morality
Lonchorcarpus laxiflorus 231

Maesopsis 35, 231
mahogany *see Khaya*
males, satellite 170–2
mammoth 18, 281, 289–90, 294
mating
  body size, effect of 172–7
  territory, effect of 172–7
  see also courtship
mating, female choice of 177–9
  elephant, African 178
  hippopotamus 178
  rhinoceros, black 178–9; Indian 178;
    white 178–9
mating gains 172–7
  elephant 174–6
  rhinoceros, black 175–7; white 176
metabolic requirements 70–1
mobility 165–7
mopane *see Colophospermum mopane*
mortality and lifespan 152–3
  elephant 153–4
  giraffe 155–9
  hippopotamus 154–5
  rhinoceros 155–9
mortality costs 168–70
mouth parts 90–1
mud wallowing
  elephant 46–7
  rhinoceros, Javan 47; Sumatran 47;
    white 48–50
musth 109, 139, 151, 167, 171

nutrients 277–8
nutritional balance 79–80

Panicum coloratum 23, 31, 36, 41, 44, 238
Panicum–Urochloa grassland 235, 237
population 200
population density 221, 224–5
  elephant 221–3
  giraffe 223–4
  hippopotamus 223
  rhinoceros 224
population regulation 246–64
  biomass 265–71
  dispersal 260–4

effect of body size 255
effect of rainfall 253–4
effect of vegetation 250–3
elephant 247–8, 257–60
hippopotamus 257, 259
interaction with vegetation 257–71,
  313–14
rhinoceros 248–53, 257, 259
patterns 313–14
population growth 212, 221
  elephant 212–4
  giraffe 214
  hippopotamus 214
  rhinoceros 214–15; white 215–20
population structure 200–1, 211
  elephant 201–3
  giraffe 204
  hippopotamus 203–4
  rhinoceros 204; white 204–11
Pterocarpus angolensis 229, 239
predation 95, 261
  human 289–92, 295
predators, response to 124
  elephant, African 124–5; Asian 125–6
  giraffe 126
  hippopotamus 126
  rhinoceros 126–31

rainfall, mean annual effect on biomass
  271–4
reproduction, male 151
  elephant 151
  giraffe 152
  hippopotamus 151
  rhinoceros 152
reproduction, female 144–51
  elephant 144–5
  giraffe 145–6
  hippopotamus 145
  rhinoceros 146–50
reproduction, maternal investment in
  190–5
reproduction, seasonality of 183–4
  elephant 183
  giraffe 184
  hippopotamus 184
  rhinoceros 184
reproduction, sex ratio of offspring
  195–9
  elephant 196
  giraffe 198
  hippopotamus 196
  rhinoceros 196–8
rhinoceros
  adolescence and puberty 140–4
  birth intervals 188
  community effects on other large
    herbivores 242

rhinoceros (*cont.*)
  conservation 300–1, 316
  courtship and mating 119–24
  diet composition 38–44
  digestion 80
  distribution 21–9
  evolutionary origins 17–19
  foraging 55–60
  groups 104–9
  infancy and juvenilehood 135–8
  males, satellite 170–2
  mating, female choice 178–9
  mating gains 174–7
  mortality and lifespan 155–9
  mortality costs 168–9
  mouth parts 90
  population growth 214–15
  population regulation interactions with
    vegetation 257, 259
  population structure 204
  predators, response to 126–31
  reproduction, male 152; female 146–50;
    seasonality of 184
  sex ratio, offspring 195–8
  temporal patterning of activities 55–60
  vegetation, impact on 234–5
rhinoceros, African black or hook-lipped
    (*Diceros bicornis*) 11
  adolescence and puberty 140–1
  courtship and mating 119–20
  dentition 11
  diet composition 39–41
  dominance 111–12
  groups 104
  infancy and juvenilehood 136–7
  mating, female choice 178–9
  mating gains 175–7
  mud wallowing 47–8
  nocturnal feeding 94
  predators, response to 126–7
  reproduction, female 146
  size 10–12
  skin color 12
  utilization of space 63–4
  water 47
rhinoceros, Indian or great one-horned
    (*Rhinoceros unicornis*) 12
  adolescence and puberty 140
  courtship and mating 119
  dentition 12
  diet composition 38
  digestion 13
  groups 104
  height 12
  infancy and juvenilehood 135–6
  mortality and lifespan 155–6
  predators, response to 126
  size 12–13

  utilization of space 63
  water 47, 48
  weight 12
rhinoceros, Javan or lesser one-horned
    (*Rhinoceros sondaicus*) 12
  dentition 12–13
  diet composition 39
  digestion 13
  dominance 111
  mud wallowing 47
Rhinoceros, Sumatran or Asian two-
    horned (*Dicerorhinus sumatrensis*) 13
  dentition 13
  diet composition 38–9
  digestion 13
  dominance 111
  mud wallowing 47
  size 13
rhinoceros, African white or square-lipped
    (*Ceratotherium simum*) 9, 10
  adolescence and puberty 141–4
  body heat 94
  copulation 121–4
  dentition 11
  diet composition 41–4
  digestion 13
  distribution 25–8
  dominance 112–16
  gestation period 120
  group 105–9
  infancy and juvenilehood 137–8
  mating, female choice 178–9
  mating gains 176
  mortality and lifespan 156–9
  mud wallowing 48–50
  population growth 215–20
  population structure 204–11
  predators, response to 127–30
  reproduction, male 152; female 146–50
  size 9–10
  skin color 12
  subspecies 10–11
  utilization of space 63–7
  vegetation, impact on 235–8
  water 48
ruminants 14, 16, 55, 72, 77, 82, 190, 310

Securinega virosa 233
sex ratio of offspring *see* reproduction
    female
Sclerocaryo birrea 232
size 6
  elephant, African 7–9
  giraffe 14
  hippopotamus 13–14
  rhinoceros, black 10–12; Indian 12–13;
    Sumatran 13; white 9–10
social and life history patterns 312–13

social organization, patterns of 101
space, utilization of
  elephant 61–2
  giraffe 62–3
  hippopotamus 62
  rhinoceros, black 63–4; Indian 63; white
    63–7
Sporobolus pyramidalis 36, 233–4
Stercula 239
subordinate *see* males, satellite

temporal patterning of activities
  elephant 53–4
  giraffe 55
  hippopotamus 54–5
  rhinoceros 55–60
Terminalia/Combretum woodland 34, 231,
    232
Themeda triandra 29, 31, 36, 42, 44, 233–8
thermoregulatory constraints 93–5
trophic adaptations 98–100
Turraea robusta 233

ungulates, comparisons with smaller
  browsing 50
  courtship and mating 131–2
  diet 51
  dominance 131
  grazing 50

group structure 131
mortality and lifespan 159
mud wallowing 51
population 25
predators, response to 131–2
space–time patterns of habitat use 67–8
water 51

vacuum zones *see* conservation
vegetation, impact on 226–39
  elephant 227–33
  giraffe 234
  hippopotamus 233–4
  rhinoceros 234–5; white 235–8

warthog 29, 51, 92, 238, 241
water, need for 45–50
waterbuck 242–3, 244, 270
weight 6
  elephant, African 8
  giraffe 14
  hippopotamus 13
  rhinoceros, Asian 13; black 10–11;
    Indian 12; white 10
Wildebeest 50–1, 68, 225, 238, 242, 244–5,
    255–6, 270, 278

Zebra 68, 82, 131, 159, 238, 240, 242
Ziziphus mucronata 234